Lecture Notes in Computer Scie.

Edited by G. Goos, J. Hartmanis and J. van Lee

T0250573

Advisory Board: W. Brauer   D. Gries   J. Sto

Springer
Berlin
Heidelberg
New York
Barcelona
Budapest
Hong Kong
London
Milan
Paris
Santa Clara
Singapore
Tokyo

Alberto Apostolico   Jotun Hein   (Eds.)

Combinatorial Pattern Matching

8th Annual Symposium, CPM 97
Aarhus, Denmark, June 30 - July 2, 1997
Proceedings

 Springer

Series Editors

Gerhard Goos, Karlsruhe University, Germany

Juris Hartmanis, Cornell University, NY, USA

Jan van Leeuwen, Utrecht University, The Netherlands

Volume Editors

Alberto Apostolico
Dipartimento di Elettronica e Informatica, University of Padova
Via Gradenigo 6/A, I-35131 Padova, Italy
E-mail: axa@artemide.dei.unipd.it
and
Department of Computer Sciences, Purdue University
West Lafayette, IN 47907, USA
E-mail: axa@cs.purdue.edu

Jotun Hein
University of Aarhus, Institute of Biological Science
Bldg. 540, Office 128
Ny Munkegade, DK-800 Aarhus C, Denmark
E-mail: jotun@pop.bio.aau.dk

Cataloging-in-Publication data applied for

Die Deutsche Bibliothek - CIP-Einheitsaufnahme

Combinatorial pattern matching : 8th annual symposium ;
proceedings / CPM 97, Aarhus, Denmark, June 30 - July 2, 1997.
Alberto Apostolico ; Jotun Hein (ed.). - Berlin ; Heidelberg ; New
York ; Barcelona ; Budapest ; Hong Kong ; London ; Milan ; Paris ;
Santa Clara ; Singapore ; Tokyo : Springer, 1997
  (Lecture notes in computer science ; Vol. 1264)
  ISBN 3-540-63220-4

CR Subject Classification (1991): F.2.2, I.5.4, I.5.0, I.7.3, H.3.3, E.4, G.2.1,
J.3

ISSN 0302-9743
ISBN 3-540-63220-4 Springer-Verlag Berlin Heidelberg New York

© Springer-Verlag Berlin Heidelberg 1997
Printed in Germany

Typesetting: Camera-ready by author
SPIN 10548911   06/3142 – 5 4 3 2 1 0   Printed on acid-free paper

Foreword

The papers contained in this volume were presented at the Eighth Annual Symposium on Combinatorial Pattern Matching (CPM 97), held June 30 – July 2, 1997, at the University of Aarhus, Denmark. They were selected from 32 papers submitted in response to a call for papers. In addition, invited lectures were given by A. Dress of the University of Bielefeld (Iterative versus Simultaneous Multiple Sequence Alignment) and J. B. Kruskal of Bell Labs (Modern Comparative Lexicostatistics).

Combinatorial Pattern Matching addresses issues of searching and matching strings and more complicated patterns such as trees, regular expressions, graphs, point sets, and arrays. The goal is to derive non-trivial combinatorial properties for such structures and then to exploit these properties in order to achieve improved performance for the corresponding computational problems.

In recent years, a steady flow of high-quality research on this subject has changed a sparse set of isolated results into a full-fledged area of algorithmics with important applications. This area is expected to grow even further due to the increasing demand for speed and efficiency that comes especially from molecular biology, but also from areas such as information retrieval, pattern recognition, compiling, data compression, and program analysis. The objective of annual CPM gatherings is to provide an international forum for research in combinatorial pattern matching.

The general organization and orientation of CPM conferences is coordinated by a steering committee composed of A. Apostolico, M. Crochemore, Z. Galil, and U. Manber.

The first six meetings were held in Paris (1990), London (1991), Tucson (1992), Padova (1993), Pacific Grove (1994), Helsinki (1995), and Laguna Beach (1996). After the first meeting, a selection of the papers appeared as a special issue of *Theoretical Computer Science*. Since the third meeting, the proceedings have appeared as volumes 644, 684, 807, 937, and 1075 of the Lecture Notes in Computer Science series.

CPM 97 was organized by Jotun Hein of the Department of Genetics and Microbial Ecology at the University of Aarhus and Christian Nørgaard Storm Pedersen of the Department of Computer Science at the University of Aarhus. The conference was supported in part by the University of Aarhus, BRICS, and the National Research Foundation of Denmark.

April 1997   Alberto Apostolico
                                                                    Jotun Hein

Program Committee

A. Amir	S. Kannan
A. Apostolico *co-chair*	R. Kosaraju
R. Baeza-Yates	R. Shamir
D. Breslauer	S. Skyum
J. Hein *co-chair*	E. Ukkonen
D. Hirschberg	K. Zhang

Additional Referees

Vineet Bafna	Stefano Lonardi
Benny Chor	Rune Lyngsø
Livio Colussi	Dalit Naor
David Eppstein	Christian N. Storm P.
Roberto Grossi	Itsik Péer
Haim Kaplan	Theis Rauhe
Richard H. Lathrop	Haim Wolfson

Table of Contents

An Improved Pattern Matching Algorithm for Strings in Terms of
Straight-Line Programs
Masamichi Miyazaki, Ayumi Shinohara, Masayuki Takeda 1

Episode Matching
*Gautam Das, Rudolf Fleischer, Leszek Gasieniec,
Dimitris Gunopulos, Juha Kärkkäinen* 12

Efficient Algorithms for Approximate String Matching with Swaps
Jee-Soo Lee, Dong Kyue Kim, Kunsoo Park, Yookun Cho 28

On the Complexity of Pattern Matching for Highly Compressed
Two-Dimensional Texts
*Piotr Berman, Marek Karpinski, Lawrence L. Larmore,
Wojciech Plandowski, Wojciech Rytter* 40

Estimating the Probability of Approximate Matches
Stefan Kurtz, Gene Myers 52

Space- and Time-Efficient Decoding with Canonical Huffman Trees
Shmuel T. Klein 65

On Weak Circular Squares in Binary Words
Aviezri S. Fraenkel, Jamie Simpson, Mike Paterson 76

An Easy Case of Sorting by Reversals
Nicholas Tran 83

External Inverse Pattern Matching
Leszek Gasieniec, Piotr Indyk, Piotr Krysta 90

Distributed Generation of Suffix Arrays
*Gonzalo Navarro, João Paulo Kitajima, Berthier A. Ribeiro-Neto,
Nivio Ziviani* 102

Direct Construction of Compact Directed Acyclic Word Graphs
Maxime Crochemore, Renaud Vérin 116

Approximation Algorithms for the Fixed-Topology Phylogenetic
Number Problem
Mary Cryan, Leslie Ann Goldberg, Cynthia A. Phillips 130

A New Algorithm for the Ordered Tree Inclusion Problem
Thorsten Richter 150

On Incremental Computation of Transitive Closure and Greedy
Alignment
Saïd Abdeddaïm 167

Aligning Coding DNA in the Presence of Frame-Shift Errors
Lars Arvestad 180

A Filter Method for the Weighted Local Similarity Search Problem
Enno Ohlebusch 191

Trie-Based Data Structures for Sequence Assembly
Ting Chen, Steven S. Skiena 206

Flexible Identification of Structural Objects in Nucleic Acid Sequences:
Palindromes, Mirror Repeats, Pseudoknots and Triple Helices
Marie-France Sagot, Alain Viari 224

Banishing Bias from Consensus Sequences
Amir Ben-Dor, Guiseppe Lancia, Jennifer Perone, R. Ravi 247

On the Nadeau-Taylor Theory of Conserved Chromosome Segments
*David Sankoff, Marie-Noelle Parent, Isabelle Marchand,
Vincent Ferretti* 262

Iterative versus Simultaneous Multiple Sequence Alignment
Andreas Dress 275

Modern Comparative Lexicostatistics
Joseph B. Kruskal 276

Author Index 277

An Improved Pattern Matching Algorithm for Strings in Terms of Straight-Line Programs

Masamichi Miyazaki, Ayumi Shinohara* and Masayuki Takeda*

{masamich, ayumi, takeda}@i.kyushu-u.ac.jp

Department of Informatics, Kyushu University 33, Fukuoka 812-81, Japan

Abstract. We show an efficient pattern-matching algorithm for strings that are succinctly described in terms of straight-line programs, in which the constants are symbols and the only operation is the concatenation. In this paper, both text T and pattern P are given by straight-line programs $\mathcal{T}$ and $\mathcal{P}$. The length of the text T (pattern P, resp.) may grow exponentially with respect to its description size $\|\mathcal{T}\| = n$ ($\|\mathcal{P}\| = m$, resp.). We show a new combinatorial property concerning with the periodic occurrences of a pattern in a text. Based on this property, we develop an $O(n^2m^2)$ time algorithm using $O(nm)$ space, which outputs a compact representation of all occurrences of P in T. This is superior to the algorithm proposed by Karpinski *et al.*[11], which runs in $O((n+m)^4 \log{(n+m)})$ time using $O((n+m)^3)$ space, and finds only one occurrence. Moreover, our algorithm is much simpler than theirs.

1 Introduction

The string pattern-matching is a task to find all occurrences of a pattern in a text. In practice the text is large and is stored in secondary storage, hence most of the time required for pattern-matching is devoted to data transmission. If the text is stored in some compressed form, the data transmission time is decreased according to the compression ratio. Text compression thus speeds up pattern-matching. Of course, the processing time (excluding I/O time) may be much longer than searching the original text. Therefore, it is important to give an efficient pattern-matching algorithm for searching a compressed text directly.

The problem of pattern-matching in compressed text is of not only practical interest but also of theoretical interest. Several researchers have studied it recently for various compression methods. For example, [1, 2, 3, 5, 6] are for the run-length coding, [4] for the LZW coding, [7, 8, 9] for the LZ77 coding.

A straight-line program is a compact representation of string. It is a context-free grammar in the Chomsky normal form that derives only one string. The length of the string represented by a straight-line program can be exponentially long with respect to the size of the straight-line program. In this sense, conversion of string into straight-line program can be viewed as a kind of text compressions.

---

* This research was supported in part by Grant-in-Aid for Scientific Research (A) from the Ministry of Education, Science, Sports and Culture, Japan, No. 07558051.

In fact, any text compressed by the LZW coding can be transformed directly into a straight-line program within a constant factor.

In this paper we concentrate on the pattern-matching problem where both text and pattern are represented in terms of straight-line programs. Karpinski *et al.*[10] showed the first polynomial-time algorithm. Later in [11] they proposed an $O((n + m)^4 \log(n + m))$ time algorithm using $O((n + m)^3)$ space, where n and m are the sizes of straight-line programs representing the text and the pattern, respectively. However, the algorithm is complicated and finds only one occurrence of the pattern. In this paper we exploit a new combinatorial property concerning with the periodic occurrences of a pattern in a text, and then present an $O(n^2m^2)$ time algorithm using $O(nm)$ space, which is based on this property. Our algorithm is simple, and outputs an $O(n)$ representation of all occurrences.

2 Preliminary

In this paper, both text and pattern are described in terms of straight-line programs. A *straight-line program* $\mathcal{R}$ is a sequence of assignments as follows:

$$X_1 = expr_1; \; X_2 = expr_2; \; ... \; ; \; X_n = expr_n,$$

where X_i are variables and $expr_i$ are expressions of the form:

- $expr_i$ is a symbol of a given alphabet Σ, or
- $expr_i = X_\ell \cdot X_r \, (\ell, r < i)$, where $\cdot$ denotes the concatenation of X_ℓ and X_r.

Denote by R the string which is derived from the last variable X_n of the program $\mathcal{R}$. The size of the straight-line program $\mathcal{R}$, denoted by $\|\mathcal{R}\|$, is the number n of assignments in $\mathcal{R}$. The length of a string w is denoted by $|w|$. We identify a variable X_i with the string represented by X_i if it is clear from the context.

Example 1. Let us consider the following straight-line program $\mathcal{R}$:

$$X_1 = \text{a}; \; X_2 = \text{b}; \; X_3 = X_1 \cdot X_2; \; X_4 = X_3 \cdot X_1; \; X_5 = X_3 \cdot X_4;$$
$$X_6 = X_5 \cdot X_5; \; X_7 = X_4 \cdot X_6; \; X_8 = X_7 \cdot X_5.$$

We can see that $R = X_8 = \text{abaababaabababaababa}$, and $\|\mathcal{R}\| = 8, |R| = 18$. The evaluation tree is shown in Fig. 1.

We define the *depth* of a variable X in a straight-line program $\mathcal{R}$ by

$$depth(X) = \begin{cases} 1 & \text{if } X = a \in \Sigma, \\ 1 + \max(depth(X_\ell), depth(X_r)) & \text{if } X = X_\ell \cdot X_r. \end{cases}$$

It corresponds to the length of the longest path from X to a leaf in the tree.

For a string w denote by $w[f..t] (1 \le f \le t \le |w|)$ the subword of w starting at f and ending at t. The *pattern matching problem for strings in terms of straight-line programs* is, given straight-line programs $\mathcal{P}$ and $\mathcal{T}$ which are the descriptions of pattern P and text T respectively, to find all occurrences of P in T. Namely, we will compute the following set:

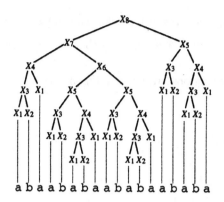

Fig. 1. Evaluation tree of $\mathcal{R}$ in Example 1.

$$Occ(T, P) = \{i : T[i..i + |P| - 1] = P\}.$$

Hereafter, we use X_i for a variable in $\mathcal{T}$ and Y_j for a variable in $\mathcal{P}$. We assume $\|T\| = n$ and $\|P\| = m$. For a set U of integers and an integer k, we denote $U \oplus k = \{i + k : i \in U\}$ and $U \ominus k = \{i - k : i \in U\}$.

3 Overview of algorithm

In this section, we give an overview of our algorithm together with its basic idea. First we consider a compact representation of the set $Occ(X, Y)$.

Suppose $X = X_\ell \cdot X_r$. We define $Occ^*(X, Y)$ to be the set of occurrences of Y in X such that Y covers the boundary between X_ℓ and X_r: (see Fig. 2):

$$Occ^*(X, Y) = \{s \in Occ(X, Y) : |X_\ell| - |Y| + 1 \le s \le |X_\ell| + 1\}.$$

For the sake of convenience, let $Occ^*(X, Y) = Occ(X, Y)$ for $X = a \in \Sigma$. Then

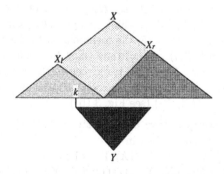

Fig. 2. $k \in Occ^*(X, Y)$, since Y covers the boundary between X_ℓ and X_r.

we have the following lemma, which is informally stated in [8].

Lemma 1. *For any X in T and any Y in $\mathcal{P}$, $Occ^\star(X,Y)$ forms a single arithmetic progression.*

We have the following observation (see Fig. 3 (a)):

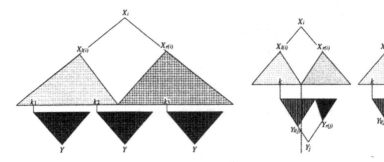

(a) $k_1, k_2, k_3 \in Occ(X_i, Y)$, while $k_1 \in Occ(X_{\ell(i)}, Y)$, $k_2 \in Occ^\star(X_i, Y)$, and $k_3 - |X_{\ell(i)}| \in Occ(X_{r(i)}, Y)$.

(b) $k \in Occ^\star(X_i, Y_j)$ if and only if either $k \in Occ^\star(X_i, Y_{\ell(j)})$ and $k + |Y_{\ell(j)}| \in Occ(X_i, Y_{r(j)})$ (left case), or $k \in Occ(X_i, Y_{\ell(j)})$ and $k + |Y_{\ell(j)}| \in Occ^\star(X_i, Y_{r(j)},)$ (right case).

Fig. 3. Decomposition of variables.

Observation 2 (decomposition of text variables).
For $X_i = X_{\ell(i)} \cdot X_{r(i)}$ in T and Y in $\mathcal{P}$,
$$Occ(X_i, Y) = Occ^\star(X_i, Y) \cup Occ(X_{\ell(i)}, Y) \cup (Occ(X_{r(i)}, Y) \oplus |X_{\ell(i)}|).$$

The above observation suggests that $Occ(X_n, Y)$ can be represented by a combination of $\{Occ^\star(X_i, Y)\}_{i=1}^n$. By Lemma 1, each $Occ^\star(X_i, Y)$ forms a single arithmetic progression, which can be stored in $O(1)$ space as a triple of the first element, the last element, and the step of the progression. Thus the desired output, a compact representation of the set $Occ(T, P) = Occ(X_n, Y_m)$ is given as a combination of $\{Occ^\star(X_i, Y_m)\}_{i=1}^n$, which occupies $O(n)$ space. Moreover, as we will show in Lemma 8 in Section 5, the membership to the set $Occ(X_i, Y_j)$ can be answered in $O(depth(X_i)) = O(n)$ time using this representation. Therefore the computation of the set $Occ(T, P)$ is reduced to the computation of each set $Occ^\star(X_i, Y_m)$, $i = 1, \ldots, n$. The next observation gives us a recursive procedure to compute the set $Occ^\star(X_i, Y_j)$ (see Fig. 3 (b)):

Observation 3 (decomposition of pattern variables).
For X_i in T and $Y_j = Y_{\ell(j)} \cdot Y_{r(j)}$ in $\mathcal{P}$,
$Occ^\star(X_i, Y_j) = Occ^\star_\ell(X_i, Y_j) \cup Occ^\star_r(X_i, Y_j)$, where
$Occ^\star_\ell(X_i, Y_j) = Occ^\star(X_i, Y_{\ell(j)}) \cap (Occ(X_i, Y_{r(j)}) \ominus |Y_{\ell(j)}|)$, and
$Occ^\star_r(X_i, Y_j) = Occ(X_i, Y_{\ell(j)}) \cap (Occ^\star(X_i, Y_{r(j)}) \ominus |Y_{\ell(j)}|)$.

The problem to be overcome is to perform the set operations, union and intersection efficiently, since each set possibly contains exponentially many elements.

Lemma 5 in the next section is a key to solving this problem. The key lemma concerns with the periodicities in strings. It guarantees that each of $Occ_\ell^\star(X_i, Y_j)$ and $Occ_r^\star(X_i, Y_j)$ forms a single arithmetic progression. This enables us to perform the union operation of these two sets in $O(1)$ time. At the same time, the key lemma gives us a basis to construct an efficient procedure of computing $Occ_\ell^\star(X_i, Y_j)$ from $Occ^\star(X_i, Y_{\ell(j)})$, assuming the function $FirstMismatch$ which returns the first position of the mismatches between X_i and $Y_{r(j)}$. We can compute the set $Occ_r^\star(X_i, Y_j)$ in the same way. In Section 5, we will explain these procedures in detail.

When computing each $Occ^\star(X_i, Y_j)$ recursively, we may often refer to the same set $Occ^\star(X_{i'}, Y_{j'})$ repeatedly for $i' < i$ and $j' < j$. We take the dynamic programming strategy. Let us consider an $n \times m$ table App where each entry $App[i, j]$ at row i and column j stores the triple representing the set $Occ^\star(X_i, Y_j)$. We compute each $App[i, j]$ in bottom-up manner, for $i = 1, \ldots, n$ and $j = 1, \ldots, m$. As we will show in Lemma 10 in Section 5, each $App[i, j]$ is computable in $O(depth(X_i) \cdot depth(Y_j))$ time. Since $depth(X_i) \leq n$ and $depth(Y_j) \leq m$ for any X_i and Y_j, we can construct the whole table App in $O(n^2 m^2)$ time. The size of the whole table is $O(nm)$, since each triple occupies $O(1)$ space. Hence we have the main theorem of this paper.

Theorem 4. *Given two straight-line programs T and P, we can compute an $O(n)$ size representation of the set $Occ(T, P)$ of all occurrences of the pattern P in the text T, in $O(n^2 m^2)$ time using $O(nm)$ work space. For this representation, the membership to the set $Occ(T, P)$ can be determined in $O(n)$ time.*

4 Key lemma

This section shows the key lemma on a property of periodic occurrences of a pattern in a text, which our algorithm based on. Let T and P be strings of a text and a pattern. At first we define the function $FirstMismatch(T, P, k)$ which returns the first (leftmost) position of mismatches, when we compare P with T at position k. Formally,

$$FirstMismatch(T, P, k) = \min\{1 \leq i \leq |P| : T[k + i - 1] \neq P[i]\},$$

for $1 \leq k \leq |T| - |P| + 1$. If there is no such i, the value of $FirstMismatch(T, P, k)$ is nil. The value is a witness of $k \notin Occ(T, P)$.

Lemma 5 (Key Lemma). *Let $T = u^p z$ $(u, z \in \Sigma^+, p \geq 0)$ and $P \in \Sigma^+$. The set $S = Occ(T, P) \cap \{1 + i|u| : i = 0, 1, \ldots, p\}$ forms a single arithmetic progression, which can be computed by at most three calls of $FirstMismatch$.*

Proof (sketch). We use the following notation in this proof: For two integers a and b, we denote by $\langle q, r \rangle = div(a, b)$ that q is the quotient and r is the remainder of the division of a by b. That is, $a = b \cdot q + r$ and $0 \leq r < b$.

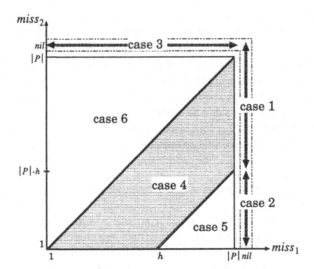

Fig. 4. Six cases depending on $miss_1$ and $miss_2$. (Cases 2 and 5 are vacant if $h > |P|$.)

Let $h = \max\{j \leq |T| - |P| + 1 : j = 1 + i|u|$ for $i = 0, 1, \ldots, p\}$. If no such j exist, $S = \phi$. The case $h \leq 1 + 2|u|$ is trivial, since S contains at most three positions. We consider the case $h \geq 1 + 3|u|$. At the beginning, we invoke the function *FirstMismatch* for two positions 1 and h as follows:

$$miss_1 = FirstMismatch(T, P, 1), \text{ and}$$
$$miss_2 = FirstMismatch(T, P, h).$$

Note that $1 \leq miss_1, miss_2 \leq |P|$, if not *nil*. It is convenient that we regard *nil* as $|P| + 1$. Depending on the values of $miss_1$ and $miss_2$, we have six cases as shown in Fig. 4.

case 1: $miss_1 = nil$ and $miss_2 = nil$ or $|P| - h + 1 < miss_2$. See Fig. 5.
Let $\langle q, r \rangle = div(|P|, |u|)$ and $\langle q', r' \rangle = div(h + miss_2 - 2, |u|)$. We can show that $P = u^q u[1..r]$ and $T = u^{q'} u[1..r']w$ for some $w \in \Sigma^+$ with $u[r' + 1] \neq w[1]$. We can show that $S = \{1 + i|u| : i \in \{0, .., t\}\}$, where $t = q' - q$ if $r' \geq r$ and $t = q' - q - 1$ otherwise. We note that such t can be directly computed by $\langle t, r'' \rangle = div(h + miss_2 - |P| - 2, |u|)$.
case 2: $miss_1 = nil$ and $miss_2 \leq |P| - h + 1$. (This is impossible if $h > |P|$).
Let $\langle q, r \rangle = div(h + miss_2 - 2, |u|)$. We can show that $P = u^q u[1..r]v$ and $T = u^q u[1..r]w$ for some $v, w \in \Sigma^+$ such that $u[r + 1] \neq v[1] = w[1]$ and v is a prefix of w. Thus we have $S = \{1\}$.
case 3: $miss_1 \neq nil$ and $miss_2 = nil$.
Let $\langle q, r \rangle = div(miss_1 - 1, |u|)$, and $\langle q', r' \rangle = div(h + miss_1 - 2, |u|)$. We can show that $r = r'$, $P = u^q u[1..r]v$ and $T = u^{q'} u[1..r]w$ for some $v, w \in \Sigma^+$ such that $u[r + 1] \neq v[1]$ and v is a prefix of w. Thus we have $S = \{h\}$.

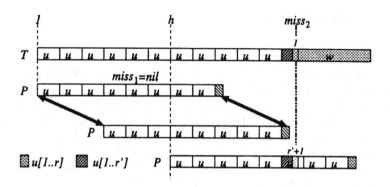

Fig. 5. Case 1, $P = u^q u[1..r]$ and $T = u^{q'} u[1..r']w$.

case 4: $miss_1 - h + 1 < miss_2 < miss_1$. See Fig. 6.

Let $\langle q, r \rangle = div(miss_1 - 1, |u|)$ and $\langle q', r' \rangle = div(h + miss_2 - 2, |u|)$. We can show that $P = u^q u[1..r]v$ and $T = u^{q'} u[1..r']w$ for some $v, w \in \Sigma^+$ such that $u[r+1] \neq v[1]$ and $u[r'+1] \neq w[1]$. If $r = r'$ and v is a prefix of w, then S is a singleton of $s = 1 + p|u| - miss_1 + miss_2$. Otherwise $S = \phi$. That is, the only candidate for the elements in S is s. We can verify whether $S = \{s\}$ or $S = \phi$ by the third call of $FirstMismatch(T, P, s)$.

case 5: $miss_2 \leq miss_1 - h + 1$. (This is impossible if $h > |P|$.)

Let $\langle q, r \rangle = div(h + miss_2 - 2, |u|)$, and $s = miss_1 - h - miss_2 + 2$. Since we can show that $P = u^q u[1..r]v$ and $T = u^q u[1..r]w$ for some $v, w \in \Sigma^+$ such that $u[r+1] \neq v[1] = w[1]$ and $v[s] \neq w[s]$, we have $S = \phi$.

case 6: $miss_1 \leq miss_2$.

Let $\langle q, r \rangle = div(miss_1 - 1, |u|)$, and $\langle q', r' \rangle = div(h + miss_1 - 2, |u|)$. Let $s = miss_2 - miss_1 + 1$. Since $r = r'$, $P = u^q u[1..r]v$ and $T = u^{q'} u[1..r]w$ for some $v, w \in \Sigma^+$ such that $u[r+1] \neq v[1]$ and $v[s] \neq w[s]$, we have $S = \phi$.

For any case, S forms a single arithmetic progression, and we can compute its representation by calling $FirstMismatch$ at most three times. $\qquad \square$

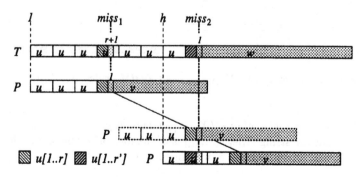

Fig. 6. Case 4, $P = u^q u[1..r]v$ and $T = u^{q'} u[1..r']w$.

5 Algorithm in detail

In this section, we explain the details on the algorithm. That is, how to compute each entry $App[i,j]$ of the table, which represents the set $Occ^*(X_i, Y_j)$. The computation is done in bottom-up manner.

If either X_i or Y_j is a symbol, we can compute the entry $App[i,j]$ in a trivial way. We show how to compute $App[i,j]$ for $X_i = X_{\ell(i)} \cdot X_{r(i)}$ and $Y_j = Y_{\ell(j)} \cdot Y_{r(j)}$, assuming that all preceding entries $App[i',j']$ for $i' < i$ and $j' < j$ are already computed. We can also assume that we know all lengths $|X_{i'}|$ and $|Y_{j'}|$. As we have explained in Section 3, the critical point is the computation of $Occ^*_\ell(X_i, Y_j) = Occ^*(X_i, Y_{\ell(j)}) \cap (Occ(X_i, Y_{r(j)}) \ominus |Y_{\ell(j)}|)$.

Lemma 6. *Independently of the cardinality of the set $Occ^*(X_i, Y_{\ell(j)})$, we can compute the set $Occ^*_\ell(X_i, Y_j)$ by using the function FirstMismatch$(X_i, Y_{r(j)}, k)$ at most three times.*

Proof. If the cardinality of the set $Occ^*(X_i, Y_{\ell(j)})$ is at most two, we can compute the set $Occ^*_\ell(X_i, Y_j)$ easily: For each $s \in Occ^*(X_i, Y_{\ell(j)})$, check whether or not $s \in Occ(X_i, Y_{r(j)}) \ominus |Y_{\ell(j)}|$ by using FirstMismatch$(X_i, Y_{r(j)}, s + |Y_{\ell(j)}|)$.

For the case that $Occ^*(X_i, Y_{\ell(j)})$ contains more than two positions, we apply Lemma 5 as follows. Let $e_{\min}$ and $e_{\max}$ be the minimum and the maximum elements in $Occ^*(X_i, Y_{\ell(j)}) \oplus |Y_{\ell(j)}|$, respectively (Fig. 7). Let d be the step of

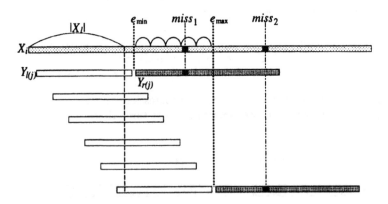

Fig. 7. *FirstMismatch$(X_i, Y_{r(j)}, e_{\min})$ and FirstMismatch$(X_i, Y_{r(j)}, e_{\max})$.*

the arithmetic progression of $Occ^*(X_i, Y_{\ell(j)})$, and let $p = (e_{\max} - e_{\min})/d$. Then we can see that the string $X_i[e_{\min}..|X_i|]$ is of the form $u^p z$, where u is the suffix of $Y_{\ell(j)}$ of length d. By Lemma 5, we can compute the set

$$S = Occ(X_i[e_{\min}..|X_i|], Y_{r(j)}) \cap \{1, 1+d, \ldots, 1+p \cdot d\}$$

by calling the function $FirstMismatch(X_i, Y_{r(j)}, k)$ at most three times. Since

$$S \oplus (e_{min} - 1)$$
$$= (Occ(X_i[e_{min}..|X_i|], Y_{r(j)}) \oplus (e_{min} - 1)) \cap \{e_{min}, e_{min} + d, \ldots, e_{min} + p \cdot d\}$$
$$= Occ(X_i, Y_{r(j)}) \cap (Occ^\star(X_i, Y_{\ell(j)}) \oplus |Y_{\ell(j)}|),$$

we have $S \oplus (e_{min} - 1 - |Y_{\ell(j)}|) = (Occ(X_i, Y_{r(j)}) \ominus |Y_{\ell(j)}|) \cap Occ^\star(X_i, Y_{\ell(j)}) = Occ_\ell^\star(X_i, Y_j)$, which is the desired set. □

We show how to realize the function $FirstMismatch(X, Y, k)$ for variables X in $\mathcal{T}$ and Y in $\mathcal{P}$ and an integer k. Remark the following recursive property:

Observation 7. For two variables X in $\mathcal{T}$ and Y with $Y = Y_\ell \cdot Y_r$ in $\mathcal{P}$,

$$FirstMismatch(X, Y, k) = \begin{cases} FirstMismatch(X, Y_\ell, k) & \text{if } k \notin Occ(X, Y_\ell), \\ |Y_\ell| + FirstMismatch(X, Y_r, k) & \text{if } k \in Occ(X, Y_\ell). \end{cases}$$

We show a pseudo-code of the function $FirstMismatch$ in Fig. 8, where the function $Match(X, Y, k)$ returns true if and only if $k \in Occ(X, Y)$. The correctness of $Match(X, Y, k)$ is directly derived from Observation 2.

Lemma 8. *The function $Match(X_i, Y_j, k)$ answers in $O(depth(X_i))$ time.*

Proof. The membership query of the form $k \in Occ^\star(X_{i'}, Y_{j'})$ can be answered in $O(1)$ time by simple calculations for any $i' < i$ and $j' < j$, since it is already computed and stored in the entry $App[i', j']$. Moreover, the number of recursive calls of $Match(X_i, Y_j, k)$ is at most $depth(X_i)$. Thus the lemma holds. □

Lemma 9. *The function $FirstMismatch(X_i, Y_j, k)$ answers in $O(depth(X_i) \cdot depth(Y_j))$ time.*

Proof. The number of recursive calls of $FirstMismatch(X_i, Y_j, k)$ is at most $depth(Y_j)$. At each call, the function $Match(X_i, Y_j, k)$ is called once. By Lemma 8, it answers in $O(depth(X_i))$ time. Thus the lemma holds. □

By Lemma 6 and Lemma 9, we have the following result.

Lemma 10. *Each entry $App[i, j]$ is computable in $O(depth(X_i) \cdot depth(Y_j))$ time.*

6 Conclusion

We have shown an improved pattern-matching algorithm for strings in terms of straight-line programs. In Table 1, we summarize the results compared to the previous ones [10, 11], from the view points of time complexity, space complexity, and pattern detection ability for multiple occurrences.

We briefly state the improvement of our algorithm compared to the one in [11]. The latter algorithm consists of two phases: At the first phase, it computes two sets: $Pref(X_i, Y_j)$ of the lengths of prefixes of Y_j that are suffixes of X_i,

---

```
function FirstMismatch(X, Y, k): integer;
/* returns the minimum s such that X[k + s − 1] ≠ Y[s] if exists,
   and nil otherwise */
begin
    if |Y| = 1 then
        if X[k] = Y then return 1 else return nil
    else        /* assume Y = Yₗ · Yᵣ */
        if Match(X, Yₗ, k) then
            return |Yₗ| + FirstMismatch(X, Yᵣ, k + |Yₗ|)
        else
            return FirstMismatch(X, Yₗ, k)
end

function Match(X, Y, k): boolean;
/* returns true iff X[k..k + |Y| − 1] = Y. */
begin
    if (k < 0) or (|X| < k + |Y|) then return false;
    if |X| = 1 then
        if Y = X then return true else return false
    else        /* assume X = Xₗ · Xᵣ */
        if k + |Y| < |Xₗ| then return Match(Xₗ, Y, k)
        else if |Xₗ| < k then return Match(Xᵣ, Y, k − |Xₗ|)
        else
            if k ∈ Occ*(X, Y) then return true
            else return false
end
```

Fig. 8. Pseudo-codes of the functions *FirstMismatch* and *Match*.

and $Suff(X_i, Y_j)$ of the lengths of suffixes of Y_j that are prefixes of X_i. At the second phases, it computes the set $Occ(X_i, Y_j)$ from $Pref(X_i, Y_j)$ and $Suff(X_i, Y_j)$ by solving certain linear Diophantine equations with using Euclid's algorithm. Each $Suff(X_i, Y_j)$ and $Pref(X_i, Y_j)$ can be stored in $O(depth(X_i) + depth(Y_j))$ space, although $Occ(X_i, Y_j)$ occupies only $O(1)$ space. On the other hand, our algorithm directly computes $Occ(X_i, Y_j)$. The property of periodic occurrences of a pattern in a text shown in the key lemma enabled the direct computation.

Our algorithm uses the first position of mismatches between text and pattern,

Table 1. Summary

algorithm	time	space	detection
KRS'95 [10]	$O((n + m)^7)$	not estimated	some one
KRS'97 [11]	$O((n+m)^4 \log(n+m))$	$O((n+m)^3)$	some one
Ours	$O(n^2 m^2)$	$O(nm)$	all

whereas the previous algorithm additionally requires text-to-text and pattern-to-pattern comparisons. This is the reason why the previous algorithm requires $(n+m) \times (n+m)$ table with $O(n+m)$ size entries, while our algorithm requires only $n \times m$ table with $O(1)$ size entries. This is also the contribution of the key lemma.

References

1. A. Amir and G. Benson. Efficient two-dimensional compressed matching. In *Proc. Data Compression Conference*, page 279, 1992.
2. A. Amir and G. Benson. Two-dimensional periodicity and its application. In *Proc. 3rd Symposium on Discrete Algorithms*, page 440, 1992.
3. A. Amir, G. Benson, and M. Farach. Optimal two-dimensional compressed matching. In *Proc. 21st International Colloquium on Automata, Languages and Programming*, 1994.
4. A. Amir, G. Benson, and M. Farach. Let sleeping files lie: Pattern matching in Z-compressed files. *Journal of Computer and System Sciences*, 52:299–307, 1996.
5. A. Amir, G. M. Landau, and U. Vishkin. Efficient pattern matching with scaling. *Journal of Algorithms*, 13(1):2–32, 1992.
6. T. Eilam-Tsoreff and U. Vishkin. Matching patterns in a string subject to multi-linear transformations. In *Proc. International Workshop on Sequences, Combinatorics, Compression, Security and Transmission*, 1988.
7. M. Farach and M. Thorup. String-matching in Lempel-Ziv compressed strings. In *27th ACM STOC*, pages 703–713, 1995.
8. L. Gąsieniec, M. Karpinski, W. Plandowski, and W. Rytter. Efficient algorithms for Lempel-Ziv encoding. In *Proc. 4th Scandinavian Workshop on Algorithm Theory*, volume 1097 of *Lecture Notes in Computer Science*, pages 392–403. Springer-Verlag, 1996.
9. L. Gąsieniec, M. Karpinski, W. Plandowski, and W. Rytter. Randomized efficient algorithms for compressed strings: the finger-print approach. In *Proc. Combinatorial Pattern Matching*, volume 1075 of *Lecture Notes in Computer Science*, pages 39–49. Springer-Verlag, 1996.
10. M. Karpinski, W. Rytter, and A. Shinohara. Pattern-matching for strings with short descriptions. In *Proc. Combinatorial Pattern Matching*, volume 637 of *Lecture Notes in Computer Science*, pages 205–214. Springer-Verlag, 1995.
11. M. Karpinski, W. Rytter, and A. Shinohara. An efficient pattern-matching algorithm for strings with short descriptions. *Nordic Journal of Computing*, 1997. (to appear).

Episode Matching

Gautam Das[1]   Rudolf Fleischer[2]   Leszek Gąsieniec[2]
Dimitris Gunopulos[3]   Juha Kärkkäinen[4]

[1] Dept. of Mathematical Sciences, The University of Memphis, Memphis TN 38152,
USA; dasg@mathsci.msci.memphis.edu
[2] Max-Planck Institut für Informatik, Im Stadtwald, Saarbrücken D-66123, Germany;
{rudolf,leszek}@mpi-sb.mpg.de
[3] IBM Almaden RC k55/B1, 650 Harry Rd, CA 95120, USA;
gunopulo@almaden.ibm.com
[4] Dept. of Computer Science, P.O. Box 26, FIN-00014 University of Helsinki,
Finland; juha.karkkainen@cs.helsinki.fi

Abstract. Given two words, text T of length n and episode P of length
m, the episode matching problem is to find all minimal length substrings
of text T that contain episode P as a subsequence. The respective optimization problem is to find the smallest number w, s.t. text T has a
subword of length w which contains episode P.
In this paper, we introduce a few efficient off-line as well as on-line algorithms for the entire problem, where by on-line algorithms we mean
algorithms which search from left to right consecutive text symbols only
once. We present two alphabet independent algorithms which work in
time $O(nm)$. The off-line algorithm operates in $O(1)$ additional space
while the on-line algorithm pays for its property with $O(m)$ additional
space. Two other on-line algorithms have subquadratic time complexity.
One of them works in time $O(nm/\log m)$ and $O(m)$ additional space.
The other one gives a time/space trade-off, i.e., it works in time $O(n +
s + nm \log \log s/ \log(s/m))$ when additional space is limited to $O(s)$.
Finally, we present two approximation algorithms for the optimization
problem. The off-line algorithm is alphabet independent, it has superlinear time complexity $O(n/\epsilon + n \log \log(n/m))$ and it uses only constant
space. The on-line algorithm works in time $O(n/\epsilon + n)$ and uses space
$O(m)$. Both approximation algorithms achieve $1 + \epsilon$ approximation ratio,
for any $\epsilon > 0$.

1 Introduction

In [6], Mannila et al. introduced the problem of finding frequent episodes in
event sequences. An *episode* is a collection of events that occur within short time
interval. Given a long sequence of events, e.g. alarms from a telecommunication
network, it can be useful to know what episodes occur frequently in the sequence.

A simplified version of this problem in string matching terms is: Given a text
T, a window width w and a frequency threshold t, find all strings (episodes) P
that satisfy the condition:

Episode Condition 1. The text T has (at least) t different substrings of length w that contain P as a subsequence.

The problem of finding frequent episodes was also considered in [5], but with a different definition of "frequent" based on minimal substrings. A substring containing P is *minimal* if no proper substring of it contains P. The corresponding episode condition is

Episode Condition 2. The text T has (at least) t different *minimal* substrings of length at most w that contain P as a subsequence.

Obviously, Condition 2 implies Condition 1, but not vice versa.

In this paper, we consider problems rising from the episode conditions when both T and P are given. We call these episode matching problems. Our main problem is the following.

Problem 1. Given a text T and an episode P, find all minimal substrings of T that contain P as a subsequence.

From the set of minimal substrings it is easy to compute the smallest w for any t or the largest t for any w that satisfies the episode condition. Only linear time is needed whether the first or the second condition is used. We will give four algorithms for this problem. In addition, we will give two *approximation* algorithms for the problem of finding a minimal w when $t = 1$:

Problem 2. Given a text T and an episode P, find the smallest w such that T has a substring of length w that contains P as a subsequence.

Some of our algorithms are alphabet independent, that is, the only character operation needed is the comparison for equality or inequality. The other algorithms may need an alphabet transformation. Let $\langle P \rangle$ denote the number of different characters in the episode P. The alphabet transformation maps characters appearing in P bijectively into the set $\{1, 2, \ldots, \langle P \rangle\}$ and all other characters into 0. Applying this transformation to P and T changes the alphabet into $\{0, \ldots, \langle P \rangle\}$ but does not change the solutions to our problems. With general alphabets allowing only equality comparisons, the transformation needs $O(nm)$ time, where $n = |T|$ and $m = |P|$. With ordered alphabets allowing order comparisons, the transformation can be done in $O(n \log m)$ time. In practice, array indexing, hashing or other techniques can be used for making the transformation in linear time.

The properties of our algorithms for Problem 1 are summarized below.

Algorithm A Alphabet independent, $O(nm)$ time, $O(1)$ additional space.

Algorithm B Alphabet independent, on-line, $O(nm)$ time, $O(m)$ additional space.

Algorithm C On-line, $O(nm/\log m)$ time, $O(m)$ additional space.

Algorithm DB On-line, $O(n + s + nm \log \log s / \log(s/m))$ time, $O(s)$ additional space.

The approximation algorithms for Problem 2, both achieving $1+\epsilon$ approximation ratio for any $\epsilon > 0$, are summarized below.

Algorithm AA Alphabet independent, $O(n/\epsilon + n \log \log(n/m))$ time, $O(1)$ additional space.

Algorithm AB On-line, $O(n/\epsilon + n)$ time, $O(m)$ additional space.

The time and space requirements for the alphabet dependent algorithms do not include the alphabet transformation but assume that it has been made. No other assumptions are made about the alphabet.

As was mentioned in the beginning, the episode conditions are a simplification of the situation described in [6]. The text T is actually a sequence of events, each of which has a type and an occurrence time. The window width w refers to a time window instead of the number of characters in a substring. Our algorithms for Problem 1 are easily modified to handle this generalization, but the approximation algorithms for Problem 2 lose either their approximation guarantee or their running time guarantee when the events are unevenly distributed in time.

Another generalization given in [6] concerns the episodes. The episode P as described above is called a *serial* episode because the characters (or events) must occur in a fixed order in the substring or time window. A *parallel* episode gives a (multi)set of characters that can occur in any order. A *general* episode specifies an arbitrary partial order on the characters. At least Algorithms A, B, AA and AB can be modified to handle general episodes. Further generalizations to the concept of episodes are described in [5].

An even more general class of patterns are regular expressions. The problem of finding minimal substrings matching a regular expression was described and solved in [1, Sect. 9.2.]. Our Algorithms A, B, AA and AB can be generalized even for regular expressions. In fact, the generalization of Algorithm B is exactly the solution given in [1].

Another related pattern matching problem is approximate string matching which looks for those substrings of the text T that can be transformed into the pattern P with at most k edit operations. When deletion is the only edit operation allowed and we choose $k = w - m$, the problem is equivalent to finding all substrings of T of length at most w that contain P as a subsequence. Our Algorithm B is closely related to the classical $O(nm)$ time dynamic programming algorithm for approximate string matching [8,9].

There are more advanced variations of the dynamic programming algorithm, including $O(nk)$ time [3,2] and $O(nm/\log n)$ time [7,13] algorithms. Due to different properties of the dynamic programming table, most of these algorithms are not directly applicable to episode matching. The general techniques in those algorithms, however, can be useful. Our algorithm D is based on the "Four Russians" technique also used in [7,13]. In [11], Ukkonen describes an automaton approach to approximate string matching and suggests that a part of the dynamic programming table could be computed with the automaton and the rest with a simpler method. This is very similar to the idea of our Algorithm DB.

There are a couple of reasons why episodes have not drawn attention before. One is that while episodes are very useful in event sequences, it is not easy to find natural applications in normal strings. The other is the apparent triviality of subsequence matching. For example, the *agrep* package [12] can find all lines or records that contain the given pattern as a subsequence, which is trivial, indeed. The new twist that makes the problem nontrivial is to look for matching substrings whose length is minimal or limited.

2 Minimal Substrings

Let $T = t_1 t_2 \ldots t_n$ be the text and $P = p_1 p_2 \ldots p_m$ the episode. A substring $T[i..j] = t_i \ldots t_j$ contains P if there exist a sequence $i \leq i_1 < i_2 < \cdots < i_m \leq j$ such that $t_{i_k} = p_k$ for all $k = 1, \ldots, m$. The sequence $i_1, \ldots, i_m$ is called an occurrence of P in $T[i..j]$. The substring $T[i..j]$ is *minimal* if $i_1 = i$ and $i_m = j$ for all occurrences of P in $T[i..j]$, i.e., no proper substring of $T[i..j]$ contains P. The following lemma gives an important property of occurrences in minimal substrings.

Lemma 3. *Let* $i_1, \ldots, i_m$ *and* $i'_1, \ldots, i'_m$ *be occurrences of an episode P in minimal substrings* $T[i_1..i_m]$ *and* $T[i'_1..i'_m]$ *with* $i_1 < i'_1$ *or* $i_m < i'_m$. *Then* $i_{j+1} \leq i'_j$ *for* $j = 1, \ldots, m-1$.

Proof. Suppose $i_{j+1} > i'_j$ for some j. Then $i'_1, \ldots, i'_j, i_{j+1}, \ldots, i_m$ is an occurrence, which contradicts $T[i_1..i_m]$ and $T[i'_1..i'_m]$ being minimal. $\qquad\square$

Let $ms(P,T)$ denote the number of minimal substrings of T that contain P. Particularly, with a single character c, $ms(c,T)$ is the number of occurrences of c in T. We will need the following results in the analysis of our algorithms.

Lemma 4. *For any* $j = 1, \ldots, m$, $ms(P,T) \leq ms(p_j, T)$.

Proof. If $i_1, \ldots, i_m$ and $i'_1, \ldots, i'_m$ are occurrences in two different minimal substrings, then by Lemma 3, $i_j \neq i'_j$. That is, the same text character cannot serve as the jth character of an occurrence in two different minimal substrings. Therefore, the number of p_j's in T must be at least $ms(P,T)$. $\qquad\square$

Lemma 5. *The sum of the lengths of the minimal substrings of T containing P is at most nm.*

Proof. Let $1 \leq i \leq n$ and $1 \leq j \leq m$. By Lemma 3, there exists at most one minimal substring with an occurrence $i_1, \ldots, i_m$ such that $i_j \leq i < i_{j+1}$ (where $i_{m+1} = i_m + 1$). Therefore, at most m different minimal substrings can contain the text character t_i. Summing over all text characters gives the result since the length of a substring is the number of characters it contains. $\qquad\square$

3 Algorithm A

The following simple algorithm finds all minimal substrings containing P.

Algorithm A
Let $i \leftarrow 0$ and repeat until T ends during the forward scan:

1. [Forward Scan] Starting from position $i + 1$ scan text T first looking for p_1, then the next p_2, and so on until p_m is found at position j.
2. [Backward Scan] Starting from position j scan text T *backwards* first looking for p_m, then the next p_{m-1}, and so on until p_1 is found at position i.
3. Report a minimal occurrence at $T[i..j]$.

The algorithm is alphabet independent and requires only $O(1)$ additional space. The running time of the algorithm is clearly $\Theta(n + s)$, where s is the total length of the minimal substrings, which by Lemma 5 is $O(nm)$.

Theorem 6. *The time complexity of Algorithm A is $O(nm)$.*

Algorithm A can be improved by removing unnecessary characters from the text during the scan. When a forward scan is looking for p_{i+1} or a backward scan is looking for p_i, all encountered characters not in $\{p_1, \ldots p_i\}$ can be removed. This makes later scans faster, although the algorithm still needs $\Theta(nm)$ time in the worst case.

Algorithm A is easily modified to handle other kinds of patterns, including general episodes and regular expressions, by doing the forward and backward scans with suitable automatons for the pattern.

4 Algorithm B

Our second algorithm for finding all minimal substrings is closely related to the basic dynamic programming algorithm for approximate string matching [8,9]. The algorithm could also be derived from a minimal substring algorithm for regular expressions [1]. Essentially the same algorithm was also described in [6], and in more detail in [10], as a part of an algorithm for finding all frequent episodes.

The algorithm computes a table $S[0..n, 0..m]$, where $S[i, j]$ is the largest value k such that $T[k..i]$ contains $P[1..j]$. Then, for every i and j such that $k = S[i, j] > S[i - 1, j]$, $T[k..i]$ is a minimal substring containing $P[1..j]$. In particular, $T[S[i, m]..i]$ is a minimal substring containing P if and only if $S[i, m] > S[i - 1, m]$. The table S can be computed by dynamic programming using the recurrence relation

$$S[i, j] = \begin{cases} S[i - 1, j - 1] & \text{if } t_i = p_j \\ S[i - 1, j] & \text{otherwise.} \end{cases}$$

with the initialization $S[0, j] = 0$ for $j = 1, \ldots, m$ and $S[i, 0] = i + 1$ for $i = 0, \ldots, n$. The recurrence relation in the basic dynamic programming algorithm for approximate string matching reduces to exactly this form when deletion is the only edit operation allowed.

In practice, the algorithm maintains a table $s[0..m]$ while scanning the text. After reading t_i, $s[j] = S[i,j]$ for all j. To quickly find the entries of s that need to be updated, the algorithm maintains for every character c the set WAITS[c] of those entries that need to be updated if the next text character is c. The resulting algorithm is as follows.

Algorithm B

```
1      for j ← 0 to m do s[j] ← 0
2      for each c ∈ Σ do WAITS[c] ← ∅
3      for i ← 1 to n do
4          s[0] ← i; WAITS[p₁] ← WAITS[p₁] ∪ {1}
5          Q ← WAITS[tᵢ]; WAITS[tᵢ] ← ∅
6          for each j ∈ Q in descending order do
7              s[j] ← s[j − 1]
8              if j = m then report minimal substring T[s[m]..i]
9                      else WAITS[pⱼ₊₁] ← WAITS[pⱼ₊₁] ∪ {j + 1}
```

Theorem 7. *Algorithm B works correctly and has time complexity $O\big(n + |\Sigma| + pms(P,T)\big)$, where*

$$pms(P,T) = \sum_{j=1}^{m} ms(P[1..j], T).$$

Proof. The following invariant holds just before executing line 5.

> For $j = 1, \ldots, m$, $s[j]$ is the largest value k such that $T[k..i-1]$ contains $P[1..j]$ and j is in WAITS[p_j] if and only if $s[j] > s[j-1]$.

The invariant can be proven by induction on m; we leave the details to the reader. The first part of the invariant shows that the value of $s[j]$ changes when t_i is processed if and only if a minimal substring containing $P[1..j]$ ends at i. The second part shows that $s[j]$ changes every time line 7 is executed. Therefore, lines 6–9 are executed exactly $pms(P,T)$ times which proves the time complexity. The algorithm works correctly because a minimal substring is reported exactly when $s[m]$ changes. □

The value $pms(P,T)$ is bounded by nm. As such, Algorithm B is alphabet dependent and requires $O(m + |\Sigma|)$ additional space. Combined with the alphabet transformation described in the introduction, the algorithm is alphabet independent, and works in $O(nm)$ time and $O(m)$ additional space. Unlike Algorithm A, Algorithm B works on-line, that is, it reads each text character only once and never needs more than $O(m)$ time to process it.

The value $pms(P,T)$ can be further analyzed using Lemma 4. For each prefix $P[1..j]$, we can choose an integer $1 \leq \ell(j) \leq j$. By Lemma 4,

$$pms(P,T) \leq \sum_{j=1}^{m} ms(p_{\ell(j)}, T) = \sum_{c \in \Sigma} |\{j \mid p_{\ell(j)} = c\}| ms(c, T) \qquad (1)$$

no matter how we choose the $\ell(j)$'s. The following examples show some of the applications of this result.

- If the character distribution of T is even over the alphabet Σ, $pms(P,T) \leq n|P|/|\Sigma|$ for all P.
- If no character appears more than k times in P, $pms(P,T) \leq k|T|$ for all T.
- If $P = $ aabcaa, $pms(P,T) \leq 2(ms(\text{a},T) + ms(\text{b},T) + ms(\text{c},T)) \leq 2|T|$ for all T. (Choose $\ell(5) = 3$, $\ell(6) = 4$ and $\ell(j) = j$ for $j = 1, \ldots, 4$.)

The value in entry $s[j]$ depends only on the prefix $P[1..j]$. When there are several episodes to be matched to the same text, we can build a trie of them and combine the computation of the entries that depend only on the common prefixes. The result can be a significant saving in computation since the entries $s[j]$ with small j are updated more often than entries with large j. Algorithm B for tries is given in detail below.

Algorithm B for tries
Input: Trie $TR = (V, E)$, V is the set of nodes, E is the set of edges
Egde $e \in E$ is from node FROM[e] to node TO[e] and is labeled by LABEL[e].

```
1      for each c ∈ Σ do WAITS[c] ← ∅
2      for each v ∈ V do SLEEPS[v] ← ∅; s[v] ← 0
3      for each e ∈ E do add e into SLEEPS[FROM[e]]
4      for i ← 1 to n do
5          s[root] ← i
6          for each e ∈ SLEEPS[root] do
7              move e from SLEEPS[root] into WAITS[LABEL[e]]
8          Q ← WAITS[tᵢ]; WAITS[tᵢ] ← ∅
9          for each e ∈ Q (in FIFO order) do
10             add e into SLEEPS[FROM[e]]
11             s[TO[e]] ← s[FROM[e]]
12             for each f ∈ SLEEPS[TO[e]] do
13                 move f from SLEEPS[TO[e]] into WAITS[LABEL[f]]
14             if TO[e] represents Pⱼ then
15                 report minimal substring T[s[TO[e]]..i] containing Pⱼ
```

The values $s[\cdot]$ are stored with the nodes of the trie. An edge e labelled with c is in the set WAITS[c] if the next occurrence of c will cause an update of $s[\text{TO}[e]]$. Otherwise, e is stored in the set SLEEPS[FROM[e]]. Let

$$pms(TR, T) = \sum_{v \in V \setminus \{root\}} ms(\text{STR}(v), T) \; ,$$

where STR(v) is the concatenation of the labels on the path from the root to the node v.

Theorem 8. *Algorithm B for tries works correctly and has time complexity* $O(n + |\Sigma| + pms(TR, T))$.

The proof is essentially the same as the proof of Theorem 7. Otherwise, too, the algorithm has the same properties as the basic algorithm, including time and space requirements with m replaced by the size of the trie.

The partial order of a general episode can be represented with a directed acyclic graph (DAG). The idea of associating a value $s[v]$ with each node v generalizes for DAG's, too. Similarly, the value $s[v]$ can be associated with each state v in an automaton that recognizes a regular expression. This is exactly the idea of the method in [1, Sect. 9.2.].

5   Algorithm C

Algorithm B performs well when the episode has a lot of variation. For example, if every character of the episode is different, Algorithm B runs in $O(n)$ time. On the other hand, if the episode has little variation, for example $P = \text{aaa}\dots$, the algorithm might require $\Theta(nm)$ time. In this section, we describe an algorithm that takes advantage of, on one hand, the good performance of Algorithm B for episodes of high variation, and on the other hand, the repetitiveness of an episode of low variation.

Algorithm C

1. Divide the episode into k distinct pieces $P_1, P_2, \dots, P_k$.
2. Build a trie TR of the pieces and use Algorithm B for tries to find the minimal substrings containing the pieces.
3. Combine the minimal substrings containing the pieces to find the minimal substrings containing the whole P.

The key to the algorithm is a right choice of the pieces. Our solution is to make each piece P_i as long as possible under the constraint

$$|P_i| \leq \log_{\langle P_i \rangle} m - \log_{\langle P_i \rangle} \log_{\langle P_i \rangle} m,$$

where $\langle P_i \rangle$ is the number of different characters in P_i. The "as long as possible" gives the lower bound

$$|P_i| > \log_{\langle P_i \rangle + 1} m - \log_{\langle P_i \rangle + 1} \log_{\langle P_i \rangle + 1} m - 1$$

for the length of a piece P_i (except possibly P_k). From this we get

$$|P_i| \log\langle P_i \rangle = \Omega(\log m), \text{ if } \langle P_i \rangle \geq 2$$
$$|P_i| = \Omega(\log m), \text{ if } \langle P_i \rangle = 1. \tag{2}$$

Step 2 of Algorithm C was already described in the previous section. One additional detail is the handling of pieces P_i with $\langle P_i \rangle = 1$. They are not included in the trie TR. Instead, their minimal occurrences are found by an $O(n)$ time algorithm that scans the text and keeps account of the positions of the last $|P_i|$ occurrences of the only character of P_i. By (2), there are at most $O(m/\log m)$ such pieces, and therefore, handling them separately takes $O(nm/\log m)$ time. The running time of the rest of Step 2 is given by the following theorem. The proof is given in the appendix.

Theorem 9. *Algorithm B for trie TR has time complexicity $O(nm/\log m)$.*

Step 3 is done with a modified Algorithm B. In Algorithm B, the value $s[j]$ is the starting position of the latest minimal substring containing the prefix $P[1..j]$. Two changes are needed in the accounting of the minimal substrings. First, it is not enough to keep only the starting position, but also the ending position is needed. Second, keeping only the latest minimal substring is not enough. Instead, the minimal substrings containing $P_1 \cdots P_j$ are stored into a queue $Q[j]$. The algorithm is given below.

Algorithm B for sequence of pieces (Step 3 of Algorithm C)

```
1       for j ← 0 to k do Q[j] ← ∅
2       for i ← 1 to n do
3           BEGIN(R) ← i; END(R) ← i − 1; APPEND(Q[0], R)
4           for each Pj with minimal substring S ending at i do
5               if not EMPTY(Q[j − 1]) then
6                   while not EMPTY(Q[j − 1]) and
                            END(TOP(Q[j − 1])) < BEGIN(S) do
7                       R ← POP(Q[j − 1])
8                   if j = k then report minimal substring T[BEGIN(R)..i]
9                   else END(R) ← i; APPEND(Q[j], R)
```

The minimal substrings ending at i (line 4) are found by Step 2 of Algorithm C. Therefore, Steps 2 and 3 are best run in parallel, so that Step 3 immediately processes any minimal substrings found by Step 2.

Theorem 10. *Step 3 of Algorithm C has time complexity $O(nm/\log m)$.*

Proof. Step 3 works in time $\Theta(n + \sum_{h=1}^{k} ms(P_h, T))$. This is clear but for the innermost loop. Every round of that loop removes an item from a queue, so the total time spend in the loop cannot be more than the time spend elsewhere adding items to the queues. We still need to show that

$$\sum_{h=1}^{k} ms(P_h, T) = O\left(\frac{nm}{\log m}\right) \ .$$

Let $P_h = P[a_h..b_h]$ and let $j_1, \ldots, j_{\langle P_h \rangle}$ be a subsequence of $a_h, \ldots, b_h$ such that $\{p_{j_1}, \ldots, p_{j_{\langle P_h \rangle}}\}$ is the set of $\langle P_h \rangle$ characters appearing in P_h. Let $T[i_{a_h}..i_{b_h}]$ be a minimal substring containing an occurrence $i_{a_h}, \ldots, i_{b_h}$ of P_h. The minimal substring will contribute 1 to the above sum. This cost will be evenly distributed among the pairs (i, j) where $i \in \{i_{j_1}, \ldots, i_{j_{\langle P_h \rangle}}\}$ and $j \in \{a_h, \ldots, b_h\}$. The number of such pairs is $\langle P_h \rangle |P_h|$. Given a pair (i, j) that gets a share of a cost, j uniquely identifies the piece P_h. By the condition $t_i = p_{j_l}$, i identifies the j_l with $i = i_{j_l}$. By Lemma 3, the minimal substring is then unique to (i, j). Therefore, no pair (i, j) can get more than one share of the cost of one minimal substring, giving an upper bound $1/\min_h(\langle P_h \rangle |P_h|)$ for the total cost assigned to a pair. By (2) and the fact that $\langle P_h \rangle |P_h| > |P_h| \log \langle P_h \rangle$ when $\langle P_h \rangle \geq 2$, the cost is $O(1/\log m)$. Since the number of different pairs (i, j) is nm, the total cost is $O(nm/\log m)$. □

The additional space requirement of Step 3 can be $\Theta(n)$ because the queues can grow large. However, from the trie TR, we can get all potential starting positions of minimal substrings containing P_j up to the current position. The queue $Q[j-1]$ needs to keep only those minimal substrings containing $P_1 \cdots P_{j-1}$ which best match the potential starting positions. By purging the unneeded minimal substrings from the queues when necessary, the additional space requirement can be kept at $O(m)$.

To summarize, Algorithm C works in $O(nm/\log m)$ time and $O(m)$ additional space. Like Algorithm B, it is an on-line algorithm and needs only $O(m)$ time to process each character.

6   Algorithms D and DB

The algorithms in the previous sections work in $O(m)$ extra space. Faster episode matching is possible if more space is available. Even $O(n)$ text scanning time can be achieved but this may need exponential space. In this section, we will describe an algorithm that works already in moderate space but can take advantage of what space is available. It is based on the "Four Russians" technique that is well-known in approximate string matching [7,13].

Algorithm B computes the full table $s[0..m]$ for every text position, even though knowing just $s[m]$ would be enough. The first algorithm of this section, Algorithm D, computes the full table $s[0..m]$ only at every hth position (for a suitably chosen h) and just $s[m]$ at other positions.

Let $s_i[j]$ denote the value of $s[j]$ at position i. Assume that $s_{kh}[0..m]$ is known for some k. The value $s_{(k+1)h}[j]$ must be one of the values in the nondecreasing sequence $\sigma_j[1..h+1] = (s_{kh}[j], s_{kh}[j-1], \ldots, s_{kh}[1], kh+1, kh+2, \ldots, (k+1)h - j + 1)$, and which it is depends only on the substring $T[kh+1..(k+1)h]$. Similarly, assuming $i \leq h \leq m$, $s_{kh+i}[m]$ must be one of the values $\sigma^i[1..i+1] = (s_{kh}[m], s_{kh}[m-1], \ldots, s_{kh}[m-i])$ depending only on the substring $T[kh+1..kh+i]$.

Algorithm D starts by building a full trie of height h, that is, a trie containing all strings of length h in the transformed alphabet $\{0, \ldots, \langle P \rangle\}$. Each leaf representing a string S stores a table $\delta_S[1..m]$ such that $s_{(k+1)h}[j] = \sigma_j[\delta_S[j]]$

for $j = 1, \ldots, m$. Similarly, each internal node at depth i representing a string S stores the value δ^S such that $s_{kh+i}[m] = \sigma^i[\delta^S]$. The size of the augmented trie is $O(mc^h)$, where $c = \langle P \rangle + 1$ is the size of the transformed alphabet. The trie can be computed in $O(mc^h)$ time.

Armed with the trie, Algorithm D then scans the text and computes the table $s[0..m]$ at every hth position and $s[m]$ at every position reporting a new minimal substring whenever $s[m]$ changes. The scan works on-line and needs $O(n + nm/h)$ time.

Algorithm D

1. Build a full trie of height h.
2. Augment the trie with the δ values.
3. Using the trie, scan the text computing $s[0..m]$ at every hth position and $s[m]$ at every position.

The algorithm works in $O(n + mc^h + nm/h)$ time and $O(mc^h)$ additional space. If enough space is available, we can choose $h = \log_c n - \log_c \log_c n$, thus getting an algorithm that works in $O(n + nm/\log_c n))$ time and $O(nm/log_c n)$ additional space. If the additional space is limited to $s = o(nm/log_c n)$, the algorithm works in $O(n + nm/\log_c(s/m))$ time.

A large alphabet size c is bad for Algorithm D while it is good for Algorithm B. Algorithm DB combines these two to achieve lesser dependence on the alphabet size and better worst case time complexity. For suitably chosen l, Algorithm D does the matching for the prefix $P[1..l]$ and Algorithm B for the suffix $P[l + 1..m]$.

Algorithm DB
for $i \leftarrow 1$ **to** n **do**

1. Execute a step of Algorithm D to compute $s_i[l]$.
2. Execute a step of Algorithm B to compute $s_i[l + 1..m]$.

Algorithm B updates the entries $s[l + 1..m]$ exactly as often as if it was computing the whole table. Its time complexity is, therefore,

$$O\left(n + \sum_{j=l+1}^{m} ms(P[1..j], T)\right) .$$

Let $c_l = \langle P[1..l] \rangle$. Using the technique of Inequality 1 to analyze the sum, we can distribute the costs evenly among the c_l different characters of $P[1..l]$. This gives $O(n(m - l)/c_l)$ as Algorithm B's contribution to the time complexity of Algorithm DB.

The combined algorithm works in $O(n + lc_l^h + nl/h + n(m - l)/c_l)$ time and $O(lc_l^h)$ additional space. With the time-optimal choice of h, the time is $O(n + nl/\log_{c_l} n + n(m - l)/c_l)$. We will choose l to minimize the time. In the worst case, we have $l \approx m/2$ and $c_l \approx \log n/\log \log n$. This gives an algorithm working in $O(n + nm \log \log n / \log n)$ time and $O(nm \log \log n / \log n)$ additional space. Similar analysis for the case where additional space is limited to $O(s)$ gives the running time $O(n + nm \log \log s / \log(s/m))$.

7 Approximation Algorithms for Problem 2

The problem considered in this section is to find the length w of the shortest substring of T that contains P. Both of the algorithms of this section return a value $\hat{w}$ with the approximation guarantee $w \leq \hat{w} \leq (1 + \epsilon)w$, for an arbitrary $\epsilon > 0$.

The first algorithm, Algorithm AA, uses the simple text scan that was also used in Algorithm A. The basic operation is SCAN(h, l) which works as follows. Starting from every hth position of the text T, the operation scans l characters forward looking for the episode P. The operation returns the smallest number of characters scanned from a starting position before finding an occurrence of P. If SCAN(h, l) returns $\hat{w}$, it is known that $\hat{w} - h < w \leq \hat{w}$. If no occurrences were found, it must be that $w > l - h + 1$. The procedure works in $O(nl/h)$ time.

In the first part of Algorithm AA, a binary search is used for finding the $j \in \{0, \ldots, \lfloor \log(n/m) \rfloor\}$ for which SCAN($2^j m, 2^{j+1} m - 1$) finds an occurrence but SCAN($2^{j-1} m, 2^j m - 1$) does not. This requires $O(n \log \log(n/m))$ time. As a result, it is known that $2^{j-1} m < w \leq 2^{j+1} m$.

Finally, Algorithm AA does SCAN($h, 2^{j+1} m + h - 1$). The return value $\hat{w}$ then satisfies

$$2^{j-1} m < w \leq \hat{w} < w + h.$$

Selecting $h = 2^{j-1} m \epsilon$, we get $w \leq \hat{w} < (1 + \epsilon)w$. This part requires $O(n/\epsilon)$ time.

Overall, Algorithm AA works in $O(n \log \log(n/m) + n/\epsilon)$ time. Like Algorithm A, it is alphabet independent, runs in $O(1)$ additional space, and is easily modified for other kinds of patterns.

The other approximation algorithm, Algorithm AB, is a simple modification of Algorithm B. Let $h = 1 + \lfloor m\epsilon \rfloor$. Line 4 of Algorithm B is replaced with

4'   **if** $(i - 1) \bmod h = 0$ **then** $s[0] \leftarrow i$; WAITS$[p_1] \leftarrow$ WAITS$[p_1] \cup \{1\}$

In other words, $s[0]$ is only changed at every hth position. These lines are the only place where new values enter the table s; otherwise values are just copied from one entry to another. Each entry $s[j]$ is the starting position of the latest minimal substring containing $P[1..j]$. With the modifications, the starting positions can be up to $h - 1$ smaller than they would be in the original algorithm. As a result, the reported minimal substrings are longer than actual minimal substrings by at most $h - 1$. Thus, the algorithm has $(1 + \epsilon)$ approximation ratio.

Algorithm AB has the running time $O(n + u)$, where u is the number of times an entry of s changes. With the modifications, only $O(n/h)$ different values are entered into the table, which means that no entry can change more than $O(n/h)$ times. Therefore, $u = O(nm/h)$ and the running time is $O(n + n/\epsilon)$.

8 Concluding Remarks

We have presented a new class of string patterns, called episodes, and given several algorithms for two different episode matching problems. Despite their simple formulation, episode matching problems can be quite nontrivial as our algorithms demonstrate. Indeed, we have just scratched the surface of episode matching problems. In addition to the various problems that can be derived from the basic episode conditions, there are the generalizations, such as occurrence times and general episodes, that rise from the applications to event sequences [6,5]. Other aspects that we have mostly ignored in this paper include average case analysis and lower bounds.

9 Acknowledgements

Discussions with Mordecai Golin, Heikki Mannila and Esko Ukkonen have been helpful in writing this paper.

References

1. A. V. Aho, J. E. Hopcroft and J. D. Ullman: *The Design and Analysis of Computer Algorithms*. Addison-Wesley, 1974.
2. Z. Galil and K. Park: An improved algorithm for approximate string matching. *SIAM J. Comp.*, **19**(6) (Dec. 1990), 989–999.
3. G. M. Landau and U. Vishkin: Fast parallel and serial approximate string matching. *J. Algorithms*, **10**(2) (June 1989), 157–169.
4. J. H. van Lint and R. M. Wilson: *A Course in Combinatorics*. Cambridge University Press, 1992.
5. H. Mannila and H. Toivonen: Discovering frequent episodes in sequences. *Proc. 2nd International Conference on Knowledge Discovery and Data Mining (KDD'96)*, 146–151. AAAI Press 1996.
6. H. Mannila, H. Toivonen and A. I. Verkamo: Discovering frequent episodes in sequences. *Proc. 1st International Conference on Knowledge Discovery and Data Mining (KDD'95)*, 210–215. AAAI Press 1995.
7. W. J. Masek and M. S. Paterson: A faster algorithm for computing string edit distances. *J. Comput. System Sci.*, **20** (1980), 18–31.
8. S. B. Needleman and C. D. Wunsch: A general method applicable to the search for similarities in the amino acid sequences of two proteins. *J. Molecular Biol.* **48** (1970), 443–453.
9. P. H. Sellers: The theory and computation of evolutionary distances: pattern recognition. *J. Algorithms*, **1**(4) (Dec. 1980), 359–373.
10. H. Toivonen: *Discovery of Frequent Patterns in Large Data Collections*. Ph.D. Thesis, Report A-1996-5, Department of Computer Science, University of Helsinki, 1996.
11. E. Ukkonen: Finding approximate patterns in strings. *J. Algorithms*, **6**(1) (May 1985), 132–137.
12. S. Wu, U. Manber: Agrep – a fast approximate pattern-matching tool. *Proc. Usenix Winter 1992 Technical Conference*, 153–162. Jan. 1992.
13. S. Wu, U. Manber and G. Myers: A subquadratic algorithm for approximate limited expression matching. *Algorithmica*, **15**(1) (Jan. 1996), 50–67.

Appendix. Proof of Theorem 9

Let $P_1, P_2, \ldots, P_k$ be strings over alphabet Σ satisfying the following conditions.

Condition 1.
$$|P_1| + |P_2| + \cdots + |P_k| \leq m$$

Condition 2. For all $i = 1, \ldots, k$,
$$\langle P_i \rangle \geq 2 \quad \text{and} \quad |P_i| \log\langle P_i \rangle - \log\log\langle P_i \rangle \leq \log m - \log\log m$$

or
$$\langle P_i \rangle = 1 \quad \text{and} \quad |P_i| \leq \log m - \log\log m.$$

The pieces of P in the trie of Algorithm C satisfy these conditions.

Let TR be the trie build from the strings $P_1, P_2, \ldots, P_k$. We shall prove that Algorithm B for trie TR has time complexity $O(nm/\log m)$, where n is the length of the text. By Theorem 8, Algorithm B for trie TR has time complexity $O(n + |\Sigma| + pms(TR, T))$. After alphabet transformation, $|\Sigma| \leq m + 1$. It is, therefore, enough to show that $pms(TR, T) = O(nm/\log m)$.

We will start by bounding the size $|TR|$ of the trie.

Lemma 11. $|TR| = O(\langle TR \rangle m/\log m)$, where $\langle TR \rangle$ is the number of different characters in TR.

Proof. Let Σ_{TR} be the set of characters in TR and let Γ be a subset of Σ_{TR}. Consider those strings P_i that contain at least one of each of the characters in Γ and no characters in $\Sigma_{TR} \setminus \Gamma$. Let m_Γ be the total length of these strings and let TR_Γ be the trie build of them. Based on Condition 2, it can be shown that

$$|TR_\Gamma| = O\left(\frac{m_\Gamma \left(1 + \log \frac{m}{m_\Gamma}\right)}{\log m}\right).$$

Clearly, $|TR| \leq \sum_{\Gamma \subseteq \Sigma_{TR}} |TR_\Gamma|$. The sum is maximized when

$$m_\Gamma = \frac{m}{2^{|\Sigma_{TR}|}} = \frac{m}{2^{\langle TR \rangle}}$$

for all Γ, giving the result. $\qquad\qquad\square$

Next, we will analyze $pms(TR, T)$ using the amortized analysis technique from Section 4. Since there is one-to-one correspondence between nodes $V \setminus \{root\}$ and edges E in the trie TR, we can redefine $pms(TR, T)$ as

$$pms(TR, T) = \sum_{e \in E} ms(\text{STR}(e), T) ,$$

where STR(e) is the concatenation of the labels on the path from the root to e including the label of e itself. Let ℓ be a relabeling of the edges such that, for all $e \in E$, $\ell(e) \in$ STR(e). Then, following (1),

$$pms(TR, T) \leq \sum_{e \in E} ms(\ell(e), T) = \sum_{c \in \Sigma} |\{e \in E \mid \ell(e) = c\}| ms(c, T) \ . \qquad (3)$$

Let L be the set of all such relabelings ℓ. For all $\Gamma \subseteq \Sigma_{TR}$, let $L_\Gamma \subseteq L$ be the set of relabelings such that $\ell(e) \in \Gamma \cup \{\text{LABEL}[e]\}$ for all $e \in E$. Let $L_\Gamma^{max} \subseteq L_\Gamma$ be the set of relabelings that maximize the number of edges e with $\ell(e) \in \Gamma$.

By Lemma 11, there exists a constant b such that $|TR| \leq b\langle TR\rangle m / \log m$ for large enough m. Let $q = \lceil bm / \log m \rceil$. Then

$$|TR| \leq \langle TR\rangle q \ . \qquad (4)$$

We will prove the following lemma at the end of the appendix.

Lemma 12. *There exists a nonempty subset Γ of Σ_{TR} and a relabeling $\ell \in L_\Gamma^{max}$ such that no character of Γ appears more than q times in the relabelled trie TR.*

Consider now the trie TR relabelled with ℓ of Lemma 12. Since $\ell \in L_\Gamma^{max}$, every path from the root to a leaf can be divided into two parts, the first having only labels in $\Sigma_{TR} \setminus \Gamma$ and the second only labels in Γ. We call the latter parts the Γ-tails of trie TR.

Removing the Γ-tails from TR and the corresponding suffixes from the strings $P_1, \ldots, P_k$, we get a trie TR' build from strings $P_1', \ldots, P_k'$. The strings $P_1', \ldots, P_k'$ satisfy Conditions 1 and 2 since $|P_i'| \leq |P_i|$ and $\langle P_i'\rangle \leq \langle P_i\rangle$. Therefore, Lemmas 11 and 12 apply to TR', too, and there exist Γ' and ℓ' with the properties stated in Lemma 12.

Applying this procedure recursively until TR is empty, we get a partition $\Gamma, \Gamma', \Gamma'', \ldots$ of Σ_{TR} and the corresponding sequence of relabelings $\ell, \ell', \ell'', \ldots$. Combining the relabelings by restricting each ℓ to the corresponding Γ-tail, we have a relabeling of TR such that no character appears more than q times in the relabelled trie. Therefore, by (3),

$$pms(TR, T) \leq q \sum_{c \in \Sigma} ms(c, T) \leq qn \ .$$

Since $q = O(m / \log m)$, this completes the proof of the theorem.

The only thing left is to prove Lemma 12.

Proof (of Lemma 12). This proof is closely related to the maximum bipartite matching problem. Let G be a bipartite graph $(X \cup Y, F)$, where X contains a vertex x_e for each edge e in TR, and Y contains q vertices $y_c^1, y_c^2, \ldots, y_c^q$ for each character $c \in \Sigma_{TR}$. Vertex x_e is connected to the vertices $y_c^1, \ldots, y_c^q$ iff e can be relabelled with c. Note that, by (4), $|X| \leq |Y|$.

A matching M is a subset of F such that no vertex is an endpoint of more than one edge in M. Matching M is complete if it matches every $x \in X$. Given a matching M, let ℓ_M be the relabeling of trie TR with $\ell_M(e) = c$ if x_e is matched to one of the vertices $y_c^1, \ldots, y_c^q$, and $\ell_M(e) = \text{LABEL}[e]$ otherwise. If M is complete, ℓ_M satisfies the conditions of the lemma when we choose $\Gamma = \Sigma_{TR}$.

Given a matching M, an alternating path is a path from a vertex $x \in X$ to a vertex $y \in Y$, where every even-numbered edge is in M and every odd-numbered edge (including the first and last) are in $F \setminus M$. An alternating path is an augmenting path if neither x nor y is matched in M. The classic result on bipartite matching is that a matching M is maximal if and only if there are no augmenting paths (see e.g. [4, Sect. 5]).

Suppose a maximal matching M is not complete. Let Y' be the set of vertices $y \in Y$ such that there is no alternating path to y from a non-matched vertex $x \in X$. All non-matched $y \in Y$ must be in Y'. Y' is nonempty, because there are non-matched vertices in Y, since $|X| \le |Y|$.

Let X' be the set of $x \in X$ that are connected to some $y \in Y'$. By definition of Y', all $x \in X'$ are matched in M. Assume there is $x' \in X'$ that is matched to $y \in Y \setminus Y'$. By definition of X', x' is also connected to some $y' \in Y'$, and by definition of $Y \setminus Y'$, there is an alternating path from a non-matched vertex $x \in X \setminus X'$ to y. But then we could extend the alternating path from y to x' to y'. This is a contradiction. Therefore every $x \in X'$ is matched to $y \in Y'$.

Let G' be the graph G restricted to $X' \cup Y'$ and let M' be the matching M restricted to G'. By the above, M' is a complete matching for G'. For any $c \in \Sigma_{TR}$, $y_c^1, \ldots, y_c^q$ have the same connections in G and are, therefore, either all in Y' or all in $Y \setminus Y'$. Let Γ be the set of $c \in \Sigma_{TR}$ with y_c^i in Y'. Then X' represents the edges of TR that can be labelled with a character in Γ. Since all $x \in X'$ are matched in M', $\ell_{M'} \in L_\Gamma^{max}$, and no $c \in \Gamma$ appears more than q times in the trie TR relabelled with $\ell_{M'}$. $\qquad \Box$

Efficient Algorithms for Approximate String Matching with Swaps

(Extended Abstract)

Jee-Soo Lee[†][‡] Dong Kyue Kim[‡] Kunsoo Park[‡*] Yookun Cho[‡]

†Department of Computer Science, Korea National Open University
Seoul 110-791, Korea
‡Department of Computer Engineering, Seoul National University
Seoul 151-742, Korea

Abstract. Most research on the edit distance problem and the k-differences problem considered the set of edit operations consisting of changes, deletions, and insertions. In this paper we include the *swap* operation that interchanges two adjacent characters into the set of allowable edit operations, and we present an $O(t \min(m, n))$-time algorithm for the extended edit distance problem, where t is the edit distance between the given strings, and an $O(kn)$-time algorithm for the extended k-differences problem. That is, we add swaps into the set of edit operations without increasing the time complexities of previous algorithms that consider only changes, deletions, and insertions for the edit distance and k-differences problems.

1 Introduction

Given two strings $A[1..m]$ and $B[1..n]$ over an alphabet Σ, the *edit distance* between A and B is the minimum number of *edit operations* needed to convert A into B. The edit distance problem is to find the edit distance between A and B. Most common edit operations are the following.

(1) *change*: replace one character of A by another single character of B.
(2) *deletion*: delete one character from A.
(3) *insertion*: insert one character into B.

These three edit operations are the ones commonly used in applications [2, 9, 13, 14], though only insertions and deletions are considered in some work [8]. A discrepancy between A and B that is corrected by an edit operation is called a *difference*.

The problem of string matching with *k-differences* (or the k-differences problem) is defined as follows: Given a pattern A of length m, a text B of length n, and an integer k, find all positions of B where A occurs with at most k differences.

---

* Email: kpark@theory.snu.ac.kr

Many algorithms have been developed for the edit distance problem and the k-differences problem [1, 3]. When the edit distance t between A and B is small, an $O(t\min(m,n))$-time algorithm due to Ukkonen [10] is the best one for the edit distance problem. When the given difference k is small, $O(kn)$-time algorithms due to Landau and Vishkin [6], Galil and Park [4], and Ukkonen and Wood [11] are best for the k-differences problem.

In this paper we consider an additional edit operation:

(4) *swap*: interchange two adjacent characters in A.

The swap operation was first considered in [7, 12] and it is a special case of a *reversal* which is one of common genome rearrangements [5]. Lowrance and Wagner [7] proposed an $O(mn)$-time algorithm for the *extended* edit distance problem including the swap operation. For the k-differences problem the swap operation has never been considered. The k-differences problem including the swap operation will be called the *extended* k-differences problem.

Ukkonen [10] considered *transpositions* in the edit distance problem. Galil and Park [4] also considered transpositions in the k-differences problem. A transposition is a correction of a difference that two adjacent characters in A corresponds to two adjacent characters in B. However, the swap operation is more general than transpositions because deletions may occur before a swap and insertions may occur after a swap, but a transposition must not accompany deletions or insertions.

Example 1: Let $A = $ abcdeefg and $B = $ ahceegif. The edit distance between A and B is four, as shown in Fig. 1 (a). The character b in A is changed to h in B, d in A is deleted, f and g in A are swapped to g and f in B, and i in B is inserted. However, if only transpositions (but not swaps) are allowed, a transposition cannot be applied to fg in A because g and f in B are not adjacent. Hence the edit distance is five, as shown in Fig. 1 (b). □

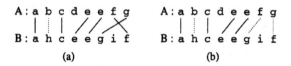

Fig. 1. The edit distances between A and B.

We present efficient algorithms for the extended edit distance problem and the extended k-differences problem. To compute the edit distance t between A and B, our algorithm takes $O(t\min(m,n))$ time, which is the same as that of Ukkonen [10]. Our algorithm for the extended k-differences problem also takes $O(kn)$ time as in [4, 6, 11]. That is, we add swaps into the set of edit operations without increasing the time complexities of [4, 6, 10, 11] that consider only changes, insertions, and deletions for the edit distance and k-differences problems.

The paper is organized as follows. In the next section we describe the tables to compute the edit distance between two strings, i.e., the D-table, the C_D-table, and the H-table. In Section 3 we compute the C_H-table for the extended edit distance problem, and in Section 4 we present an efficient algorithm for the extended k-differences problem.

2 Preliminaries

In this section we describe well-known algorithms for computing edit distances.

2.1 D-table and C_D-table

We will describe the D-table and the C_D-table for the edit distance problem between two strings A and B when the set of edit operations consists of change, deletion, and insertion.

Wagner and Fischer [13] devised an algorithm that takes $O(mn)$ time to compute the D-table. Let $D(i, j)$, $0 \leq i \leq m$ and $0 \leq j \leq n$, be the edit distance between $A[1..i]$ and $B[1..j]$. An entry $D(i, j)$ of the D-table is determined by the three entries $D(i-1, j-1)$, $D(i-1, j)$, and $D(i, j-1)$. The recurrence for the D-table is as follows: For all $1 \leq i \leq m$ and $1 \leq j \leq n$,

$$D(i, j) = \min\{D(i-1, j-1) + \delta_{ij}, D(i-1, j) + 1, D(i, j-1) + 1\}$$

where $\delta_{ij} = 0$ if $A[i] = B[j]$; $\delta_{ij} = 1$ otherwise.

Ukkonen [10] proved that $D(i, j) = D(i-1, j-1)$ or $D(i, j) = D(i-1, j-1) + 1$ in the D-table and proposed the C_D-table which provided a more compact way to store the information of the D-table.

Let D-diagonal d be the entries $D(i, j)$ such that $d = j - i$. For each D-diagonal we store only the positions where the value increase. For a D-diagonal d and a difference e, the entry $C_D(e, d)$ of the C_D-table is the largest column j such that $D(j - d, j) = e$. In other words, the entries of value e on D-diagonal d end at column $C_D(e, d)$. Note that $C_D(e, d) - d$ is the row of the last entry on D-diagonal d whose value is e.

Algorithm *Make-C_D* $(A[1..m], B[1..n])$
Initialize the C_D-table and set $A[m+1] = \#_a$, $B[n+1] = \#_b$, $e \leftarrow 0$
while $C_D(e, m-n) < n$ do
    for $d \leftarrow -e$ to e do
        $c \leftarrow \max\{C_D(e-1, d-1) + 1, C_D(e-1, d) + 1, C_D(e-1, d+1)\}$
        $c' \leftarrow \min\{c, m+d, n\}$
        while $A[c'+1-d] = B[c'+1]$ do $c' \leftarrow c'+1$ od
        $C_D(e, d) \leftarrow c'$
    od
    $e \leftarrow e+1$
od

Fig. 2. Algorithm *Make-C_D* for constructing the C_D-table.

The C_D-table is computed by the algorithm *Make-C_D* in Fig. 2, which is essentially Ukkonen's algorithm [4, 10, 11]. We add special characters $\#_a$ and $\#_b$ (not in Σ) at the end of A and B, respectively, to simplify codes.

2.2 H-table

In this subsection we describe the H-table for the extended edit distance problem [7].

Wagner [12] showed that the extended edit distance problem is in general NP-complete and most of the restricted cases can be solved in polynomial time. Lowrance and Wagner [7] considered a restricted case of the problem, i.e., the cost of two swaps is at least as large as the sum of costs of an insertion and a deletion, and computed the edit distance in $O(mn)$ time by constructing the H-table. When we count the number of edit operations, the restriction above holds, and thus we can use the H-table. Let $H(i, j)$, $0 \leq i \leq m$ and $0 \leq j \leq n$, be the edit distance between $A[1..i]$ and $B[1..j]$ when swaps are added to the set of edit operations.

For $1 \leq i \leq m$ and $1 \leq j \leq n$, let p_{ij} be the largest position less than i such that $A[p_{ij}] = B[j]$ and q_{ij} be the largest position less than j such that $A[i] = B[q_{ij}]$. Such positions p_{ij} and q_{ij} are called the *last-positions* for $H(i, j)$.

To compute an entry $H(i, j)$ of the H-table, we need two last-positions p_{ij} and q_{ij} [7]. When we apply a swap at $H(i, j)$, we should do the following:

(1) delete the characters from $p_{ij} + 1$ to $i - 1$ in A,
(2) swap $A[p_{ij}] (= B[j])$ and $A[i] (= B[q_{ij}])$, and
(3) insert the characters from $q_{ij} + 1$ to $j - 1$ in B.

Let *d-length* α_{ij} (resp. *i-length* β_{ij}) be the number of deleted (resp. inserted) characters to apply a swap at $H(i, j)$, i.e., $\alpha_{ij} = i - p_{ij} - 1$ and $\beta_{ij} = j - q_{ij} - 1$. Then the *swap-cost* $s(i, j)$ is as follows:

$$s(i, j) = H(p_{ij} - 1, q_{ij} - 1) + \alpha_{ij} + \beta_{ij} + 1.$$

The recurrence used in the H-table is

$$H(i, j) = \min\{H(i - 1, j - 1) + \delta_{ij}, H(i - 1, j) + 1, H(i, j - 1) + 1, s(i, j)\}.$$

If $s(i, j)$ is less than $H(i - 1, j - 1) + \delta_{ij}$, $H(i - 1, j) + 1$, and $H(i, j - 1) + 1$, then we say that a swap *occurs* at $H(i, j)$.

3 The Extended Edit Distance Problem

In this section we present a more efficient algorithm for the extended edit distance problem.

3.1 Properties of the H-table

We present some properties of the H-table when every edit operation has a unit cost. Lemma 1 shows that we need not consider any swaps that accompany both deletions and insertions during the computation of the H-table.

Lemma 1. *Let α_{ij} be the d-length and β_{ij} be the i-length for $H(i,j)$. If $\alpha_{ij} > 0$ and $\beta_{ij} > 0$ then $s(i,j) \geq H(i-1,j-1)+1$.*

By Lemma 1 there are three types of swaps occurring at $H(i,j)$:

(1) *transposition*: the case when $p_{ij} = i-1$ and $q_{ij} = j-1$. $A[i-1]$ and $A[i]$ are swapped to $B[j-1]$ and $B[j]$.

(2) *d-swap*: the case when $p_{ij} < i-1$ and $q_{ij} = j-1$. After the characters $A[p_{ij}+1..i-1]$ are deleted, $A[p_{ij}]$ and $A[i]$ are swapped to $B[j-1]$ and $B[j]$.

(3) *i-swap*: the case when $p_{ij} = i-1$ and $q_{ij} < j-1$. $A[i-1]$ and $A[i]$ are swapped to $B[q_{ij}]$ and $B[j]$ and then the characters $B[q_{ij}+1..j-1]$ are inserted between them.

Example 2: Fig. 3 shows three types of swaps. The characters c and e in A are swapped to e and c in B.   □

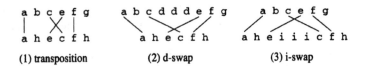

(1) transposition   (2) d-swap   (3) i-swap

Fig. 3. Three types of swaps.

Lemma 2. $H(i,j) = H(i-1,j-1)$ *or* $H(i,j) = H(i-1,j-1)+1$ *for every* $i \geq 1$ *and* $j \geq 1$.

Lemma 2 implies that we can make the C_H-table for a H-table as we constructed the C_D-table for a D-table.

3.2 Construction of the C_H-table

We will construct the C_H-table which is an extension of the C_D-table. The C_H-table is to the H-table what the C_D-table is to the D-table. Let H-diagonal d be the entries $H(i,j)$ such that $d = j-i$. For a H-diagonal d and a difference e, the entry $C_H(e,d)$ of the C_H-table is the largest column j such that $H(j-d,j) = e$.

We define two kinds of diagonals in the C_H-table. Let A-*diagonal* x be the entries $C_H(e,d)$ such that $e+d = x$ and B-*diagonal* y be the entries $C_H(e,d)$ such that $e-d = y$. See Fig 5. Since the value of an entry in the C_H-table is a position of B, if $C_H(e,d) = j$ then we say that the position j of B appears on B-*diagonal* $e-d$. If $i = C_H(e,d)-d$ then we say that the position i of A appears on A-*diagonal* $e+d$.

When we compute an entry of the C_H-table, the main difficulty is to find correct last-positions since all positions of A and B may not appear in the C_H-table. Galil and Park [4] proposed a property of A-diagonals in Lemma 3. Based on Lemma 3, we will maintain some information of last-positions for each A-diagonal and B-diagonal.

Lemma 3. *The positions of A (resp. B) that appear on the same A-diagonal (resp. B-diagonal) are nondecreasing in the C_H-table.*

We describe how to compute the C_H-table. Consider the computation of an entry $C_H(e, d)$. Assume by induction that $C_H(e - 1, d - 1)$, $C_H(e - 1, d)$, and $C_H(e - 1, d + 1)$ were computed correctly. This means that in the H-table the entries of value $e - 1$ reach column $C_H(e - 1, d - 1)$ on H-diagonal $d - 1$, $C_H(e - 1, d)$ on H-diagonal d, and $C_H(e - 1, d - 1)$ on H-diagonal $d - 1$. Let $c_i = C_H(e-1, d-1)+1$, $c_t = C_H(e-1, d)+1$, and $c_d = C_H(e-1, d+1)$, and let c be the maximum of c_i, c_t, and c_d. $H(c - d, c)$ gets value e from one of the last entries of value $e-1$ on H-diagonal $d-1$, d, and $d+1$ by one of insertion, change, and deletion, respectively. The entries of value e on H-diagonal d continue to the smallest column $c' \geq c$ such that $A[c' - d + 1] \neq B[c' + 1]$, and thus $C_H(e, d) = c'$.

Now we consider the swap operation. If $c' = c$ (i.e., $A[c - d + 1] \neq B[c + 1]$) then $H(c - d + 1, c + 1)$ would get value $e + 1$ without swaps. However, if a swap occurs at $H(c - d + 1, c + 1)$, the value of $H(c - d + 1, c + 1)$ can be still e. In this case, the entries of value e on H-diagonal d continue to the smallest column $c'' > c'$ such that $A[c'' - d + 1] \neq B[c'' + 1]$, and thus $C_H(e, d) = c''$. Hence, in the computation of $C_H(e, d)$ we must check whether any swaps occur at $H(c - d + 1, c + 1)$ or not. Let $u = c - d + 1$ and $v = c + 1$. The positions u and v are called the *swap-positions* to be considered for the computation of $C_H(e, d)$. There are three types of swaps occurring at $H(u, v)$:

(1) The case when $c = c_t$: Characters $A[u-1]$ and $B[v-1]$ have been changed to get $H(u-1, v-1)$. Hence we need to consider an occurrence of a *transposition* at $H(u, v)$.
(2) The case when $c = c_d$: $A[u - 1]$ has been deleted to get $H(u - 1, v - 1)$, and thus we need to consider an occurrence of a *d-swap* at $H(u, v)$.
(3) The case when $c = c_i$: $B[v - 1]$ has been inserted to get $H(u - 1, v - 1)$, and thus we need to consider an occurrence of an *i-swap* at $H(u, v)$.

To determine $C_H(e, d)$, we need to compute the swap-cost $s(u, v)$ for the three cases above. By Lemma 4, however, we need only compute swap-cost $s(u, v)$ when $c = c_t$.

Lemma 4. *Let $c_t = C_H(e-1, d)+1$, $c_d = C_H(e-1, d+1)$, $c_i = C_H(e-1, d-1)+1$ and $c = \max\{c_t, c_d, c_i\}$. Let u and v be the swap-positions for $C_H(e, d)$.*

(1) If a d-swap occurs at $H(u, v)$ then $c = c_d = c_t$.
(2) If an i-swap occurs at $H(u, v)$ then $c = c_i = c_t$.

We can easily find a transposition occurring at $H(u, v)$. Since a transposition must come from H-diagonal d, it occurs at $H(u, v)$ if $A[u] = B[v-1]$, $A[u-1] = B[v]$, and $c = c_t$ [4, 10]. In the rest of this section we will describe the cases of d-swaps and i-swaps in the computation of $C_H(e, d)$.

Definition 5. A position j (resp. $j - d$) appearing on B-diagonal $e - d$ (resp. A-diagonal $e + d$) (i.e., $C_H(e, d) = j$) is change-dominated if $j = C_H(e-1, d)+1$ and $j > \max\{C_H(e-1, d-1)+1, C_H(e-1, d+1)\}$ in the C_H-table.

Recall that $c_t = C_H(e-1, d) + 1$, $c_i = C_H(e-1, d-1) + 1$, and $c_d = C_H(e-1, d+1)$ in the computation of $C_H(e, d)$. If positions $j - d$ and j are change-dominated on A-diagonal $e + d$ and B-diagonal $e - d$, entry $H(j - d, j)$ gets value e from the last entry of value $e - 1$ on H-diagonal d by a change, and cannot get value e by any of a deletion, an insertion, a swap, and a match.

Definition 6.

(1) A position p in A is d-effective on A-diagonal $e + d$ if every position from p to $u - 2$ in A appears on A-diagonal $e + d$.

(2) A position q in B is i-effective on B-diagonal $e - d$ if every position from q to $v - 2$ in B appears on B-diagonal $e - d$.

Lemma 7.

(1) A position p in A is d-effective on A-diagonal $e + d$ if and only if $C_H(e - \alpha, d + \alpha) = c$, where $\alpha = u - 1 - p$.

(2) A position q in B is i-effective on B-diagonal $e - d$ if and only if $C_H(e - \beta, d + \beta) = q$, where $\beta = v - 1 - q$.

We describe how to find last-positions p_{uv} and q_{uv} for $H(u, v)$ in the C_H-table.

Lemma 8. *Consider a position p in A and a position q in B.*

(1) If $A[p] = B[v]$ and p is d-effective on A-diagonal $e + d$ then p is equal to the last-position p_{uv} in A.

(2) If $A[u] = B[q]$ and q is i-effective on B-diagonal $e - d$ then q is equal to the last-position q_{uv} in B.

If we know last-positions p_{uv} or q_{uv}, we determine whether we can apply a d-swap or an i-swap in the computation of $C_H(e, d)$ by Lemma 9.

Lemma 9. *Suppose that $c = c_t$ and $A[u] \neq B[v]$.*

(1) A d-swap occurs at $H(u, v)$ if and only if $A[u] = B[v - 1]$ and p_{uv} is d-effective and change-dominated on A-diagonal $e + d$, where p_{uv} is the last-position in A.

(2) An i-swap occurs at $H(u, v)$ if and only if $A[u - 1] = B[v]$ and q_{uv} is i-effective and change-dominated on B-diagonal $e - d$, where q_{uv} is the last-position in B.

Algorithm *Make-C_H* $(A[1..m+1], B[1..n+1])$
1 Initialize the C_H-table.
2 Clear arrays LA and LB, and set $e \leftarrow 0$
3 **while** $C_H(e, m-n) < n$ **do**
4   **for** $d \leftarrow -e$ **to** e **do**
5          $c_t \leftarrow C_H(e-1, d) + 1, c_d \leftarrow C_H(e-1, d+1), c_i \leftarrow C_H(e-1, d-1) + 1$
6          $c \leftarrow \max\{c_t, c_d, c_i\}$
7   **if** $c = c_t$ **then** // consider swap operations
8              $u \leftarrow c - d + 1, v \leftarrow c + 1$
9              $p \leftarrow LA[e+d], q \leftarrow LB[e-d]$
10             $\alpha \leftarrow \min\{e, u-1-p\}, \beta \leftarrow \min\{e, v-1-q\}$
11   **if** ($A[u] = B[v-1]$ and $A[u-1] = B[v]$) or
12   ($A[p] = B[v]$ and $C_H(e-\alpha, d+\alpha) = c$ and $A[u] = B[v-1]$) or
13   ($A[u] = B[q]$ and $C_H(e-\beta, d+\beta) = q$ and $A[u-1] = B[v]$)
14   **then** $c \leftarrow c + 1$ **fi**
15   **fi**
16          $c' \leftarrow \min\{c, m+d, n\}$
17   **while** $A[c'+1-d] = B[c'+1]$ **do** $c' \leftarrow c' + 1$ **od**
18          $C_H(e, d) \leftarrow c'$
19   **if** $c' = c_t$ and $c' > \max\{c_d, c_i\}$ **then** // update arrays LA and LB
20             $LA[e+d] \leftarrow c' - d$
21             $LB[e-d] \leftarrow c'$
22   **fi**
23   **od**
24      $e \leftarrow e + 1$
25 **od**

Fig. 4. Algorithm *Make-C_H* for constructing the C_H-table.

In our algorithm we maintain two arrays LA and LB of size $(2t + 1)$ each, where t is the edit distance between A and B. Array LA will be used for applying d-swaps and array LB for i-swaps. In the computation of entry $C_H(e, d)$, we define arrays LA and LB as follows. An element $LA[x]$ is p if and only if p is the largest position less than $c_d - d = C_H(e-1, d+1) - d$ in A such that p is change-dominated on A-diagonal $x = e + d$. An element $LB[y]$ is q if and only if q is the largest position less than $c_i = C_H(e-1, d-1) + 1$ in B such that q is change-dominated on B-diagonal $y = e - d$. Since the position corresponding to $C_H(e-1, d+1)$ (resp. $C_H(e-1, d-1)$) is $c_d - d - 1$ of A (resp. $c_i - 1$ of B), the largest position that appears on A-diagonal $e + d$ (resp. B-diagonal $e - d$) must be less than $c_d - d$ (resp. c_i) in the computation of entry $C_H(e, d)$.

Let $LA[e+d] = p$ and $LB[e-d] = q$ in the computation of $C_H(e, d)$. We will show that a d-swap cannot occur at $H(u, v)$ when $p \neq p_{uv}$. Because we want to check an occurrence of a d-swap rather than a transposition, $p_{uv} \neq u - 1$. Suppose that $p < p_{uv} < u - 1$. If $c_d \neq c$, a d-swap cannot occur at $H(u, v)$. Hence we may regard $c_d - d$ as $u - 1$. By definition of array LA, p is the largest position less than $u - 1$ such that p is change-dominated on A-diagonal $e + d$. Since $p < p_{uv} < u - 1$, p_{uv} is not change-dominated on A-diagonal $e + d$. Hence a d-swap does not occur at $H(u, v)$ by Lemma 9.

Next, suppose that $p_{uv} < p$. Since p appears on A-diagonal $e + d$ by Definition 5, there must exist an integer $1 \leq j < e$ such that $C_H(e - j, d + j) = p + d + j$ in the C_H-table. Moreover, the value of the previous entry $C_H(e - j - 1, d + j + 1)$ on the same A-diagonal is less than $p + d + j$. Then the position of A corresponding to $C_H(e - j - 1, d + j + 1)$ is less than $p + d + j - (d + j + 1) = p - 1$. Hence $p - 1$ does not appear on A-diagonal $e + d$ by Lemma 3, and thus p_{uv} is not d-effective. Therefore, again by Lemma 9, a d-swap does not occur at $H(u, v)$.

We now show that if $A[p] \neq B[v]$ or p is not d-effective then a d-swap cannot occur at $H(u, v)$. If $A[p] \neq B[v]$ then $p \neq p_{uv}$ by definition of p_{uv}. Hence a d-swap does not occur at $H(u, v)$ when $A[p] \neq B[v]$. When p is not d-effective, p may or may not be p_{uv}. If $p \neq p_{uv}$, a d-swap does not occur at $H(u, v)$. If $p = p_{uv}$ then p_{uv} is not d-effective, and thus a d-swap does not occur at $H(u, v)$ by Lemma 9.

However, if $A[p] = B[v]$ and p is d-effective then we have $p = p_{uv}$ by Lemma 8. By this, p_{uv} is also d-effective. Since $LA[e + d] = p = p_{uv}$, p_{uv} is change-dominated. Hence, by Lemma 9, if $A[u] = B[v - 1]$ then we have found a d-swap occurring at $H(u, v)$. Similarly, if $A[u] = B[q]$ and q is i-effective then we have $q = q_{uv}$ by Lemma 8. If $A[u - 1] = B[v]$ then we have found an i-swap occurring at $H(u, v)$.

We present algorithm *Make-C_H* for constructing the C_H-table in Fig. 4. Algorithm *Make-C_H* works as algorithm *Make-C_D* does except the parts for swap operations. We initialize the C_H-table and arrays LA and LB. For each H-diagonal d and a difference e, we compute c. If $c = c_t$, then we determine whether a swap occurs or not by Lemma 4. We set $\alpha = \min\{e, u - 1 - p\}$ and $\beta = \min\{e, v - 1 - q\}$ because $u - 1 - p$ or $v - 1 - q$ may be larger than e, in which case no swaps occur at $H(u, v)$ because swap-cost $s(u, v) > e$. If one of a transposition (at line 11), a d-swap (at line 12), and an i-swap (at line 13) occurs, we increase c. Then we follow the procedure of *Make-C_D* and compute $C_H(e, d)$. After the computation of $C_H(e, d) = c'$, we update arrays LA and LB. If $c' = c_t$ and $c' > \max\{c_d, c_i\}$, then positions $c' - d$ and c' are change-dominated on A-diagonal $e + d$ and B-diagonal $e - d$, respectively. Hence we set $LA[e + d] = c' - d$ and $LB[e - d] = c'$ (at lines 20 and 21).

Example 3: Let $A =$ abcdddefg and $B =$ ahecfh. The edit distance between A and B is six, as shown in Fig. 3 (2). Fig. 5 shows the H-table and the C_H-table of A and B. Consider the computation of $C_H(5, -3)$. We have $c_t = C_H(4, -3) + 1 = 3$, $c_d = C_H(4, -2) = 3$, and $c_i = C_H(4, -4) + 1 = 2$. Since $c = \max\{c_t, c_d, c_i\} = 3$, $u = c - d + 1 = 7$ and $v = c + 1 = 4$.

- *The last-position $p_{7,4}$:* $C_H(5, -3)$ is on A-diagonal $e + d = 2$. Consider the computation of $C_H(2, 0)$. Because $C_H(2, 0) - 0 = C_H(1, 0) + 1 = 3$, and $\max\{C_H(1, -1) + 1 = 2, C_H(1, 1) = 2\} < 3$, position 3 of A is change-dominated on A-diagonal 2. Hence $LA[2]$ has position $p = 3$. Let $\alpha = u - 1 - p = 3$. Since $C_H(e - \alpha, d + \alpha) = C_H(2, 0) = 3 = c$, p is d-effective on A-diagonal 2 by Lemma 7. Because $A[3] = B[4] = c$ and p is d-effective, we can find last-position $p_{7,4} = p = 3$ by Lemma 8.

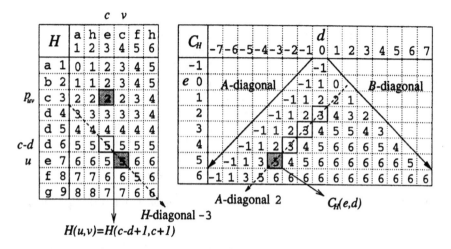

Fig. 5. The H-table and the C_H-table of strings **abcdddefg** and **ahecfh**.

- *An occurrence of a d-swap:* Since $LA[2] = p = p_{7,4} = 3$, $p_{7,4}$ is change-dominated and d-effective. Because $A[7] = B[3] = $ **e**, we have found a d-swap occurring at $H(7,4)$ by Lemma 9. Since swap-cost $s(7,4) = H(2,2) + (7 - 3 - 1) + (4 - 3 - 1) + 1 = 5$, we have $H(7,4) = s(7,4) = 5$.
- *The computation of $C_H(5,-3)$:* Since we found a d-swap, we increase c. Then $c' = \min\{c = 4, m + d = 6, n = 6\} = 4$. Because there is a continuing match at column 5 on H-diagonal -3 (i.e., $A[8] = B[5] = $ **f**), we increase c'. Since there is a mismatch at column 6 (i.e., $A[9] \neq B[6]$), the value of $C_H(5,-3)$ becomes 5.
- *Updating LA and LB:* After we compute $C_H(5,-3) = c' = 5$, we check the conditions to update LA and LB. Since $c' > c_t = c_d = 3$ in this case, we need not update elements $LA[2]$ and $LB[8]$.

$\square$

Theorem 10. *Algorithm Make-C_H solves the extended edit distance problem in $O(t \min(m,n))$ time, where t is the edit distance between A and B.*

4 The Extended k-differences Problem

In this section we present an algorithm for the *extended k-differences* problem. Landau and Vishkin [6], Galil and Park [4], and Ukkonen and Wood [11] proposed $O(kn)$-time algorithms for the k-differences problem when the set of edit operations consists of changes, deletions and insertions. Here we do not mention preprocessing because preprocessings are all different in [4, 6, 11] and preprocessing time is absorbed into $O(kn)$ when n is sufficiently larger than m.

We can apply the method used in Section 3 to any of the three algorithms. The algorithms can proceed A-diagonal by A-diagonal since the maximal difference k is given. There are $(n - m + 1 + k)$ A-diagonals in the C_H-table, but we

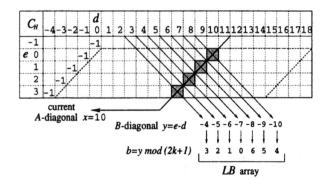

Fig. 6. The C_H-table and array LB for the extended k-differences problem.

need one variable LA only for all A-diagonals because the C_H-table is computed A-diagonal by A-diagonal.

We maintain array LB of size $(2k+1)$. When we compute an entry $C_H(e, x - e)$ on A-diagonal x, we need the element $LB[y]$ such that $y = 2e - x$. Since $0 \le e \le k$, we should keep the positions of B associated with the entries on B-diagonal y such that $-x \le y \le 2k - x$. Since $(2k + 1)$ B-diagonals are needed for each A-diagonal, let $b = y \bmod (2k+1)$. Then we can use $LB[b]$ as the array LB in the algorithm for the extended k-differences problem. After the computation of $C_H(k, x - k)$ on each A-diagonal x, we do not need $LB[2k - x]$ any more and need $LB[-(x + 1)]$. Let $b' = (2k - x) \bmod (2k + 1) = (-x - 1) \bmod (2k + 1)$. Hence we reset variable LA and element $LB[b']$ in order to use for the next A-diagonal $x + 1$ and B-diagonal $-x - 1$.

Example 4: Fig. 6 shows the C_H-table when $k = 3$ and there are 18 A-diagonals. The dotted parallelogram is the region where the entries of the C_H-table exist. Consider the computation of the entries on A-diagonal $x = 10$. Since $-10 = -x \le y \le 2k - x = -4$, we maintain array LB for B-diagonals in the range of $[-10, -4]$. After the computation of $C_H(3, 7)$, we reset LA and $LB[3]$. □

Theorem 11. *We can solve the extended k-differences problem in $O(kn)$ time, not including preprocessing.*

References

1. R. Baeza-Yates and G. Navarro, A faster algorithm for approximate string matching, In *Proc. of the 7th Symp. on Combinatorial Pattern Matching, Springer-Verlag Lecture Note in Comp. Sci.* 1075 (1996), 1–23.
2. D. Eppstein, Z. Galil, R. Giancarlo, and G. Italiano, Sparse dynamic programming I: linear cost functions, *J. Assoc. Comput. Mach.* 39 (1992), 519–545.
3. Z. Galil and R. Giancarlo, Data structures and algorithms for approximate string matching, *J. Complexity* 4 (1988), 33–72.

4. Z. Galil and K. Park, An improved algorithm for approximate string matching, *SIAM J. Comput.* 19 (1990), 989–999.

5. S. Hannenhalli and P.A. Pevzner, Transforming men into mice (polynomial algorithm for genomic distance problem), *IEEE Symp. Found. Computer Science* (1995), 581–592.

6. G.M. Landau and U. Vishkin, Fast parallel and serial approximate string matching, *J. Algorithms* 10 (1989), 157–169.

7. R. Lowrance and R.A. Wagner, An extension of the string-to-string correction problem, *J. Assoc. Comput. Mach.* 22 (1975), 177–183.

8. E.W. Myers, An $O(ND)$ difference algorithm and its variations, *Algorithmica* 1 (1986), 251–266.

9. D. Sankoff and J.B. Kruskal, *Time Warps, String Edits, and Macromolecules: The Theory and Practice of Sequence Comparison*, New York, Addison-Wesley (1983).

10. E. Ukkonen, Algorithms for approximate string matching, *Inform. and Control* 64 (1985), 100–118.

11. E. Ukkonen and D. Wood, Approximate string matching with suffix automata, *Algorithmica* 10 (1993), 353–364.

12. R.A. Wagner, On the complexity of the extended string-to-string correction problem, *ACM Symp. Theory of Computing* (1975), 218–223.

13. R.A. Wagner and M.J. Fischer, The string-to-string correction problem, *J. Assoc. Comput. Mach.* 21 (1974), 168–173.

14. S. Wu and U. Manber, Fast text searching allowing errors, *Comm. ACM* 35 (1992), 83–91.

On the Complexity of Pattern Matching for Highly Compressed Two-Dimensional Texts

Piotr Berman [*1] and Marek Karpinski [**2] and Lawrence L. Larmore [***3] and Wojciech Plandowski [†4] and Wojciech Rytter [‡45]

[1] Dept. of Computer Science & Eng., Pensylvania State University, University Park, PA16802, USA
[2] Dept. of Computer Science, University of Bonn, D-53117, Bonn, Germany.
[3] Department of Computer Science, University of Nevada, Las Vegas, NV 89154-4019.
[4] Instytut Informatyki, Uniwersytet Warszawski, Banacha 2, 02–097 Warsaw.
[5] Department of Computer Science, University of Liverpool, U.K.

Abstract

We consider the complexity of problems related to 2-dimensional texts (2d-texts) described succinctly. In a succinct description, larger rectangular sub-texts are defined in terms of smaller parts in a way similar to that of Lempel-Ziv compression for 1-dimensional texts, or in shortly described strings as in [9], or in hierarchical graphs described by context-free graph grammars. A given 2d-text T with many internal repetitions can have a **hierarchical description** (denoted $Compress(T)$) which is up to exponentially smaller and which can be the only part of the input for a pattern-matching algorithm which gives information about T. Such a hierarchical description is given in terms of a straight-line program, see [9] or, equivalently, a 2-dimensional grammar.

We consider **compressed pattern-matching**, where the input consists of a 2d-pattern P and of a hierarchical description of a 2d-text T, and **fully compressed pattern-matching**, where the input consists of hierarchical descriptions of both the pattern P and the text T. For 1-dimensional strings there exist polynomial-time deterministic algorithms for these problems, for similar types of succinct text descriptions [2, 6, 8, 9]. We show that the complexity dramatically increases in a 2-dimensional setting. For example, compressed 2d-matching is $\mathcal{NP}$-complete, fully compressed 2d-matching is Σ_2^P-complete, and testing a given occurrence of a two dimensional compressed pattern is co-$\mathcal{NP}$-complete.

On the other hand, we give efficient algorithms for the related problems of randomized equality testing and testing for a given occurrence of an uncompressed pattern.

We also show the surprising fact that the compressed size of a subrectangle of a compressed 2d-text can grow exponentially, unlike the one dimensional case.

---

 * Email:berman@cse.psu.edu

 ** This research was partially supported by the DFG Grant KA 673/4-1 Email:marek@cs.uni-bonn.de.

 *** Research partially supported by National Science Foundation grant CCR-9503441. Part of this work was done while the author was visiting Institut Informatik V, Universität Bonn, Germany. Email:larmore@cs.unlv.edu.

 † Supported by the grant KBN 8T11C01208. Email:wojtekpl@mimuw.edu.pl.

 ‡ Supported by the grant KBN 8T11C01208. Email: rytter@mimuw.edu.pl.

1   Introduction

We consider algorithms for problems dealing with *highly compressed* 2d-texts, *i.e.*, two dimensional arrays with entries from some finite alphabet. A 2d-text T is represented hierarchically in a succinct way, denoted $Compress(T)$. Texts are as much as exponentially compressed. Our main problem is the **Fully Compressed Matching Problem**:

   Instance: $Compress(P)$ and $Compress(T)$.
   Question: does P occur in T?

where P and T are rectangular 2d-texts. The **Compressed Matching Problem** is essentially the same, the only difference being that $P = Compress(P)$, in other words, the pattern is uncompressed. Our results show that an attempt to deal with exponentially compressed 2d-texts should fail algorithmically. The size of the problem is $n+m$, where $n = |Compress(T)|$ and $m = |Compress(P)|$. Let N be the total uncompressed size of the problem. Note that in general N can be exponential with respect to n, and any algorithm which decompresses T takes exponential time in the worst case.

We also consider the problems of **Pattern Checking**, that is, testing an occurrence of a pattern at a given position. This problem has also its *compressed* and *fully compressed* versions. The hierarchical description of a 2d-text is in terms of *straight-line programs* (SLP's for short), or equivalently, two dimensional context-free grammars generating single objects with the following two operations:

   $A \leftarrow BC$, which concatenates 2d-texts B and C (both of equal height)
   $A \leftarrow B \ominus C$, which puts the 2d-text B on top of C (both of equal width)

An SLP of size n consists of n statements of the above form, where the result of the last statement is the compressed 2d-text. The only constants in our SLP's are symbols of an alphabet, interpreted as 1×1 images. We view SLP's as compressed (descriptions of) images.

The complexity of basic string problems for one dimensional texts is polynomial, see [4, 6, 8, 10]. Surprisingly, the complexity jumps if we pass to two dimensions. The compressed size of a subrectangle of a compressed 2d-text A can be exponential with respect to the compressed size of A, though such a situation cannot occur in the 1-dimensional case. This phenomenon appears to be responsible for the increase in the time complexity.

Theorem 1. *For each n there exists an SLP of size n describing a text image A_n and a subrectangle B_n of A_n such that the smallest SLP describing B_n has exponential size.*

Proof. The proof is omitted in this version.   □

Example. Hilbert's curve can be viewed as an image which exponentially compressible in terms of SLP's. An SLP which describes the nth Hilbert's curve, H_n, uses six (terminal) symbols ⌐ , ⌐ , ⌐ , ⌐ , ⊓ , ⊟ , and 12 variables ⬚$_i$, ⬚$_i$, ⬚$_i$, ⬚$_i$, ⬚$_i$, ⬚$_i$, ⬚$_i$, ⬚$_i$, ⬚$_i$, ⬚$_i$, ⬚$_i$, ⬚$_i$, for each $0 \le i \le n$. A variable with index i represents a text square of size $2^i \times 2^i$ containing part of a curve. The dots in the boxes show the places where the curve enters and leaves the box.

The 2d-text $T = $ ⬚$_3$ describing the 3rd Hilbert's curve is shown in Figure 1. It is composed of four smaller square 2d-texts ⬚$_2$, ⬚$_2$, ⬚$_2$, ⬚$_2$ according to the composition rule in Figure 1. T consists of 64 (terminal) symbols.

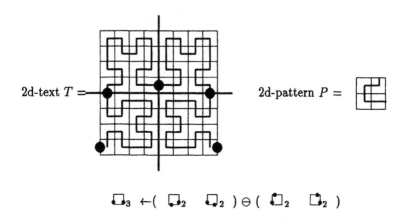

2d-text $T = $ 2d-pattern $P = $

⬚$_3$ ←(⬚$_2$ ⬚$_2$) ⊖ (⬚$_2$ ⬚$_2$)

Fig. 1. An example of a 2d-text T and a pattern P. The pattern occurs twice in T. The black dots are not part of T.

The 1×1 text squares are described as follows.

$$\text{⬚}_0 \leftarrow \text{⊓}, \quad \text{⬚}_0 \leftarrow \text{⊓}, \quad \text{⬚}_0 \leftarrow \text{⊟}, \quad \text{⬚}_0 \leftarrow \text{⊟},$$
$$\text{⬚}_0 \leftarrow \text{⌐}, \quad \text{⬚}_0 \leftarrow \text{⌐}, \quad \text{⬚}_0 \leftarrow \text{⌐}, \quad \text{⬚}_0 \leftarrow \text{⌐},$$
$$\text{⬚}_0 \leftarrow \text{⌐}, \quad \text{⬚}_0 \leftarrow \text{⌐}, \quad \text{⬚}_0 \leftarrow \text{⌐}, \quad \text{⬚}_0 \leftarrow \text{⌐},$$

The text squares for variables indexed by $i \ge 1$ are rotations of text squares for the variables ⬚$_i$, ⬚$_i$, ⬚$_i$. These variables are composed according to the rules:

$$\text{⬚}_i \leftarrow \text{⬚}_{i-1}\ \text{⬚}_{i-1} \ominus \text{⬚}_{i-1}\ \text{⬚}_{i-1},$$
$$\text{⬚}_i \leftarrow \text{⬚}_{i-1}\ \text{⬚}_{i-1} \ominus \text{⬚}_{i-1}\ \text{⬚}_{i-1},$$
$$\text{⬚}_i \leftarrow \text{⬚}_{i-1}\ \text{⬚}_{i-1} \ominus \text{⬚}_{i-1}\ \text{⬚}_{i-1}.$$

2   Equality testing in randomized polynomial time

We reduce equality of two 2d-texts $\mathcal{A}$ and $\mathcal{B}$ to equality of two polynomials $Poly_A(x,y)$ and $Poly_B(x,y)$. Assume that the symbols are integers in some small range. For an $n \times n$ 2d-text $\mathcal{Z}$ define its corresponding polynomial

$$Poly_{\mathcal{Z}}(x,y) = \sum_{i,j=1}^{n} \mathcal{Z}_{i,j} x^i y^j.$$

Observation.
Let $\mathcal{A}$ and $\mathcal{B}$ are two 2d-texts. Then $\mathcal{A} = \mathcal{B} \Leftrightarrow Poly_A \equiv Poly_B$.

Fact 2 *Let $\mathcal{A}, \mathcal{B}, \mathcal{C}$ be 2d-texts corresponding to variables A, B, C in some SLP.*

1. *If $A \leftarrow B \ominus C$ then $Poly_A(x,y) = Poly_C(x,y) \cdot x^{height(\mathcal{B})} + Poly_B(x,y)$.*
2. *If $A \leftarrow BC$ then $Poly_A(x,y) = Poly_C(x,y) \cdot y^{width(\mathcal{B})} + Poly_B(x,y)$.*
3. *for given values (x_0, y_0) of arguments the value $Poly_A(x_0, y_0)$ mod k can be computed in time polynomial w.r.t. the compressed size of $\mathcal{A}$ and the number of bits of k.*
4. *degree($Poly_A$) = $height(\mathcal{A}) \cdot width(\mathcal{A})$.*

The following result is a version of theorems given by Schwartz and by Zippel [13].

Lemma 3.
Let $\mathcal{P}$ be a nonzero polynomial of degree at most d. Assume that we assign to each variable in $\mathcal{P}$ a random value from a set Ω of integers of cardinality R. Then

$$Prob\{\mathcal{P}(\bar{x}) \neq 0 \} \geq 1 - \frac{d}{R}.$$

Theorem 4. There exists a polynomial time randomized algorithm for testing equality of two 2d-texts, given their hierarchical compressed representation.

Proof. Let n be the total size of compressed description of $\mathcal{A}$, $\mathcal{B}$. Denote
$$deg = \max\{degree(\mathcal{A}), \ degree(\mathcal{B})\}.$$
The value of deg corresponds to maximum size of 2d-texts and we have $deg \leq c^n$, for a constant c.

We can test equality of $\mathcal{A}$ and $\mathcal{B}$ in a randomized way, due to Lemma 3, by selecting random values x_0, y_0 of variables in the range $[1 \ldots 2 \cdot deg]$ and testing $y_1 = y_2$, where $y_1 = Poly_A(x_0, y_0)$ and $y_2 = Poly_B(x_0, y_0)$.

However there is one technical difficulty, the numbers y_1, y_2 are exponential w.r.t. deg and can need exponential number of bits, then, obviously, we are not able to compute them in polynomial time. Hence instead of computing the exact values of y_1, y_2 we choose a random prime number p from a suitable interval and compute values y_1, y_2 *modulo* p. We refer the reader to Theorem 7.5 and the discussion in Example 7.1 in [13], for details about randomized testing of the equality of two number using prime numbers and modular arithmetic with exponentially smaller number of bits than the numbers to be tested.

If the computed values $y_1 \bmod p$ and $y_2 \bmod p$ are different, then the polynomials are different. Otherwise, by Lemma 3, the polynomials are identical with high probability. The computation of $y_1 \bmod p$ and $y_2 \bmod p$, where p is a prime number with polynomially many bits, can be done in polynomial time w.r.t. n. This completes the proof. □

Open Problem: We designed a fast **randomized** algorithm for the equality of two compressed 2d-texts, and we conjecture that there is a polynomial time **deterministic** algorithm.

3 Compressed two dimensional pattern-matching

Recall that the compressed matching problem is to find, given an uncompressed pattern and compressed text, whether the pattern occurs within the text.

In our constructions we will use, as a building block, rectangles filled with one kind of letter only, say a. We will use $[a]_j^i$ to denote such an $i \times j$ 2d-text. It is easy to see that $[a]_j^i$ can be compressed to an SLP of size $O(\log(i) + \log(j))$. We will use $I, J, \ldots, P, Q, \ldots$ for uncompressed 2d-texts, and $\mathcal{I}, \mathcal{J}, \ldots, \mathcal{P}, \mathcal{Q}, \ldots$ for compressed ones. Given a compressed 2d-text $\mathcal{R}$ (uncompressed 2d-text R), we use $\mathcal{R}_{i,j}$ ($R_{i,j}$) to denote the symbol at position (i, j); if the position (i, j) is out of range, we will have $\mathcal{R}_{i,j} = \bot$. We will number the rows and columns starting from 0. We also use the convention that given a number m, $\tilde{m}$ is a 0-1 vector $(a_0, \ldots, a_{k-1})$ such that $m = \sum_{i=0}^{k-1} 2^i a_i$. The length of $\tilde{m}$ should be clear from context. Let $Positions(P) = \{(i, j) : P_{i,j} \neq \bot\}$ and $Positions(\mathcal{P}) = \{(i, j) : \mathcal{P}_{i,j} \neq \bot\}$.

First we consider the **Point test problem**: compute the symbol $\mathcal{I}_{i,j}$ for given $\mathcal{I}$, i and j.

Lemma 5. There exists an $O(n|P|)$ time algorithm for the point test problem, where n is the size of $\mathcal{T}$.

Theorem 6. *The compressed matching problem for two dimensional 2d-texts is $\mathcal{NP}$-complete.*

Proof. To see that compressed matching is in $\mathcal{NP}$, we express this problem as the following property of pattern P and 2d-text $\mathcal{R}$:
$$\exists(i, j)\{\forall(k, l) \in Positions(P)\ P_{k,l} = \mathcal{R}_{i+k,j+l}\}.$$
The equality inside the braces can be tested in polynomial time (Lemma 5), hence we have expressed the problem in the normal form for $\mathcal{NP}$.

To show $\mathcal{NP}$ hardness, we will use a reduction from 3SAT. Consider a set of clauses $C_0, \ldots, C_{k-1}$, where each clause is a Boolean function of some three variables from the set $\{x_0, \ldots, x_{n-1}\}$. The 3SAT question is whether there exists m such that $0 \leq m \leq 2^n - 1$ and $C_i(\tilde{m}) = 1$ for $i = 0, \ldots, k - 1$.

Define a $k \times 2^n$ 2d-text A by: $A_{i,m} = C_i(\tilde{m})$. Then the 3SAT question is equivalent to the following: does A contain a column consisting of k 1's (i.e. the pattern $[1]^1_m$)? We will reduce 3SAT to the compressed matching problem by showing how to compress A to an SLP with $O(kn)$ statements. It suffices to show that we can compress any row of A to an SLP with $O(n)$ statements, because we combine the compressed rows using k "$\ominus$" operations.

Consider a row R of A corresponding to a clause $C(x_h, x_i, x_j)$ where $h < i < j$. Define $\imath(v_0, \ldots, v_{n-1}) = v_h + 2v_i + 4v_j$, then $R_m = C(\imath(\tilde{m}))$. We will show how to compress the string I over $\Sigma = \{0, 1, 2, 3, 4, 5, 6, 7\}$ defined by $I_m = \imath(\tilde{m})$, for $0 \le m < 2^n$. Then obtain an SLP for R from the SLP $\mathcal{I}$ for I by replacing each constant $a \in \Sigma$ with $C(\tilde{a})$. We omit an easy proof of the following fact.

Fact 7 I can be expressed as

$$(((0^{2^h} 1^{2^h})^{2^{i-h-1}} (2^{2^h} 3^{2^h})^{2^{i-h-1}})^{2^{j-i-1}} ((4^{2^h} 5^{2^h})^{2^{i-h-1}} (6^{2^h} 7^{2^h})^{2^{i-h-1}})^{2^{j-i-1}})^{2^{n-j-1}}$$

To compress I, write a constant length SLP that computes all subexpressions of I, then replace each statement of the form $K \leftarrow L^{2^i}$ with i statements $L \leftarrow LL$ followed by $K \leftarrow L$. This results in an SLP with $O(n)$ statements. $\qquad\square$

4 Fully compressed two dimensional pattern-matching

Recall that the fully compressed matching problem is to determine, given a pattern and a text that are both compressed, whether the pattern occurs within the text. We prove that this problem is Σ^P_2-complete, see [14] for the definition of the class Σ^P_2.

Theorem 8 (main result). *The fully compressed matching problem for 2d-texts is Σ^P_2-complete.*

Given compressed pattern $\mathcal{P}$ and compressed 2d-text $\mathcal{I}$, the positive answer to the fully compressed two dimensional pattern matching question is equivalent to the following:

$$\exists (i, j) \forall (k, l) \in Positions(\mathcal{P}) \; \{\mathcal{P}_{k,l} = \mathcal{I}_{i+k, j+l}\}$$

By Lemma 5, the equality in this formula can be checked in polynomial time, hence the problem can be formulated in the normal form of Σ^P_2 problems. This proof of Σ^P_2-hardness requires two lemmas.

Lemma 9. *There exists a log-SPACE function f such that for any 3CNF formula F, $f(F) = (u, v, t)$, where u and v are vectors of non-negative integers, t is an integer and*

$$\forall x \quad F(x) \equiv \exists y \; ux + vy = t.$$

where the quantifiers range over 0-1 vectors of appropriate length.

Proof. Assume that F has n variables, a clauses with three literals, b clauses with two literals and c clauses with one literal. Vector u will consists of n numbers and vector v of $7a + 3b$ numbers. We will describe each of these numbers, (and t as well) using the identity $\tilde{d} = d^0 \ldots d^{(a+b+c-1)}$, where $d^{(k)}$ is *the fragment of d corresponding to clause C_k*. The fragments corresponding to a clause with l literals will have length $2l$. We describe in detail the case of a clause with three literals, the other cases being similar, only simpler.

Assume that clause C_k contains three variables, x_h, x_i, x_j. The fragments of u_h, u_i, and u_j corresponding to C_k are 000100, 000010 and 000001 respectively, while for $l \notin \{h, i, j\}$ we have $u_l^{(k)} = 000000$.

There are 7 truth assignments for (x_h, x_i, x_j) that satisfy $C(k)$, for each one we have an entry in vector v; if v_l is the entry corresponding to a truth assignment (b_0, b_1, b_2) for C_k, then $v_l^{(k)} = 100(1 - b_0)(1 - b_1)(1 - b_2)$. Moreover, for $k' \neq k$ we have $v_l^{(k')} = 0 \ldots 0$. Finally, $t^{(k)} = 100111$.

Consider now x such that $F(x)$ is true. Then the fragment of $\widehat{ux}$ corresponding to a clause C_k is $000b_0b_1b_2$, where (b_0, b_1, b_2) is a truth assignment satisfying C_k (note that x satisfies all the clauses of F).

Let v_l be the entry of v corresponding to this truth assignment, and $v_{l_1}, \ldots, v_{l_6}$ be the entries corresponding to other truth assignments that may satisfy C_k. We set y_l to 1 and $y_{l_1}, \ldots, y_{l_6}$ to 0; it is easy to see that the fragment of $ux + vy$ corresponding to C_k is 100111, the same as the corresponding fragment of t. Since this is true for every fragment of t, we have $ux + vy = t$.

Now suppose that there exists y such that $ux + vy = t$. If for every clause C_k exactly one of the entries corresponding to the truth assignments that satisfy C_k has coefficient 1 in the vector y, and if the addition is performed without carries, then each C_k is satisfied. It can be proved by induction that this is indeed the case (note that in our string representations of numbers we write the least significant bit first). We leave the details are left to the reader.

Finally, the method of creating (u, v, t) is so regular that it can be carried out by a deterministic log-SPACE Turing machine. $\qquad\Box$

Define the Σ_2(Subset Sum) problem as follows: given (u, v, t) where u and v are vectors of positive integers and t is an integer; the question is whether $\exists x \forall y \; ux + vy \neq t$, where the quantifiers range over 0-1 vectors of appropriate length.

Lemma 10. *The Σ_2(Subset Sum) problem is $\Sigma_2^\mathcal{P}$-complete.*

Proof. Consider now an arbitrary property L of binary strings that belongs to $\Sigma_2^\mathcal{P}$. In its normal form, L is represented as

$$L(x) \equiv \exists y_1 \forall y_2 \; P(x, y_1, y_2)$$

where P is a polynomial time predicate.

Because $\mathcal{P} \subseteq \mathcal{NP} \cap \text{co-}\mathcal{NP}$, the predicate P can be represented as

$$P(x, y_1, y_2) \equiv \neg(\exists y_3 F(x, y_1, y_2, y_3))$$

where F is a 3CNF formula (computed using space which is logarithmic in the size of x in unary).

Let "$\cdot$" denote concatenation of vectors. By the previous lemma,

$$F(x, y_1, y_2, y_3) \equiv \exists y_4\ u(x \cdot y_1 \cdot y_2 \cdot y_3) + vy_4 = t$$

where (u, v, t) can be computed in logarithmic space from F. Define $\bar{u}, \bar{v}, \bar{w}$ and $\bar{t}$ so that $u(x \cdot y_1 \cdot y_2 \cdot y_3) + vy_3 = \bar{w}x + \bar{u}y_1\bar{v}(y_2 \cdot y_3 \cdot y_4)$ and $\bar{t} = t - \bar{w}x$. By substitution and the De Morgan laws, we have

$$
\begin{aligned}
L(x) &\equiv\ \exists y_1 \forall y_2 \neg(\exists y_3 \exists y_4\ u(x \cdot y_1 \cdot y_2 \cdot y_3) + vy_4 = t) \\
&\equiv\ \exists y_1 \forall y_2 \forall y_3 \forall y_4\ u(x \cdot y_1 \cdot y_2 \cdot y_3) + vy_4 \neq t \\
&\equiv\ \exists y_1 \forall y_2 \forall y_3 \forall y_4\ \bar{w}x + \bar{u}y_1\bar{v}(y_2 \cdot y_3 \cdot y_4) \neq t \\
&\equiv\ \exists y_1 \forall (y_2 \cdot y_3 \cdot y_4)\ \bar{u}y_1\bar{v}(y_2 \cdot y_3 \cdot y_4) \neq \bar{t}
\end{aligned}
$$

Because the last of the above statements is an instance of Σ_2(Subset Sum), we have shown that L can be reduced to Σ_2(Subset Sum). $\qquad\square$

To prove that fully compressed two dimensional pattern matching is Σ_2^P complete, it suffices to show how to translate an instance of Σ_2(Subset Sum).

Consider an instance given by (u, v, t). Recall the definition of T^w from our proof of co-$\mathcal{NP}$ completeness.

Let U be the 2d-text T^u and let V be the 2d-text T^v with all rows reversed. Recall that dimensions of U and V are $2^n \times (1+r)$ and $2^m \times (1+s)$ respectively, where m and n are the lengths of u and v, while r and s are their sums. We define the pattern and the test as follows:

$$
\begin{array}{lll}
P \leftarrow 1 \ominus [0]^1_{2^n + 2^m} & S_0 \leftarrow [0]^{2^n}_{s-t} U & S_1 \leftarrow V[1]^{2^m}_{r-t} \\
S_2 \leftarrow [0]^{2^n}_{1+r+s-t} & T \leftarrow R_1 \ominus R_2 \ominus R_2 &
\end{array}
$$

The subrectangles S_i's are *stripes* of the text T. Observe first that T contains P if and only if there exists a column of T, say c, that contains P. Because the length of P equals the sum of heights of S_1 and S_2 plus 1, P can start anywhere in the upper stripe S_0 but only there. Because P starts with 1, it must start within U, so $c = s - t + a$ for some $a \geq 0$. Therefore column c consists of column a of U, column $s - t + a$ of V and zeros at the bottom—we can easily exclude the case when this column crosses the middle stripe S_1 through the subrectangle consisting of 1's only.

Now, column a of U is column a of T^u, so a 1 exists in this column if and only if for some $x < 2^n$ we have $u\tilde{x} = a$. Moreover, column $s - t + a$ of V is column $s - (s - t - a) = t - a$ of T^v, and we have all 0's in this column if and only if $vy \neq t - a$ for every $y < 2^m$. Summarizing, P occurs in T if and only if there exists x with the following property: for $a = ux$ the equality $vy = t - a \equiv a + vy = t \equiv ux + vy = t$ holds for no y. Therefore the positive answer to our pattern matching problem is equivalent to the positive answer to the original Σ_2(Subset Sum) problem. This concludes the proof of Theorem 8.

5 Fully compressed pattern checking

The problem of fully compressed pattern checking at a given location is to check, given pattern $\mathcal{P}$ and text $\mathcal{R}$ that are both compressed and a position within the text, whether $\mathcal{P}$ occurs within $\mathcal{R}$ at this particular place.

Theorem 11. *Fully compressed pattern checking for d-texts is co-$\mathcal{NP}$-complete.*

Proof. We can use Lemma 5 to express this problem in the normal form of co-$\mathcal{NP}$:

$$\forall (k,l) \in Positions(\mathcal{P}) \;\; \mathcal{P}_{k,l} = \mathcal{R}_{k+i,l+j}.$$

To prove co-NP hardness, we will reduce co-(Subset Sum) to our problem. An instance of co-(Subset Sum) is a vector of integer weights $w = (w_0, \ldots, w_{n-1})$ and a target integer value t; the question is whether $\forall m < 2^n \; w\tilde{m} \neq t$. (Here $w\tilde{m}$ stands for the inner product; because $\tilde{m}$ is a 0-1 vector, $w\tilde{m}$ is a sum of a subset of the terms of w.) We can transform this question to a pattern checking question in a natural manner. Let $s = 1 + \sum_{i=0}^{n-1} w_i$, and let the 2d-text T^w consists of 0's and 1's, with $T^w_{m,i} = 1$ if and only if $w\tilde{m} = i$.

Then our co-(Subset Sum) question is whether column t of T^w consists of 0's only. In terms of the pattern checking problem, we specify the text T^w, the pattern $[0]^1_{2^n}$ and the position $(t, 0)$.

To finish the proof, we need to compress T^w. Observe that row m of T^w contains exactly one 1, at position $w\tilde{m}$. Moreover, for $m < 2^{n-1}$ we have $w(m + 2^{n-1}) = w(\tilde{m}) + w_{n-1}$. Therefore when we split T^w into upper and lower halves (each with 2^{n-1} rows), the pattern of 1's is very similar, the only difference being that in the lower half (with higher row numbers) the 1's are shifted by w_{n-1} to the right. Moreover, if we remove the last w_{n-1} zeros from each row in the upper half, we obtain $T^{w(n-1)}$, a 2d-text defined just as T^w, but where $w(n-1) = (w_0, \ldots, w_{n-2})$. Proceeding inductively, we compute T^w:

$$T^{w(0)} \leftarrow 1$$
$$\textbf{for } i \leftarrow 0 \textbf{ to } n-1 \textbf{ do}$$
$$U \leftarrow T^{w(i)}[0]^{2^i}_{w_i}; \;\; L \leftarrow [0]^{2^i}_{w_i} T^{w(i)}; \;\; T^{w(i+1)} \leftarrow U \ominus L$$

To obtain an SLP for T^w, we combine $3n + 1$ statements of the above program with SLP's that compute auxiliary 2d-texts $[0]^{2^i}_{w_i}$. The resulting SLP has $O(n^2 + b)$ statements, where b is the total number of bits in the binary representations of the numbers in vector w. $\qquad\square$

6 Compressed pattern checking

Recall that the compressed pattern checking problem is to check whether an uncompressed pattern P occurs at a position (x, y) of a 2d-text T given by an SLP $\mathcal{T}$. Let n be the size of $\mathcal{T}$ and N be the size of T. The compressed pattern checking problem can be solved easily in polynomial time by using an algorithm for point test problem $m \cdot k$ times.

By Lemma 5 there is an algorithm which solves the compressed pattern checking problem in $O(n|P|)$ time. We improve this by replacing n by $\log N \log m$. This is similar to the approach of [6]. If the 2d-text is not very highly compressed then $\log(N)$ is close to $\log(n)$.

The idea behind the algorithm is to consider point tests in groups, called a *queries*. Denote by $\mathcal{V}$ a text which is generated by a variable V. A *query* is a triple $(V, p, \mathcal{R})$ where V is a variable in the SLP $\mathcal{T}$, p is a position inside $\mathcal{V}$ and $\mathcal{R}$ is a subrectangle of the pattern P. Denote by $\mathcal{R}'$ the subrectangle of $\mathcal{V}$ which is placed at position p in $\mathcal{V}$ and is of the same shape as the rectangle $\mathcal{R}$. We require that $\mathcal{R}'$ abut one of the sides of the rectangle $\mathcal{V}$. An *answer* to the query tells whether $\mathcal{R}' = \mathcal{R}$.

Queries are answered by replacing them by equivalent "simpler" queries. We say that a query $(V, p, \mathcal{R})$ is *simpler* than a query $(V', p', \mathcal{R}')$ if $|\mathcal{V}| < |\mathcal{V}'|$. A query which contains a variable V is called a V-*query*. An *atomic query* is a query $(V, p, \mathcal{R})$ such that $\mathcal{V}$ is a 1×1 square, which can be answered in $O(1)$ time.

The queries are divided into three classes: *strip queries*, *edge queries*, and *corner queries*. Let $(V, p, \mathcal{R})$ be a query. Let $\mathcal{R}'$ be the rectangle of the same shape as $\mathcal{R}$ which is positioned at p in $\mathcal{V}$. $(V, p, \mathcal{R})$ is a *corner* query if $\mathcal{R}$ contains at least one side of the pattern or $\mathcal{R}$ is a corner subrectangle of the pattern and $\mathcal{R}'$ is a corner subrectangle of $\mathcal{V}$. $(V, p, \mathcal{R})$ is an *edge* query if $\mathcal{R}'$ contains one side of $\mathcal{V}$. There are four types of edge queries depending on which side of $\mathcal{V}$ is contained in $\mathcal{R}'$. We call these *down*, *left*, *right* and *up* queries. $(V, p, \mathcal{R})$ is a *strip* query if $\mathcal{R}$ is a strip of the pattern and $\mathcal{R}'$ is a strip of $\mathcal{V}$.

The algorithm CHECKING for the checking problem uses two procedures, *Split*(V, Q) and *Remove_Edge_Queries*(V, Q), where V is a variable in $\mathcal{T}$ and Q is a set of queries.

Algorithm CHECKING
{ input: an SLP $\mathcal{T}$, a pattern P and a position p }
{ output: true iff P occurs at p in a text described by $\mathcal{T}$ }
begin
    $V_1, V_2, \ldots, V_n$: = sequence of variables in $\mathcal{T}$,
             in descending order w.r.t. the sizes of their texts
    $Q := \{(V_1, p, P)\}$
    for $i := 1$ **to** n **do**
        $Q := Remove\_Edge\_Queries(V_i, Q)$ $Q := Split(V_i, Q)$

{there are now only atomic queries in Q}
answer all atomic queries in Q

end

The procedure *Compress_Edge_Queries*(V, Q) deals only with edge V-queries in Q. Its aim is to eliminate, for each type of edge query separately, all edge V-queries except the query which contains the largest subrectangle of the pattern. We describe how this procedure works for left-edge queries. Let $(V, (0,0), \mathcal{R})$ be a left-edge query and $\mathcal{R}$ be of maximal size among all left-edge V-queries in Q. Let $(V, (0,0), \mathcal{R}')$ be any other left-edge V-query. Then the rectangle of shape $\mathcal{R}'$ positioned at $(0,0)$ in $\mathcal{V}$ is a subrectangle of the rectangle of shape $\mathcal{R}$ positioned at $(0,0)$ in $\mathcal{V}$.

Hence, to answer both queries it is enough to answer the query $(V, (0,0), \mathcal{R})$ and to check whether the text $\mathcal{R}'$ occurs in $\mathcal{R}$ at $(0,0)$. Before removing each edge query equality of appropriate rectangles is checked and if the rectangles do not match then the procedure stops and the algorithm returns false.

Assume that $A:=BC$ or $A:=B \ominus C$ is an assignment for A. The *Split*(A, Q) procedure replaces A-queries in Q by equivalent B-queries and C-queries. Let $(A, p, \mathcal{R})$ be an A-query in Q. Consider a rectangle R of shape $\mathcal{R}$ positioned at p in $\mathcal{A}$.

Then division of A into B and C according to the assignment for A causes that either R to be wholly contained in B or C, or to be divided into two smaller rectangles one of which is in B and the other in C. In the latter case the split of a query is called a *division* of the query.

Fact 12 *The total number of all divisions of queries during the work of the algorithm is exactly $|P| - 1$.*

For each variable, edge and corner queries are stored in a list. The data structure for storing strip queries is more sophisticated. For each variable it is a 2-3-tree [1] in which keys are positions of strip rectangles in the variable. Recall that 2-3 trees provide operations *split* and *join* in $O(\log s)$ time where s is the number of elements in the tree.

Fact 13 *In each step of algorithm* CHECKING *the set Q contains at most four corner queries and m strip queries.*

Implementation of the *Split* operation, if it is not a division, requires merging 2-3 trees and this may result in a large number of splits of 2-3 trees. Fortunately, it is possible to prove the following lemma, using arguments similar to those of [6].

Lemma 14. *The number of splits of 2-3 trees in algorithm* CHECKING *is $O(m \log N)$.*

Theorem 15. *The algorithm* CHECKING *works in $O(|P|+n+(m \log N)(\log m))$ time.*

Proof. By Fact 12, the total cost of all divisions is $O(|P|)$. The total cost of all *Splits* which are not divisions is determined by the number of all corner queries and all edge queries which survive after the *Remove_Edge_Queries* operation during the execution of the algorithm and the number of splits of 2-3 trees. This gives, by Lemma 14, $O(n + (m \log N)(\log m))$. □

References

1. A.V. Aho, J.E. Hopcroft, and J.D. Ullman, *The design and analysis of computer algorithms*, Addison-Wesley, Reading, Mass., 1974.
2. A. Amir, G. Benson and M. Farach, *Let sleeping files lie: pattern-matching in Z-compressed files*, in *SODA'94*.
3. A. Amir, G. Benson, *Efficient two dimensional compressed matching*, Proc. of the *2nd IEEE Data Compression Conference* 279-288 (1992).
4. A. Amir, G. Benson and M. Farach, *Optimal two-dimensional compressed matching*, in *ICALP'94* pp.215-225.
5. M. Crochemore and W. Rytter, *Text Algorithms*, Oxford University Press, New York (1994).
6. M. Farach and M. Thorup, *String matching in Lempel-Ziv compressed strings*, in STOC'95, pp. 703-712.
7. M.R. Garey and D.S. Johnson, *Computers and Intractability: A Guide to the Theory of NP-Completeness*. W.H. Freeman (1979).
8. L. Gąsieniec, M. Karpiński, W. Plandowski and W. Rytter, *Efficient Algorithms for Compressed Strings*. in proceedings of the SWAT'96 (1996).
9. M. Karpinski, W. Rytter and A. Shinohara, *Pattern-matching for strings with short description*, in *Combinatorial Pattern Matching*, 1995.
10. D. Knuth, *The Art of Computing, Vol. II: Seminumerical Algorithms. Second edition*. Addison-Wesley, 1981.
11. A. Lempel and J. Ziv, *On the complexity of finite sequences*, *IEEE Trans. on Inf. Theory* 22, 75-81 (1976).
12. A. Lempel and J. Ziv, *Compression of two-dimensional images sequences*, *Combinatorial algorithms on words* (ed. A. Apostolico, Z.Galil) Springer Verlag (1985) 141-156.
13. R. Motwani, P. Raghavan, Randomized algorithms, Cambridge University Press 1995.
14. Papadimitriou, Ch. H., *Computational complexity*, Addison Wesley, Reading, Massachusetts, 1994.
15. W. Plandowski, *Testing equivalence of morphisms on context-free languages*, ESA'94, Lecture Notes in Computer Science 855, Springer-Verlag, 460-470 (1994).
16. J. Storer, *Data compression: methods and theory*, Computer Science Press, Rockville, Maryland, 1988.
17. R.E. Zippel, Probabilistic algorithms for sparse polynomials, in EUROSAM 79, Lecture Notes in Comp. Science 72, 216-226 (1979).
18. J. Ziv and A. Lempel, *A universal algorithm for sequential data compression*, *IEEE Trans. on Inf. Theory* vo. IT-23(3), 337-343, 1977.

Estimating the Probability
of Approximate Matches

Stefan Kurtz[1]* and Gene Myers[2]**

[1] Technische Fakultät, Universität Bielefeld, Postfach 100 131, D-33501 Bielefeld, Germany, E-mail: kurtz@techfak.uni-bielefeld.de
[2] Department of Computer Science, The University of Arizona, Tucson, Arizona 85721, USA, E-mail: gene@cs.arizona.edu

1 Introduction

While considerable effort and some progress has been made on developing an analytic formula for the probability of an approximate match, such work has not achieved fruition [4,6,2,1]. Therefore, we consider here the development of an unbiased estimation procedure for determining said probability given a specific string $P \in \Sigma^*$ and a specific cost function δ for weighting edit operations. Problems of this type are of general interest, see for example a recent paper [5] giving an unbiased estimator for counting the words of a fixed length in a regular language. We were further motivated by a particular application arising in the pattern matching system *Anrep* designed by us for use in genomic sequence analysis [8,11]. *Anrep* accomplishes a search for a complex pattern by backtracking over subprocedures that find approximate matches. The subpatterns are searched in an order that attempts to minimize the expected running time of the search. Determining this optimal backtrack order requires a reasonably accurate estimate of the probability with which one will find an approximate match to each subpattern. Given that the probabilities involved are frequently 10^{-6} or less, the simple expedient of measuring match frequency over a random text of several thousand characters has been less than satisfactory. The unbiased estimator herein is shown to give good results in a matter of a thousand samples even for small probability patterns. Thus it is expected to improve the performance of *Anrep* and may have utility in estimating the significance of similarity searches.

Proceeding formally, suppose we are given

- a pattern string $P = p_1 p_2 \ldots p_m \in \Sigma^*$,
- an integer cost function δ, and
- a match threshold $k \in \mathbb{N}_0$.

* Research done while visiting the University of Arizona, partially supported by NLM grant LM-4960.
** Partially supported by NLM grant LM-4960.

Let $A_k(P,T)$ denote the event that P can be aligned to a *prefix* of text T with cost k or less. We seek $Pr[A_k(P,T)]$ under the distributional assumption that T is generated by independent, uniform Bernoulli trials over the alphabet Σ. Our results depend on δ satisfying the following two conditions:

$$\delta(a \rightarrow \varepsilon) + \delta(\varepsilon \rightarrow b) \geq \delta(a \rightarrow b) \geq 0 \tag{1}$$

$$\delta(\varepsilon \rightarrow b), \delta(a \rightarrow \varepsilon) > 0 \tag{2}$$

where $a \rightarrow \varepsilon$ denotes the deletion of the character a, $a \rightarrow b$ denotes the replacement of the character a by the character b, and $\varepsilon \rightarrow b$ denotes the insertion of the character b. These conditions are generally met by the scoring schemes required in most applications, and if not, one can usually transform the scoring scheme to one satisfying them. Note that δ may otherwise be any cost function over the integers.

Let the condensed k-neighborhood of P, $CN_k(P)$, be the set of all strings approximately matching P within threshold k, less those that have another such string as a prefix. Because the occurrence of each string in $CN_k(P)$ as a prefix of T is an independent event, it immediately follows that

$$Pr[A_k(P,T)] = \sum_{v \in CN_k(P)} \frac{1}{|\Sigma|^{|v|}}$$

Unfortunately, the size of $CN_k(P)$ grows exponentially with the threshold value, quickly rendering direct computation by enumerating $CN_k(P)$ hopelessly inefficient. Thus we turn to developing a Monte-Carlo algorithm that estimates $Pr[A_k(P,T)]$ by sampling a subset of $CN_k(P)$. The difficulty immediately encountered is that a direct procedure for sampling $CN_k(P)$ appears formidable if not impossible. So instead we turn to sampling a surrogate space, $S_k(P)$, of edit scripts of cost k or less that convert P into strings it approximately matches. The set $S_k(P)$ can be any set of edit scripts provided that it is efficiently enumerable and *complete* in that for every $v \in CN_k(P)$, there is at least one edit script $s \in S_k(P)$ that when applied to P generates $s(P) = v$.

The difficulty to be overcome in sampling the surrogate space $S_k(P)$ is that there is not a one-to-one correspondence between its edit scripts and the strings in $CN_k(P)$. Some strings in $CN_k(P)$ may be generated by several distinct edit scripts in $S_k(P)$ and some edit scripts in $S_k(P)$ may not even generate strings in $CN_k(P)$. This bias is removed by considering the random variable $X : S_k(P) \rightarrow \mathbb{R}_0^+$ defined by:

$$X(s) = \begin{cases} \frac{1}{|\Sigma|^{|v|}g(v)} & \text{if } v \in CN_k(P) \\ 0 & \text{otherwise} \end{cases}$$

where $v = s(P)$, and $g(v)$, called the cluster size of v, is the number of edit scripts in $S_k(P)$ that generate the string v. That X removes the bias is the claim of our central lemma:

Lemma 1. *Suppose $S_k(P)$ is uniformly distributed. Then the expected value $E[X]$ of X is $Pr[A_k(P,T)]/|S_k(P)|$.*

Proof.

$$
\begin{aligned}
E[X] &= \sum_{s \in S_k(P)} \frac{X(s)}{|S_k(P)|} \\
&= \frac{1}{|S_k(P)|} \cdot \sum_{s \in S_k(P)} \left\{ \frac{1}{|\Sigma|^{|v|} g(v)} \mid s(P) \in CN_k(P), v = s(P) \right\} \\
&= \frac{1}{|S_k(P)|} \cdot \sum_{v \in CN_k(P)} \left\{ \frac{1}{|\Sigma|^{|v|} g(v)} \mid s \in S_k(P), s(P) = v \right\} \\
&= \frac{1}{|S_k(P)|} \cdot \sum_{v \in CN_k(P)} \frac{g(v)}{|\Sigma|^{|v|} g(v)} \\
&= \frac{1}{|S_k(P)|} \cdot \sum_{v \in CN_k(P)} \frac{1}{|\Sigma|^{|v|}} \\
&= \frac{Pr[A_k(P,T)]}{|S_k(P)|} \qquad \square
\end{aligned}
$$

Our unbiased estimation procedure thus consists of

- selecting a suitably large set of samples from $S_k(P)$ with uniform probability,
- computing the average of $X(s)$ over the samples s, and
- multiplying the average by $|S_k(P)|$.

While the procedure, in outline, is quite simple, the remainder of this paper addresses how to define $S_k(P)$ and how to solve the algorithmic sub-problems involved in an efficient realization with respect to this definition. Section 2 introduces as our choice for $S_k(P)$ the set of what we call the *condensed, canonical edit scripts*. Our choice attempts to keep small, both (*i*) the number of edit scripts for which $X(s) = 0$, and (*ii*) the size of $g(v)$. Doing so improves the convergence of the estimator as it places $S_k(P)$ and $CN_k(P)$ in closer correspondence. The remaining sections present dynamic programming algorithms for the following subtasks:

Section 3: Determining the size of $S_k(P)$ in $O(|\Sigma|mk)$ time.
Section 4: Selecting an edit script from $S_k(P)$ with uniform probability in $O(|\Sigma|(m+k))$ time.
Section 5: Deciding if $s \in S_k(P)$ generates a string $s(P) \in CN_k(P)$ in $O(m \cdot \min(m,k))$ time.
Section 6: Computing the cluster size $g(v)$ in $O(\Delta \cdot m \cdot \min(m,k))$ time, where $\Delta \in [1, k+1]$ is depending on δ and P.

Altogether, our Monte-Carlo algorithm achieves a running time of $O(|\Sigma|mk + t(|\Sigma|m + |\Sigma|k + \Delta \cdot m \cdot \min(m,k)))$ and requires $O(mk)$ space where t is the number of samples collected. If δ is the unit cost function that scores all mismatches, insertions, and deletions as 1, the running time further improves to simply $O(tmk)$. The paper concludes with Section 7 that presents empirical results demonstrating the accuracy of the estimates, the rate of convergence of the sampling process, and the speed of the procedure.

In the context of scanning a text, our estimator above gives the probability that a match *begins* at a specific position in the text. However, these events are not independent. One match may significantly condition a match at the next position. To estimate the number of such "clumps", one needs to compute the probability that a match begins at a position and there is no overlapping match to the left of it. Addressing this issue is beyond the scope of this paper. The current result is still of value as it overestimates this probability and in the context of our *Anrep* application these probabilities are small enough that the clump size is almost always one, so the overestimation is slight.

We close the introduction by noting that our treatment is extensible to patterns that are network expressions [11] (regular expression without Kleene closure), and to models of the text where characters are generated by a weighted Bernoulli process. We do not directly treat these extensions in this paper, as they complicate the treatment and obscure the basic ideas.

2 Condensed, Canonical Edit Scripts

Approximate matches are typically characterized as an alignment or trace between P and a string that it matches. In our context we must turn to the equivalent operational view of an *edit script* that transforms P into the string it matches as originally introduced, for example, in the seminal work of Wagner and Fischer [13]. There are three kinds of edit operations: $a \to \varepsilon$ denotes the *deletion* of the character a, $a \to b$ denotes the *replacement* of the character a by the character b, and $\varepsilon \to b$ denotes the *insertion* of the character b. An edit script for $P = p_1 p_2 \ldots p_m$ is a list $s = [\alpha_1 \to \beta_1, \alpha_2 \to \beta_2, \ldots, \alpha_r \to \beta_r]$ of edit operations for which $r \geq m$ and $P = \alpha_1 \cdot \alpha_2 \cdot \ldots \cdot \alpha_r$ when one interprets ε as denoting the empty string. P is viewed as the *source* string and the application of the edit script s results in the *target* string $s(P) = \beta_1 \cdot \beta_2 \cdot \ldots \cdot \beta_r$. We further say that s *generates* the string $s(P)$. The correspondence between edit scripts and alignments is immediate given the observation that each edit operation corresponds to an alignment column.

The underlying *cost function* δ assigns a non-negative integer cost, $\delta(\alpha \to \beta)$, to each edit operation $\alpha \to \beta$. Recall that in addition to integrality, we are assuming that (1) and (2), given in the introduction, hold. The *cost* $\delta(s)$ of an edit script $s = [\alpha_1 \to \beta_1, \alpha_2 \to \beta_2, \ldots, \alpha_r \to \beta_r]$ is the sum of the costs of its edit operations, i.e., $\delta(s) = \sum_{i=1}^{r} \delta(\alpha_i \to \beta_i)$. We say that s is a (P, l) edit script if s is an edit script for P of cost l. The *edit distance* between P and $v \in \Sigma^*$, denoted by $\delta(P, v)$, is the minimal cost over all edit scripts for P that generate v. An edit script s for P is *optimal* if $\delta(s) = \delta(P, s(P))$.

Given P and $k \geq 0$, the *k-neighborhood* of P is the set of all strings matching P with cost k or less, i.e., $N_k(P) = \{v \in \Sigma^* \mid \delta(P, v) \leq k\}$. In order to remove dependent events, we restrict our attention to the *condensed* *k-neighborhood*, $CN_k(P)$, which is defined by

$$CN_k(P) = \{v \in N_k(P) \mid \forall \text{ proper prefixes } v' \text{ of } v, v' \notin N_k(P)\}$$

Note that $\varepsilon \in N_k(P)$ implies $Pr[A_k(P,T)] = 1$. Hence we do not need to estimate $Pr[A_k(P,T)]$ whenever k is greater or equal to the cost of deleting every character in P. This implies that $k = O(m)$ in the cases of interest, assuming δ is a constant.

Recall that we will effectively be sampling $CN_k(P)$ by sampling the surrogate space $S_k(P)$ and applying the random variable X to each sample. We could simply choose to let $S_k(P)$ be the set of *all* (P,l) edit scripts for $l \leq k$. However, the convergence rate of the estimator improves, the tighter the correspondence between the set of sampled edit scripts and the condensed *k*-neighborhood. We develop a much smaller but still complete set of edit scripts in a progression of three observations below.

First, we remove easily detectable non-optimal or redundant edit scripts. An edit script is called *canonical* if it does not contain a sublist of the form: $[a \to \varepsilon, \varepsilon \to b]$, $[\varepsilon \to b, a \to \varepsilon]$, $[a \to \varepsilon, a \to b]$, or $[\varepsilon \to b, a \to b]$ for some $a, b \in \Sigma$. Such sublists are called *forbidden*.[1] Intuitively, canonical edit scripts can be characterized by the following properties:

– A replacement is preferred over a deletion/insertion or insertion/deletion combination.
– If the source string contains a substring of the same character, then only the rightmost instances in the substring are deleted (if any).
– If the edit script generates a substring of the same character in the target string, then only the rightmost instances are generated by an insertion (if any).

Let $C_l(P)$ be the set of canonical (P,l) edit scripts. Provided that δ satisfies (1), limiting our attention to canonical edit scripts is conservative as every non-canonical edit script s can be transformed into a canonical edit script s' such that $s(P) = s'(P)$ and $\delta(s') \leq \delta(s)$ holds. Thus we have $\bigcup_{l \leq k} \bigcup_{s \in C_l(P)} \{s(P)\} = N_k(P)$.

The second observation is that any edit script ending with an insertion, generates a shorter word at lesser cost if the final insertion is dropped from the edit script. Thus, any such edit script cannot generate a string in $CN_k(P)$ as it does not contain strings that have proper prefixes matching P at lesser cost.

---

[1] One could generalize the idea of a canonical edit script to those that do not contain forbidden sublists of length 3, or 4, etc. The marginal rate at which these higher-order schemes eliminate redundant edit scripts diminishes rapidly, while the cost of sampling them grows exponentially in time. Thus we present only a second order scheme here, both because it is most practical and also for the sake of keeping the exposition simple. However, it is worth noting that if one goes to an *m*-order scheme, then $CN_k(P)$ and canonical edit scripts correspond one-to-one.

Let a *condensed, canonical edit script* be a canonical edit script that does not end with an insertion. Let $\mathcal{CC}_l(P)$ be the set of condensed, canonical (P, l) edit scripts. We may restrict our attention to $\bigcup_{l \leq k} \mathcal{CC}_l(P)$ as it is complete w.r.t. $CN_k(P)$, i.e., $\bigcup_{l \leq k} \bigcup_{s \in \mathcal{CC}_l(P)} \{s(P)\} \supseteq CN_k(P)$.

Finally, observe that while the edit distance between a string $v \in CN_k(P)$ and P may be less than k, there must be a lower bound as one could replace the last non-deletion operation in an edit script generating v optimally with a deletion operation for some bounded increase in cost. Indeed, carrying this train of thinking to its furthest degree, one arrives at the following lemma:

Lemma 2. *For any* $v \in CN_k(P)$, $\delta(P, v) > k - \Delta$ *where*

$$\Delta = \max\Big\{ \delta(p_i \to \varepsilon) - \delta(p_i \to b) \mid i \in [1, m], b \in \Sigma,$$

$$\textstyle\sum_{r=i+1}^{m} \delta(p_i \to \varepsilon) \leq k - \delta(p_i \to b)\Big\}$$

Proof. Let $v \in CN_k(P)$. We have $v = wb$ for some $w \in \Sigma^*$ and some $b \in \Sigma$. Let s be an optimal edit script for P generating v. Then $s \neq []$, $\delta(s) = \delta(P, v) \leq k$, and the last edit operation in s is not an insertion. It is easy to show that there is an $i \in [1, m]$ and $b \in \Sigma$ such that $s = s' \cdot [p_i \to b] \cdot s''$ where s' is an edit script for $p_1 p_2 \ldots p_{i-1}$ generating w, and s'' is a maximal suffix of s consisting of deletions only. We have $\delta(s'') + \delta(p_i \to b) \leq \delta(s) \leq k$, and therefore $\sum_{r=i+1}^{m} \delta(p_r \to \varepsilon) = \delta(s'') \leq k - \delta(p_i \to b)$, which implies $\Delta \geq \delta(p_i \to \varepsilon) - \delta(p_i \to b)$. Since $s' \cdot [p_i \to \varepsilon] \cdot s''$ is an edit script for P generating w, we get $\delta(s') + \delta(p_i \to \varepsilon) + \delta(s'') \geq \delta(P, w)$. Hence

$$
\begin{aligned}
\delta(P, v) &= \delta(s') + \delta(p_i \to b) + \delta(s'') \\
&\geq \delta(P, w) - \delta(p_i \to \varepsilon) + \delta(p_i \to b) \\
&> k - (\delta(p_i \to \varepsilon) - \delta(p_i \to b)) \\
&\geq k - \Delta
\end{aligned}
$$

$\square$

To conclude, let

$$\mathcal{CC}(P) = \bigcup_{l=\Phi}^{k} \mathcal{CC}_l(P)$$

where $\Phi = k - \Delta + 1$. This set of edit scripts is our choice in this paper for $S_k(P)$. Note that $\mathcal{CC}(P)$ is an appropriate choice for the sampling surrogate as it is complete w.r.t. $CN_k(P)$, i.e., $\bigcup_{s \in \mathcal{CC}(P)} \{s(P)\} \supseteq CN_k(P)$.

3 Counting Edit Scripts

We first consider how to enumerate canonical edit scripts. To do so, we split the set of all canonical edit scripts into three classes according to the last edit operation in each edit script. This gives us the control we need to avoid enumerating edit scripts containing a forbidden sublist.

Definition 3. For each $i \in [0, m]$, each $l \in [0, k]$, and each $b \in \Sigma$ define:

$$D(l, i) = \{s \in C_l(p_1 p_2 \ldots p_i) \mid s \text{ ends with a deletion}\}$$
$$R(l, i) = \{s \in C_l(p_1 p_2 \ldots p_i) \mid s = [] \text{ or } s \text{ ends with a replacement}\}$$
$$I_b(l, i) = \{s \in C_l(p_1 p_2 \ldots p_i) \mid s \text{ ends with the insertion } \varepsilon \to b\}$$
$$I(l, i) = \{s \in C_l(p_1 p_2 \ldots p_i) \mid s \text{ ends with an insertion}\}$$

If $l < 0$, we define $D(l, i) = R(l, i) = I_b(l, i) = I(l, i) = \emptyset$.

From the definitions it follows that $\mathcal{CC}(P) = \bigcup_{l=\Phi}^{k} D(l, m) \cup R(l, m)$. It remains to develop recurrences describing each class of edit scripts. Note that the class of edit scripts ending with an insertion are further decomposed according to the character b inserted. This is required so that, in the recurrences below, we can avoid composing edit scripts that follow the insertion of b with a substitution of b (see the equation for $R(l, i)$ below).

Lemma 4. For $i, l \geq 0$, the following recurrences hold:

$$D(l, i) = (D(l', i-1) \cup R(l', i-1)) \cdot [p_i \to \varepsilon] \text{ where } l' = l - \delta(p_i \to \varepsilon)$$
$$R(l, i) = \bigcup_{b \in \Sigma} ((D(l^b, i-1) \text{ if } i > 1 \text{ and } p_{i-1} \neq p_i) \cup$$
$$(I(l^b, i-1) - I_b(l^b, i-1)) \cup$$
$$R(l^b, i-1)) \cdot [p_i \to b] \text{ where } l^b = l - \delta(p_i \to b)$$
$$I_b(l, i) = (I(l', i) \cup R(l', i)) \cdot [\varepsilon \to b] \text{ where } l' = l - \delta(\varepsilon \to b)$$
$$I(l, i) = \bigcup_{b \in \Sigma} I_b(l, i).$$

subject to the boundary conditions: $D(l, 0) = \emptyset$ and $R(l, 0) = (if \, l = 0 \text{ then } \{[]\} \text{ else } \emptyset)$. The notation $X \cdot a$ where X is a set of edit scripts and a is an edit operation, denotes the set $\{x \cdot a \mid x \in X\}$. The symbol "$\cdot$" denotes the concatenation of edit scripts.

With the recurrences above it is now a simple exercise to reformulate them to *count* the number of edit scripts, as opposed to *enumerating* them. We compute three $(k+1) \times (m+1)$-tables: $ND(l, i) = |D(l, i)|$, $NR(l, i) = |R(l, i)|$, and $NI(l, i) = |I(l, i)|$ for $l \in [0, k]$ and $i \in [0, m]$. Note that we make the small optimization of not *storing* the number of edit scripts in $I_b(l, i)$ for each $b \in \Sigma$, as it requires only constant time to compute these numbers on demand given the other entries. This saves us a factor of $|\Sigma|$ space without any penalty in time. We now have

$$|\mathcal{CC}(P)| = \sum_{l=\Phi}^{k} ND(l, m) + NR(l, m)$$

Thus it follows:

Lemma 5. $|\mathcal{CC}(P)|$ can be evaluated in $O(|\Sigma| mk)$ time and $O(mk)$ space. This further improves to $O(mk)$ time when δ is the unit cost function.

4 Uniformly Sampling Edit Scripts

We begin by noting that each edit script in $CC(P)$ can be obtained by tracing back through the recurrences given in the previous section, thereby producing an edit script from *right to left*. In order to sample $CC(P)$ uniformly, one must select a traceback branch with probability proportional to the number of possibilities the chosen branch can generate. That is, if there are r branch possibilities and branch i leads to C_i possible finishes, then branch i should be chosen with probability $C_i / \sum_{l=1}^{r} C_l$.

As a concrete example, suppose that we have thus far chosen a suffix s for which $D(l, i) \cdot s \subseteq CC(P)$. We wish to uniformly sample a finishing prefix from $D(l, i)$. Following the recurrence for $D(l, i)$, we prepend $[p_i \to \varepsilon]$ to s and decide to either complete the result with an edit script from $D(l', i-1)$ with probability $ND(l', i-1)/ND(l, i)$, or with an edit script from $R(l', i-1)$ with probability $NR(l', i-1)/ND(l, i)$ where $l' = l - \delta(p_i \to \varepsilon)$. As one further example, the sampling process starts with a decision to either generate the edit script from $D(l, m)$ with probability $ND(l, m)/|CC(P)|$, or from $R(l, m)$ with probability $NR(l, m)/|CC(P)|$ where $l \in [\Phi, k]$.

Recall that insertions have cost 1 or more (see (2)). Thus the longest possible edit script in $CC(P)$ is of length not greater than $m + k$. Moreover, from the structure of the recurrences above it is easy to see that at each traceback point there are never more than $O(|\Sigma|)$ branches. Thus it follows:

Lemma 6. *Given the dynamic programming tables for counting edit scripts described in Section 3, an edit script $s \in CC(P)$ can be selected with uniform probability in time $O(|\Sigma|(m + k))$. This further improves to $O(m + k)$ time when δ is the unit cost function.*

5 Deciding Condensed Neighborhood Membership

Suppose we have selected s from $CC(P)$. We next wish to know if $v = s(P)$ is in $CN_k(P)$. To do so, consider performing a standard sequence comparison (cf. [13]) between P and v. That is, consider computing the $(m+1) \times (n+1)$ table $E(i, j) = \delta(p_1 p_2 \ldots p_i, v_1 v_2 \ldots v_j)$ where $n = |v|$. Observe that $E(m, n) = \delta(P, v) = \delta(P, s(P)) \leq \delta(s) \leq k$, confirming that v is in $N_k(P)$. If v is in $CN_k(P)$, then it must further be true that no prefix of v is in the k-neighborhood. That is, for all $j \in [0, n-1]$, $\delta(P, v_1 v_2 \ldots v_j) = E(m, j) > k$ must hold. Checking this condition requires just $O(n)$ additional time beyond the $O(mn)$ time spent computing E.

Whenever $k < m$, the time complexity can be further improved by observing that we need only consider the band of E of width k about the main diagonal. This follows because all entries outside this band must be greater than k as more than k insertions or deletions are required to edit the associated prefix of P into that of v. The computation of E restricted to this band consumes only $O(mk)$ time. Furthermore, only $O(k)$ space is required as each row can be computed from the preceding row. See [12,9] for more details. Recalling that $n \leq m + k \in O(m)$, we obtain the following result:

Lemma 7. $v \in CN_k(P)$ *can be decided in* $O(m \cdot \min(k, m))$ *time and in* $O(\min(k, m))$ *space.*

6 Computing Cluster Sizes

To this point an edit script s has been chosen uniformly from $CC(P)$ and we have determined that $v = s(P)$ is in $CN_k(P)$. It remains to compute the cluster size $g(v)$, i.e., the number of edit scripts in $CC(P)$ that generate v. As in Section 3, we first consider how to construct the set of canonical edit scripts for P generating v. Throughout this section, let $r = k - E(m, n)$.

Definition 8. For each $i \in [0, m]$, $j \in [0, n]$, and $l \in [E(i, j), E(i, j) + r]$ let $C(l, i, j)$ be the set of all canonical $(p_1 p_2 \ldots p_i, l)$ edit scripts generating $v_1 v_2 \ldots v_j$. Define

$$D(l, i, j) = \{s \in C(l, i, j) \mid s \text{ ends with a deletion}\}$$
$$R(l, i, j) = \{s \in C(l, i, j) \mid s = [\,] \text{ or } s \text{ ends with a replacement}\}$$
$$I(l, i, j) = \{s \in C(l, i, j) \mid s \text{ ends with an insertion}\}$$

For $l < E(i, j)$, we define $D(l, i, j) = R(l, i, j) = I(l, i, j) = \emptyset$.

If we compare Definition 8 with Definition 3, we recognize an additional argument j, accounting for the fact that we want the edit scripts generating a *fixed* string v.

Lemma 9. *For* $l \in [E(i, j), E(i, j) + r]$, *the following recurrences hold:*

$$D(l, i, j) = (D(l', i-1, j) \cup R(l', i-1, j)) \cdot [p_i \rightarrow \varepsilon] \text{ where } l' = l - \delta(p_i \rightarrow \varepsilon)$$
$$R(l, i, j) = ((D(l', i-1, j-1) \quad \text{if } i > 1 \text{ and } p_{i-1} \neq p_i) \cup$$
$$(I(l', i-1, j-1) \quad \text{if } j > 1 \text{ and } v_{j-1} \neq v_j) \cup$$
$$R(l', i-1, j-1)) \cdot [p_i \rightarrow v_j] \text{ where } l' = l - \delta(p_i \rightarrow v_j)$$
$$I(l, i, j) = (I(l', i, j-1) \cup R(l', i, j-1)) \cdot [\varepsilon \rightarrow v_j] \text{ where } l' = l - \delta(\varepsilon \rightarrow v_j)$$

subject to the boundary conditions:

$$D(l, 0, j) = \emptyset$$
$$R(l, 0, j) = \text{if } l = 0 \text{ and } j = 0 \text{ then } \{[\,]\} \text{ else } \emptyset$$
$$R(l, i, 0) = \text{if } l = 0 \text{ and } i = 0 \text{ then } \{[\,]\} \text{ else } \emptyset$$
$$I(l, i, 0) = \emptyset$$

To compute the cluster size of v, one evaluates three $(r + 1) \times (m + 1) \times (n + 1)$-tables: $ND(l, i, j) = |D(l, i, j)|$, $NR(l, i, j) = |R(l, i, j)|$, and $NI(l, i, j) = |I(l, i, j)|$, for $i \in [0, m]$, $j \in [0, n]$, and $l \in [E(i, j), E(i, j) + r]$. It immediately follows that $g(v) = \sum_{l = E(m, n)}^{k} ND(l, m, n) + NR(l, m, n)$.

Recurrences for computing each entry in the three tables in constant time can simply be derived from the recurrences of Lemma 9. Note that whenever $k < m$, it suffices to evaluate the table entries in the band of width k around the main diagonal, as described in Section 5. Thus $O(r \cdot \min(m, k) \cdot n)$ values are to be computed. At any time, only $O(r \cdot \min(m, k))$ of these need to be stored. Recall that $n \leq m + k \in O(m)$ and $r < \Delta$. Hence we can conclude with:

Lemma 10. *The cluster size $g(v)$ can be computed in $O(\Delta \cdot \min(m, k) \cdot m)$ time and $O(\Delta \cdot \min(m, k))$ space.*

Note that for the unit cost function we have $k < m$ and $\Delta = 1$, i.e., the cluster size can be computed in $O(km)$ time and $O(k)$ space.

7 Experimental Results

We implemented our estimator in C, and performed experiments to demonstrate its accuracy, its convergence, and its speed. In the first three experiments we used an alphabet of size four and the unit cost function. For this cost function our implementation runs in $O(tmk)$ time and $O(mk)$ space. We achieved a considerable speedup in practice by using the failure function of the Knuth-Morris-Pratt algorithm [7]: whenever a column in table E is computed such that the minimal entry is k, the remaining columns need not be evaluated. Instead, using the precomputed failure function, one can decide $v \in CN_k(P)$ and compute $g(v)$ in $O(k)$ additional time. For the details of this technique, see [10, page 352].

In the first experiment, we employed the neighborhood construction algorithm of [10], to compute the "real" probability $Pr = Pr[A_k(P, T)]$ for $k = 2$ and 10,000 random patterns of length $m = 20$. We applied our procedure to the same threshold value and the same random patterns, and compared the resulting estimation Pr_e^t with Pr after each trial $t \in [1, 2000]$, by evaluating the deviation $a(t) = 100 \cdot |Pr - Pr_e^t|/Pr$. Figure 1 shows the probability that $a(t) > d$ for $d = 10\%, 20\%, \ldots, 50\%$ and $t \in [1, 2000]$. It reveals that after 1,000 trials we can expect our procedure to compute a very good estimation of the real probability. The average of $a(t)$ over all 10,000 random patterns was 6.99% after 1,000 trials, the median was 8.97%, and the standard deviation was 4.93%.

In the second experiment, we chose a fixed random pattern of length $m = 50$ and $k = 10$. For $r \in [8, 16]$ we evaluated $c(2^r) = 100 \cdot |\log_2(Pr_e^{2^r}/Pr_e^{2^{r-1}})|$, where Pr_e^t is the estimated probability after t trials. Figure 2 shows the probability that $c(2^r) > d$ for $d = 10\%, 20\%, \ldots, 50\%$ in 1,000 runs of our procedure. (In each run, the random number generator used for selecting random edit scripts, was started with a different seed.) One recognizes that the oscillation of the estimated probability becomes smaller, the larger the number of trials. For instance, the average of $c(2^{16})$ was 29.6% over all 1,000 runs, the median was 21.6%, and the standard deviation was 28.0%. For $C(2^{15})$ the corresponding values were 35.7%, 26.8%, and 35.7%. To get an idea of the size of the set we sampled from, we computed the estimation $1.23 \cdot 10^{18}$ for $|CN_k(P)|$, by dividing $|CC(P)|$ by the average cluster size obtained in the successful trials. These numbers show that 2^{16} trials are not a large effort compared to the alternative of enumerating the entire condensed k-neighborhood.

In the third experiment, we ran 10,000 trials with 100 random patterns of length $m \in \{10, 20, \ldots, 50\}$ and error rates $m/k \in \{10\%, 20\%\}$. The average running times (in seconds) on a DEC Alpha $200^{4/233}$ are shown in the following table:

m	10	20	30	40	50
10%	0.2	0.6	1.2	1.9	2.8
20%	0.3	1.0	2.0	3.2	4.7

In the fourth and fifth experiment, we used the 20-character alphabet of amino-acids and the PAM120 [3] scoring function σ with score -8 for insertions and deletions. We estimated the probability $Pr_\sigma[A_k(P, T)]$ that a pattern string P of length m approximately matches some prefix v of a random string T such that the length-relative score $\sigma(P, v)/|s|$ is greater or equal to k', where $k' = 0.8$ and $|s|$ is the length of the shortest optimal edit script for P generating v. Recall that PAM120-scores are to be maximized, i.e., $\sigma(P, v)$ is the *maximal* score of any edit script for P generating v. Since σ has negative replacement scores, we transformed it into a cost function δ defined by $\delta(\alpha \to \beta) = -\sigma(\alpha \to \beta) + x$ where $x = \max_{a,b \in \Sigma} \sigma(a \to b) = 12$. δ is an integer cost function satisfying (1) and (2). Moreover, one can show that $\sigma(P, v)/|s| \geq k'$ if and only if $\delta(P, v) \leq (x - k') \cdot |s|$. δ assigns cost 20 to deletions and insertions. The maximal cost for replacements is 20, and the average is 14.01. Hence, in an optimal edit script deletions and insertions are rare, i.e., m is a good approximation for $|s|$. Therefore, we can expect a good estimation for $Pr_\sigma[A_k(P, T)]$ if we run our procedure with δ and $k = (x - k') \cdot m = 11.2 \cdot m$.

In the fourth experiment, we again evaluated $c(2^r)$ for $r \in [8, 16]$. This time we chose a random pattern of length $m = 20$ and $k = 11.2 \cdot m = 224$. The size of the sampled set $CN_k(P)$ was estimated to be $2.03 \cdot 10^{22}$. Figure 3 shows the probability that $c(2^r) > d$ for $d = 10\%, 20\%, \ldots, 50\%$ in 1,000 runs of our procedure. As in Figure 2, the oscillation of the estimated probability decreases, with the number of trials becoming larger. After about 2^{14} trials one can recognize the convergence of the estimation value. For instance, the average of $C(2^{16})$ was 12.1% over all 1,000 runs, the median was 8.4%, and the standard deviation was 15.8%. For $C(2^{14})$ the corresponding values are 14.2%, 9.8%, and 15.5%.

In the fifth experiment, we ran 10,000 trials with 100 different random patterns of length $m \in \{10, 20, \ldots, 50\}$ and $k = 11.2 \cdot m$. The average running times (in seconds) on a DEC Alpha $200^{4/233}$ are shown in the following table:

m	10	20	30	40	50
	2.7	6.7	12.2	18.7	26.8

We did not perform any experiments verifying the accuracy of our estimator for the 20-character alphabet and σ. This is because for interesting values of m and k the condensed k-neighborhood was too large to be enumerated completely.

Acknowledgements

We wish to thank Will Evans for discussions on the convergence properties of our estimator.

References

1. L. Allison and C.S. Wallace. The Posterior Probability Distribution of Alignments and Its Application to Parameter Estimation of Evolutionary Trees and to Optimization of Multiple Alignments. *Journal of Molecular Evolution*, 39:418–430, 1994.
2. W.I. Chang and J. Lampe. Theoretical and Empirical Comparisons of Approximate String Matching Algorithms. In Proc. CPM92, LNCS **644**, pages 175–184, 1992.
3. M.O. Dayhoff, R.M. Schwartz, and B.C. Orcutt. A Model of Evolutionary Change in Proteins. Matrices for Detecting Distant Relationships. *Atlas of Protein Sequence and Structure*, 5:345–358, 1978.
4. W.M. Fitch. Random Sequences. *Journal of Molecular Biology*, 163:171–176, 1983.
5. S. Kannan, Z. Sweedyk, and S. Mahaney. Counting and Random Generation of Strings in Regular Languages. In *Proceedings of the Sixth Annual ACM-SIAM Symposium on Discrete Algorithms*, pages 551–557, 1995.
6. S. Karlin and S.F. Altschul. Methods for Assessing the Statistical Significance of Molecular Sequence Features by using General Scoring Schemes. *Proc. Nat. Acad. Sci. U.S.A.*, 87:2264–2268, 1990.
7. D.E. Knuth, J.H. Morris, and V.R. Pratt. Fast Pattern Matching in Strings. *SIAM Journal on Computing*, 6(2):323–350, 1977.
8. G. Mehldau and E.W. Myers. A System for Pattern Matching Applications on Biosequences. *CABIOS*, 9(3):299–314, 1993.
9. E.W. Myers. An $O(ND)$ Differences Algorithm. *Algorithmica*, 2(1):251–266, 1986.
10. E.W. Myers. A Sublinear Algorithm for Approximate Keyword Searching. *Algorithmica*, 12(4/5):345–374, 1994.
11. E.W. Myers. Approximate Matching of Network Expressions with Spacers. *Journal of Computational Biology*, 3(1):33–51, 1996.
12. E. Ukkonen. Algorithms for Approximate String Matching. *Information and Control*, 64:100–118, 1985.
13. R.A. Wagner and M.J. Fischer. The String to String Correction Problem. *Journal of the ACM*, 21(1):168–173, 1974.

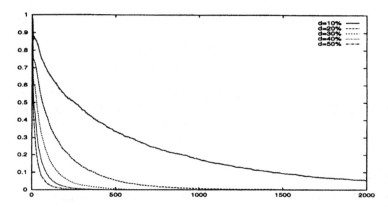

Fig. 1. $Pr[a(t) > d]$ for 10,000 patterns ($|\Sigma| = 4, m = 20, k = 2$)

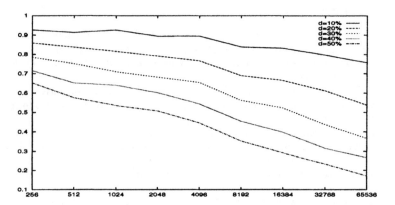

Fig. 2. $Pr[c(2^r) > d]$ for a fixed pattern and 1,000 runs ($|\Sigma| = 4, m = 50, k = 10$)

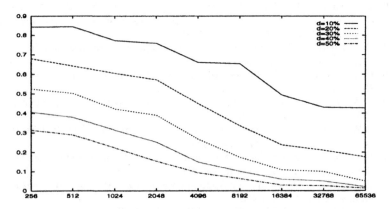

Fig. 3. $Pr[c(2^r) > d]$ for a fixed pattern and 1,000 runs ($|\Sigma| = 20, m = 20, k = 224$)

Space- and Time-Efficient Decoding
with Canonical Huffman Trees*

EXTENDED ABSTRACT

Shmuel T. Klein

Department of Mathematics and Computer Science
Bar Ilan University, Ramat-Gan 52900, Israel
Tel: (972–3) 531 8865   Fax: (972–3) 535 3325
tomi@cs.biu.ac.il

Abstract. A new data structure is investigated, which allows fast decoding of texts encoded by canonical Huffman codes. The storage requirements are much lower than for conventional Huffman trees, $O(\log^2 n)$ for trees of depth $O(\log n)$, and decoding is faster, because a part of the bit-comparisons necessary for the decoding may be saved. Empirical results on large real-life distributions show a reduction of up to 50% and more in the number of bit operations.

1   Introduction

Huffman coding [12] is still one of the best known and most popular data compression methods. While for certain applications, such as data transmission over a communication channel, both coding and decoding ought to be fast, for other applications, like the storage of the various files in a large static full text information retrieval system, compression and decompression are not symmetrical tasks. Compression is done only once, while building the system, whereas decompression is needed during the processing of every query and directly affects response time. There is thus a special focus on fast decoding techniques (see e.g., [13]).

The data structures needed for the decoding of a Huffman encoded file (a Huffman tree or lookup table) are generally considered negligible overhead relative to large texts. However, not all texts are large, and if Huffman coding is applied in connection with a Markov model [1], the required Huffman forest may become itself a storage problem. Moreover, the "alphabet" to be encoded is not necessarily small, and may, e.g., consist of all the different words in the text, so that Huffman trees with thousands and even millions of nodes are not uncommon [20]. We try, in this paper, to reduce the necessary internal memory space

* Partially supported by Grant 8560195 of the Israeli Ministry of Science and Arts

by devising efficient ways to encode these trees. In addition, the new suggested data structure also allows a speed-up of the decompression process, by reducing the number of necessary bit comparisons.

The manipulation of individual bits is indeed the main cause for the slow decoding of Huffman encoded text. A method based on large tables constructed in a pre-processing stage is suggested in [4], with the help of which the entire decoding process can be performed using only byte oriented commands (see also [23]). However, the internal memory required for the storage of these tables may be very large.

In the next section, we recall the necessary definitions of canonical Huffman trees as they are used in the sequel. Section 3 then presents the new suggested data structure, and Section 4 brings some experimental results.

2 Canonical Huffman codes

0	0 0 0
1	0 0 1 0
2	0 0 1 1
3	0 1 0 0
4	0 1 0 1 0
5	0 1 0 1 1
6	0 1 1 0 0
7	0 1 1 0 1
8	0 1 1 1 0 0
9	0 1 1 1 0 1
10	0 1 1 1 1 0
11	0 1 1 1 1 1
12	1 0 0 0 0 0
13	1 0 0 0 0 1
14	1 0 0 0 1 0
15	1 0 0 0 1 1
16	1 0 0 1 0 0 0
17	1 0 0 1 0 0 1
18	1 0 0 1 0 1 0
19	1 0 0 1 0 1 1
...	...
29	1 0 1 0 1 0 1
30	1 0 1 0 1 1 0
31	1 0 1 0 1 1 1 0
32	1 0 1 0 1 1 1 1
33	1 0 1 1 0 0 0 0
...	...
61	1 1 0 0 1 1 0 0
62	1 1 0 0 1 1 0 1
63	1 1 0 0 1 1 1 0 0
64	1 1 0 0 1 1 1 0 1
...	...
124	1 1 1 0 1 1 0 0 1
125	1 1 1 0 1 1 0 1 0
126	1 1 1 0 1 1 0 1 1 0
127	1 1 1 0 1 1 0 1 1 1
...	...
198	1 1 1 1 1 1 1 1 1 0
199	1 1 1 1 1 1 1 1 1 1

Figure 1: *Canonical Huffman code for Zipf-200*

For a given probability distribution, there might be quite a large number of different Huffman trees, since interchanging the left and right subtrees of any internal node will result in a different tree whenever the two subtrees are different in structure, but the weighted average path length is not affected by such an interchange. There are often also other optimal trees, which cannot be obtained via Huffman's algorithm. One may thus choose one of the trees that has some additional properties. The preferred choice for many applications is the *canonical* tree, defined by Schwartz and Kallick [22], and recommended by many others (see, e.g., [13, 24]).

Denote by $(p_1, \ldots, p_n)$ the given probability distribution, where we assume that $p_1 \geq p_2 \geq \cdots \geq p_n$, and let ℓ_i be the length in bits of the codeword assigned by Huffman's procedure to the element with probability p_i, i.e., ℓ_i is the depth of the leaf corresponding to p_i in the Huffman tree. A tree is called canonical if, when scanning its leaves from left to right, they appear in non-decreasing order of their depth (or equivalently, in non-increasing order, as in [19]). The idea is that Huffman's algorithm is only used to generate the lengths $\{\ell_i\}$ of the codewords, rather than the codewords themselves; the latter are easily obtained as follows: the i-th codeword consists of the first ℓ_i bits immediately to the right of the "binary point" in the infinite binary expansion of $\sum_{j=1}^{i-1} 2^{-\ell_j}$, for $i = 1, \ldots, n$ [10]. Many properties of canonical codes are mentioned in [13, 2].

The following will be used as a running example in this paper. Consider the probability distribution implied by Zipf's law, defined by the weights $p_i = 1/(i\, H_n)$, for $1 \le i \le n$, where $H_n = \sum_{j=1}^{n}(1/j)$ is the n-th harmonic number. This law is believed to govern the distribution of the most common words in a large natural language text [25]. A canonical code can be represented by the string $\langle n_1, n_2, \ldots, n_k \rangle$, called a *source*, where k denotes, here and below, the length of the longest codeword (the depth of the tree), and n_i is the number of codewords of length i, $i = 1, \ldots, k$. The source corresponding to Zipf's distribution for $n = 200$ is $\langle 0, 0, 1, 3, 4, 8, 15, 32, 63, 74 \rangle$. The code is depicted in Figure 1.

We shall assume, for the ease of description in this extended abstract, that the source has no "holes", i.e., there are no three integers $i < j < \ell$ such that $n_i \ne 0, n_\ell \ne 0$, but $n_j = 0$. This is true for many, but not all, real-life distributions.

One of the properties of canonical codes is that the set of codewords having the same length are the binary representations of consecutive integers. For example, in our case, the codewords of length 9 bits are the binary integers in the range from 110011100 to 111011010. This fact can be exploited to enable efficient decoding with relatively small overhead: once a codeword of ℓ bits is detected, one can get its relative index within the sequence of codewords of length ℓ by simple subtraction.

The following information is thus needed: let $m = \min\{i \mid n_i > 0\}$ be the length of the shortest codeword, and let $base(i)$ be the integer value of the first codeword of length i. We then have

$$base(m) = 0$$
$$base(i) = 2\,(base(i-1) + n_{i-1}) \qquad \text{for } m < i \le k.$$

Let $B_s(k)$ denote the standard s-bit binary representation of the integer k (with leading zeros, if necessary). Then the j-th codeword of length i, for $j = 0, 1, \ldots, n_i - 1$, is $B_i(base(i) + j)$. Let $seq(i)$ be the sequential index of the first codeword of length i:

$$seq(m) = 0$$
$$seq(i) = seq(i-1) + n_{i-1} \qquad \text{for } m < i \le k.$$

Suppose now that we have detected a codeword w of length ℓ. If $I(w)$ is the integer value of the binary string w (i.e., $w = B_\ell(I(w))$), then $I(w) - base(\ell)$ is the relative index of w within the block of codewords of length ℓ. Thus $seq(\ell) + I(w) - base(\ell)$ is the relative index of w within the full list of codewords. This can be rewritten as $I(w) - diff(\ell)$, for $diff(\ell) = base(\ell) - seq(\ell)$. Thus all one needs is the list of integers $diff(\ell)$. Table 1 gives the values of n_i, $base(i)$, $seq(i)$ and $diff(i)$ for our example.

i	n_i	$base(i)$	$seq(i)$	$diff(i)$
3	1	0	0	0
4	3	2	1	1
5	4	10	4	6
6	8	28	8	20
7	15	72	16	56
8	32	174	31	143
9	63	412	63	349
10	74	950	126	824

Table 1: *Decode values for canonical Huffman code for Zipf-200*

We suggest in the next section a new representation of canonical Huffman codes, which not only is space-efficient, but may also speed up the decoding process, by permitting, at times, the decoding of more than a single bit in one iteration. Similar ideas, based on tables rather than on trees, were recently suggested in [19].

3 Skeleton trees for fast decoding

The following small example, using the data above, shows how such savings are possible. Suppose that while decoding, we detect that the next codeword starts with 1101. This information should be enough to decide that the following codeword ought to be of length 9 bits. We should thus be able, after having detected the first 4 bits of this codeword, to read the following 5 bits as a block, without having to check after each bit if the end of a codeword has been reached. Our goal is to construct an efficient data-structure, that permits similar decisions as soon as they are possible. The fourth bit was the earliest possible in the above example, since there are also codewords of length 8 starting with 110.

3.1 Decoding with sk-trees

The suggested solution is a binary tree, called below an *sk-tree* (for skeleton-tree), the structure of which is induced by the underlying Huffman tree, but which has generally significantly fewer nodes. The tree will be traversed like a regular Huffman tree. That is, we start with a pointer to the root of the tree, and another pointer to the first bit of the encoded binary sequence. This sequence is scanned, and after having read a zero (resp., a 1), we proceed to the left (resp., right) son of the current node. In a regular Huffman tree, the leaves correspond to full codewords that have been scanned, so the decoding algorithm just outputs the corresponding item, resets the tree-pointer to the root and proceeds with scanning the binary string. In our case, however, we visit the tree only up to the depth necessary to identify the length of the current codeword. The leaves of the sk-tree then contain the lengths of the corresponding codewords.

```
{
    tree_pointer  ←—  root
    i  ←—  1
    start  ←—  1
    while i < length_of_string
    {
        if string [i] = 0     tree_pointer  ←—  left (tree_pointer)
        else                  tree_pointer  ←—  right (tree_pointer)
        if value (tree_pointer) > 0
        {
            codeword  ←—  string [start ⋯ (start + value (tree_pointer) − 1)]
            output  ←—  table [I(codeword) − diff [value (tree_pointer)]]
            tree_pointer  ←—  root
            start  ←—  start + value (tree_pointer)
            i  ←—  start
        }
        else          i  ←—  i + 1
    }
}
```

Figure 2: *Decoding procedure using sk-tree*

The formal decoding process using an sk-tree is depicted in Figure 2. The variable *start* points to the index of the the bit st the beginning of the current codeword in the encoded string, which is stored in the vector *string* []. Each node of the sk-tree consists of three fields: a *left* and a *right* pointer, which are not null if the node is not a leaf, and a *value*-field, which is zero for internal nodes, but contains the length in bits of the current codeword, if the node is a leaf. In an actual implementation, we can use the fact that any internal node has either zero or two sons, and store the *value*-field and the *right*-field in the same space, with *left* = *null* serving as flag for the use of the *right* pointer. The procedure also uses two tables: *table* [j], $0 \leq j < n$, giving the j-th element (in non-increasing order of frequency) of the encoded alphabet; and *diff* [i] defined above, for i varying from m to k, that is from the length of the shortest to the length of the longest codeword.

The procedure passes from one level in the tree to the one below according to the bits of the encoded string. Once a leaf is reached, the next *codeword* can be read in one operation. Note that not all the bits of the input vector are individually scanned, which yields possible time savings.

Figure 3 shows the sk-tree corresponding to Zipf's distribution for $n = 200$. The tree is tilted by $45°$, so that left (right) sons are indicated by arrows pointing down (to the right). The framed leaves correspond to the last codewords of the indicated length. The sk-tree of our example consists of only 49 nodes, as opposed to 399 nodes of the original Huffman tree.

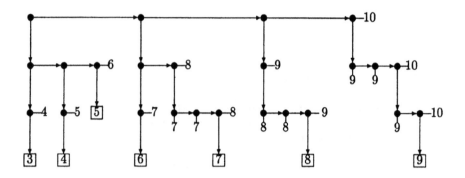

Figure 3: *sk-tree for Zipf-200 distribution*

3.2 Construction of sk-trees

While traversing a standard canonical Huffman tree to decode a given codeword, one may stop as soon as one gets to the root of any full subtree of depth h, for $h \geq 1$, i.e., a subtree of depth h that has 2^h leaves, since at this stage it is known that exactly h more bits are needed to complete the codeword. One way to look at sk-trees is therefore as standard Huffman trees from which all full subtrees of depth $h \geq 1$ have been pruned. A more direct and much more efficient construction is as follows.

The one-to-one correspondence between the codewords and the paths from the root to the leaves in a Huffman tree can be extended to define, for any binary string $S = s_1 \cdots s_e$, the path $P(S)$ induced by it in a tree with given root r_0. This path will consist of $e + 1$ nodes r_i, $0 \leq i \leq e$, where for $i > 0$, r_i is the left (resp. right) son of r_{i-1}, if $s_i = 0$ (resp. if $s_i = 1$). For example, in Figure 3, $P(111)$ consists of the four nodes represented as bullets in the top line. The skeleton of the sk-tree will consist of the paths corresponding to the last codeword of every length. Let these codewords be denoted by L_i, $m \leq i \leq k$; they are, for our example, 000, 0100, 01101, 100011, etc. The idea is that $P(L_i)$ serves as "demarcation line": any node to the left (resp. right) of $P(L_i)$, i.e., a left (resp. right) son of one of the nodes in $P(L_i)$, corresponds to a prefix of a codeword with length $\leq i$ (resp. $> i$).

As a first approximation, the construction procedure thus takes the tree obtained by $\bigcup_{i=m}^{k-1} P(L_i)$ (there is clearly no need to include the longest codeword L_k, which is always a string of k 1's), and adjoins the missing sons to turn it into a complete tree in which each internal node has both a left and a right son. The label on such a new leaf is set equal to the label of the closest leaf following it in an inorder traversal. In other words, when creating the path for L_i, one first follows a few nodes in the already existing tree, then one branches off creating new nodes; as to the labeling, the missing right son of any node in the path will be labeled $i + 1$ (basing ourselves on the assumption that there are no holes), but only the missing left sons of any *new* node in the path will be labeled i.

A closer look then implies the following refinement. Suppose a codeword L_i has a zero in its rightmost position, i.e., $L_i = \alpha 0$ for some string α of length $i - 1$. Then the first codeword of length $i + 1$ is $\alpha 10$. It follows that only when getting to the i-th bit one can decide if the length of the current codeword is i or $i + 1$. But if L_i terminates in a string of 1's, $L_i = \beta 01^a$, with $a > 0$ and $|\beta| + a = i - 1$, then the first codeword of length $i + 1$ is $\beta 10^{a+1}$, so the length of the codeword can be deduced already after having read the bit following β. It follows that one does not always need the full string L_i in the sk-tree, but only its prefix up to and not including the rightmost zero. Let $L_i^* = \beta$ denote this prefix. The revised version of the above procedure starts with the tree obtained by $\bigcup_{i=m}^{k-1} P(L_i^*)$. The nodes of this tree are depicted as bullets in Figure 3. For each path $P(L_i^*)$ there is a leaf in the tree, and the left son of this leaf is the new terminal node, represented in Figure 3 by a box containing the number i. The additional leaves are then filled in as explained above.

3.3   Space complexity

To evaluate the size of the sk-tree, we count the number of nodes added by path $P(L_i^*)$, for $m \leq i < k$. Since the codewords in a canonical code, when ordered by their corresponding frequencies, are also alphabetically sorted, it suffices to compare L_i to L_{i-1}. Let $\gamma(m) = 0$, and for $i > m$, let $\gamma(i)$ be the longest common prefix of L_i and L_{i-1}, e.g., $\gamma(7)$ is the string 10 in our example. Then the number of nodes in the sk-tree is given by:

$$size = 2 \left(\sum_{i=m}^{k-1} \max(0, |L_i^*| - |\gamma(i)|) \right) - 1, \tag{1}$$

since the summation alone is the number of internal nodes (the bullets in Figure 3).

The maximum function comes to prevent an extreme case in which the difference might be negative. For example, if $L_6 = 010001$ and $L_7 = 0101111$, the the longest common prefix is $\gamma(7) = 010$, but since we consider only the bits up to and not including the rightmost zero, we have $L_7^* = 01$. In this case, indeed, no new nodes are added for $P(L_7^*)$.

An immediate bound on the number of nodes in the sk-tree is $O(\min(n, k^2))$, since on the one hand, there are up to $k - 1$ paths $P(L_i^*)$ of lengths $\leq k - 2$, but on the other hand, it cannot exceed the number of nodes in the underlying Huffman tree, which is $2n - 1$. To get a tighter bound, consider the nodes in the upper levels of the sk-tree belonging to the full binary tree F with $k - 1$ leaves and having the same root as the sk-tree. The depth of F is $d = \lceil \log_2(k - 1) \rceil$, and all its leaves are at level d or $d - 1$. The tree F is the part of the sk-tree where some of the paths $P(L_i^*)$ must be overlapping, so we account for the nodes in F and for those below separately. There are at most $2k - 1$ nodes in F; there are at most $k - 1$ disjoint paths below it, with path $P(L_i^*)$ extending at most $i - 2 - \lfloor \log_2(k - 1) \rfloor$ nodes below F, for $\log_2(k - 1) < i \leq k$. This yields as bound for the number of nodes in the sk-tree:

$$2k + 2 \left(\sum_{i=1}^{k-2-\lfloor \log_2(k-1) \rfloor} i \right) = 2k + (k - 2 - \lfloor \log_2(k-1) \rfloor)(k - 1 - \lfloor \log_2(k-1) \rfloor).$$

There are no savings in the worst case, e.g., when there is only one codeword of each length (except for the longest, for which there are always at least two). More generally, if the depth of the Huffman tree is $\Omega(n)$, the savings might not be significant. But such trees are optimal only for some very skewed distributions. In many applications, like for most distributions of characters or character pairs or words in most natural languages, the depth of the Huffman tree is $O(\log n)$, and for large n, even the constant c, if the depth is $c \log_2 n$, must be quite small. For suppose the Huffman tree has a leaf on depth d. Then by [14, Theorem 1], the probability of the element corresponding to this leaf is $p < 1/F_{d+1}$, where F_j is the j-th Fibonacci number, and we get from [15, Exercise 1.2.1–4], that $p < (1/\phi)^{d-1}$, where $\phi = (1 + \sqrt{5})/2$ is the golden ratio. Thus if $d > c \log_2 n$, we have

$$p < \left(\frac{1}{\phi} \right)^{c \log_2 n} = n^{-c \log_2(1/\phi)} = n^{-0.693c}.$$

To give a numeric example, in Section 4 below one of the Huffman trees corresponds to the different words in English, with $n = 289,101$ leaves. The probability for a tree of this size to have a leaf at level $3 \log_2 n$ is less than 4.4×10^{-12}, which means that even if the word with this probability appears only once, the text must be at least 4400 billion words long, enough to fill about 35,000 CD-Roms! But even if the original Huffman tree would be deeper, it is sometimes convenient to impose an upper limit of $B = O(\log n)$ on the depth, which often implies only a negligible loss in compression efficiency [8]. In any case, given a logarithmic bound on the depth, the size of the sk-tree is about

$$\log n \, (\log n - \log \log n).$$

3.4 Time complexity

When decoding is based on a standard Huffman tree, the average number of comparisons per codeword is the sum, taken over all the leaves i, of the depth of i in the tree times the probability to get to i. A similar sum holds for sk-trees, with the difference that a leaf does not correspond to a single element, but to several consecutive codewords of the same length. Let w be the prefix of a codeword corresponding to a leaf of the sk-tree labeled ℓ, $\ell \geq |w|$, and denote $t = \ell - |w|$. Then the 2^t codewords corresponding to this leaf of the sk-tree are $w0^t, \ldots, w1^t$, and correspond, using the notations of Section 2, to indices in the range from $I(w0^t) - \text{diff}(\ell)$ to $I(w1^t) - \text{diff}(\ell)$. The average number of comparisons per codeword using the sk-tree can thus be evaluated as:

$$\sum_{i \in \{\text{leaves in sk-tree}\}} \left(d(i) \sum_{j=I(w(i)0^{label(i)-d(i)})-\text{diff}(label(i))}^{I(w(i)1^{label(i)-d(i)})-\text{diff}(label(i))} \text{Prob}(j) \right), \quad (2)$$

where $w(i)$ is the binary string corresponding to the leaf i, $d(i) = |w(i)|$ is the depth of i in the tree, and $\text{Prob}(j)$ is the probability of the element with index j.

As an approximation, we assume that the probability of an element on level i in the tree is 2^{-i}. This corresponds to a *dyadic* probability distribution, where all the probabilities are integral powers of $\frac{1}{2}$. There cannot be too great a difference between the actual probability distribution and this dyadic one, since they both yield the same Huffman tree (see [17] for bounds on the "distance" between such distributions). Given this model, eqn. (2) becomes

$$\sum_{i \in \{\text{leaves in sk-tree}\}} \left(d(i)\, 2^{-d(i)} \right).$$

A similar sum, but taken over all the leaves of the original Huffman tree gives the average codeword length. Therefore the savings will depend on both the size and the shape of the sk-tree.

4 Experimental Results

To test the effectiveness of the use of sk-trees, the following real-life distributions were used. The data for French was collected from the *Trésor de la Langue Française*, a database of 680 MB of French language texts (115 million words) of the 17th–20th centuries [3]; for English, the source are 500 MB (87 million words) of the *Wall Street Journal* [21]; and for Hebrew, a part of the *Responsa Retrieval Project*, 100 MB of Hebrew and Aramaic texts (15 million words) written over the past ten centuries [5]. The first set of alphabets consists of the bigrams in the three languages (the source for English for this distribution was [11]); for the next set, the elements to be encoded are the different *words*, which yields very large "alphabets"; and the final set contains the distribution of trigrams in French. For completeness, the Zipf-200 distribution used in the above examples was also added.

Table 2 displays the results. The first three columns give some statistics about the various distributions: the depth k of the Huffman tree, the size n of the encoded alphabet, and the weighted average length of a codeword, measured in bits, which equals the average number of comparisons if the standard Huffman tree is used. The next two columns bring the number of nodes in the sk-tree, as given in eqn. (1), and the average number of comparisons per codeword when decoding is based on the sk-tree, as given in eqn. (2). The final two columns show the relative savings in space and time, measured in percent. We see that for large distributions, roughly half of the comparisons may be saved, and that the cost of storing the sk-tree is only several percent of the cost for the full Huffman tree.

Source	k	total number of elements	average codeword length	number of nodes in sk-tree	average number of comparisons	relative savings	
						space	time
Zipf–200	10	200	6.024	49	3.990	87.7%	33.7%
English	13	371	7.445	67	4.200	91.0%	43.6%
bigrams French	29	2192	7.784	285	4.620	93.5%	40.6%
Hebrew	24	743	8.037	127	4.183	91.4%	48.0%
English	26	289101	11.202	425	5.726	99.93%	48.9%
words French	27	439191	10.473	443	5.581	99.95%	46.7%
Hebrew	24	296933	13.033	345	5.694	99.94%	56.3%
trigrams French	28	25781	10.546	381	5.026	99.3%	52.3%

Table 2: *Time and Space requirements for real-life distributions*

5 Future work

All the above examples satisfied the assumption that their sources had no holes. We still have to address the problem of dealing also with other distributions. For example, the distribution of one of the alphabets used for compressing a set of sparse bitmaps in [6] is $\langle 1, 0, 0, 1, 7, 0, 1, 28, 0, 46, 59, 114 \rangle$.

Another direction to explore is the time/space tradeoffs obtained by pruning the sk-tree at some internal node: one would thus get to leaves at which it is not yet possible to deduce the length of the current codeword, but at which some partial information is already available. For example, in Figure 3, if the bits already processed were 111 (corresponding to the internal node in the rightmost upper corner), we know already that the length of the current codeword is either 9 or 10.

References

1. BOOKSTEIN A., KLEIN S.T., Compression, Information Theory and Grammars: A Unified Approach, *ACM Trans. on Information Systems* **8** (1990) 27–49.
2. BOOKSTEIN A., KLEIN S.T., Is Huffman coding dead?, *Computing* **50** (1993) 279–296.
3. BOOKSTEIN A., KLEIN S.T., ZIFF D.A., A systematic approach to compressing a full text retrieval system, *Information Processing & Management* **28** (1992) 795–806.
4. CHOUEKA Y., KLEIN S.T., PERL Y., Efficient Variants of Huffman Codes in High Level Languages, *Proc. 8-th ACM-SIGIR Conf.*, Montreal (1985) 122–130.
5. FRAENKEL A.S., All about the Responsa Retrieval Project you always wanted to know but were afraid to ask, Expanded Summary, *Jurimetrics J.* **16** (1976) 149–156.

6. FRAENKEL A.S., KLEIN S.T., Novel Compression of Sparse Bit-Strings, in *Combinatorial Algorithms on Words*, NATO ASI Series Vol **F12**, Springer Verlag, Berlin (1985) 169–183.

7. FRAENKEL A.S., KLEIN S.T., Bidirectional Huffman Coding, *The Computer Journal* **33** (1990) 296–307.

8. FRAENKEL A.S., KLEIN S.T., Bounding the Depth of Search Trees, *The Computer Journal* **36** (1993) 668–678.

9. FERGUSON T.J., RABINOWITZ J.H., Self-synchronizing Huffman codes, *IEEE Trans. on Information Theory*, IT-**30** (1984) 687–693.

10. GILBERT E.N., MOORE E.F., Variable-length binary encodings, *The Bell System Technical Journal* **38** (1959) 933–968.

11. HEAPS H.S., *Information Retrieval, Computational and Theoretical Aspects*, Academic Press, New York (1978).

12. HUFFMAN D., A method for the construction of minimum redundancy codes, *Proc. of the IRE* **40** (1952) 1098–1101.

13. HIRSCHBERG D.S., LELEWER D.A., Efficient decoding of prefix codes, *Comm. of the ACM* **33** (1990) 449–459.

14. KATONA G.H.O., NEMETZ T.O.H., Huffman codes and self-information, *IEEE Trans. on Inf. Th.* IT-**11** (1965) 284–292.

15. KNUTH D.E., *The Art of Computer Programming, Vol I, Fundamental Algorithms*, Addison-Wesley, Reading, MA (1973).

16. LELEWER D.A., HIRSCHBERG D.S., Data compression, *ACM Computing Surveys* **19** (1987) 261–296.

17. LONGO G., GALASSO G., An application of informational divergence to Huffman codes, *IEEE Trans. on Inf. Th.* IT-**28** (1982) 36–43.

18. MOFFAT A., BELL T., In-situ generation of compressed inverted files, *J. ASIS* **46** (1995) 537–550.

19. MOFFAT A., TURPIN A., On the implementation of minimum redundancy prefix codes, *Proc. Data Compression Conference DCC-96*, Snowbird, Utah (1996) 182–191.

20. MOFFAT A., TURPIN A., KATAJAINEN J., Space-efficient construction of optimal prefix codes, *Proc. Data Compression Conference DCC-95*, Snowbird, Utah (1995) 192–201.

21. MOFFAT A., ZOBEL J., SHARMAN N.,Text compression for dynamic document databases, to appear in *IEEE Transactions on Knowledge and Data Engineering*. Preliminary version in *Proc. Data Compression Conference DCC-94*, Snowbird, Utah (1994) 126–135.

22. SCHWARTZ E.S., KALLICK B., Generating a canonical prefix encoding, *Comm. of the ACM* **7** (1964) 166–169.

23. SIEMINSKI, A., Fast decoding of the Huffman codes, *Information Processing Letters* **26** (1988) 237–241.

24. WITTEN I.H., MOFFAT A., BELL T.C., *Managing Gigabytes: Compressing and Indexing Documents and Images*, Van Nostrand Reinhold, New York (1994).

25. ZIPF G.K., *The Psycho-Biology of Language*, Boston, Houghton (1935).

On Weak Circular Squares in Binary Words

Aviezri S. Fraenkel[1], Jamie Simpson[2] and Mike Paterson[3]

[1] Department of Applied Mathematics and Computer Science, Weizmann Institute
of Science, Rehovot 76100, Israel
http://www.wisdom.weizmann.ac.il/~fraenkel   fraenkel@wisdom.weizmann.ac.il
[2] School of Mathematics, Curtin University, Perth WA 6001, Australia;
http://www.cs.curtin.edu.au/~simpson   simpson@cs.curtin.edu.au
[3] Department of Computer Science, University of Warwick, Coventry CV4 7AL, UK
http://www.dcs.warwick.ac.uk/   msp@dcs.warwick.ac.uk.

Abstract. A *weak square* in a binary word is a pair of adjacent non-empty blocks of the same length, having the same number of 1s. A *weak circular square* is a weak square which is possibly wrapped around the word: the tail protruding from the right end of the word reappears at the left end. Two weak circular squares are *equivalent* if they have the same length and contain the same number of ones. We prove that the longest word with only k inequivalent weak circular squares contains $4k + 2$ bits and has the form $(01)^{2k+1}$ or its complement. Possible connections to tandem repeats in the human genome are pointed out.

1 Introduction

A *binary word* or *binary sequence* is a finite or infinite concatenation of 0s and 1s. A *weak square* in a binary word w is a pair of adjacent nonempty blocks in w, each having the same number of 1s and each having the same number of 0s. We say that two weak squares are *equivalent* if they have the same length and contain the same number of 1s. Otherwise they are *inequivalent*. Equivalence is clearly an equivalence relation. The *conjugates* of a binary word are its cyclic permutations. Thus the conjugates of 0010 are 0010, 0100, 1000 and 0001.

A *weak circular square* in a binary word is a weak square in any of the word's conjugates. For example, the word w=0110110 contains 4 weak circular squares 0110, 011011, 11 and 00. The last occurs in the conjugate 1001101 (and in some others). Equivalence and inequivalence of weak circular squares are defined as for weak squares. The word w also contains the weak circular square 1001 but this is equivalent to 0110. Unless otherwise specified, a *word* means a binary word, and its digits are *bits*. In this note we are interested in the longest word which contains only k weak circular squares. This is equivalent to finding a word of given length such that the number of inequivalent weak squares appearing in its set of conjugates is minimal. We do not allow weak circular squares whose total length is greater than that of the word, that is, we do not allow a weak circular square to overlap itself.

We are interested in finding the longest word which contains only k inequivalent weak circular squares. Our main result, Theorem 2, states that such words contain $4k + 2$ bits, and have the form $(01)^{2k+1}$ or its complement, which is also its reverse.

We will prove Theorem 2 by showing, in Theorem 1, that if a word w contains n 1s, then for every $k \in \mathbb{Z}^+$, $2k \leq n$, one of the conjugates of w must contain a weak square with $2k$ 1s, that is, w contains a weak circular square, each half of which contains k 1s. Theorem 2 is based on Lemma 1, whose proof is the only place where we use circularity in an essential way.

For the non-circular case we formulate the following

Conjecture. A longest word containing only k inequivalent weak squares contains $4k + 3$ bits and has one of the forms: $(01)^{2k+1}0$, $0^{2k+1}10^{2k+1}$ or their complements.

2 Background

A *square* is a pair of adjacent nonempty blocks which are identical. Two squares are *equivalent* if one can be translated onto the other. Otherwise they are *inequivalent*. It has been shown many times that there exist infinite *ternary* sequences which are *squarefree*, i.e., they contain no square. See, e.g., Thue [1912], Morse and Hedlund [1944], Hawkins and Mientka [1956], Leech [1957], Novikov and Adjan [1968], Pleasants [1970], Burris and Nelson [1971/72], del Junco [1977], Ehrenfeucht and Rozenberg [1983]. (Currie [1993] wrote: "One reason for this sequence of rediscoveries is that nonrepetitive sequences have been used to construct counterexamples in many areas of mathematics: ergodic theory, formal language theory, universal algebra and group theory, for example...".) Actually, Thue [1912] showed more: there exists a doubly infinite squarefree ternary sequence which also avoids the two triples $a_1 a_3 a_1$ and $a_2 a_3 a_2$. See Berstel [1992, §4.2] for an exposition of the full result, and Berstel [1995] for an English translation of Thue's papers.

Every binary word of length 4, however, must already contain a square. Let $g(k)$ be the length of a longest binary word containing at most k inequivalent squares. Then clearly $g(0) = 3$; in fact, the only squarefree words of length 3 are 010 and its 1-complement 101. A computer disclosed that $g(1) = 7$; the only binary words of length 7 with only 1 square are

$$0001000, \qquad 0100010, \qquad 0111011$$

and their 1-complements and the reverse of 0111011 and its 1-complement. Further, $g(2) = 18$; the only binary words of length 18 which contain only 2 distinct squares are 010011000111001101 and its 1-complement (which is also its reverse).

Entringer, Jackson and Schatz [1974] proved that $g(5) = \infty$, i.e., there exists an arbitrarily long binary sequence with only 5 inequivalent squares. Fraenkel and Simpson [1995] showed that already $g(3) = \infty$, by constructing an arbitrarily long binary sequence whose only squares are 0^2, 1^2 and $(01)^2$. This is best possible, since $g(2)$ is finite.

A word is *strongly squarefree* if it has no weak square. Erdös [1961] has raised the question of the maximum length of a word on a symbols which is strongly squarefree. Pleasants [1970] constructed an arbitrarily long word on $a = 5$ symbols which is strongly squarefree, and showed that for $a \leq 3$, every sufficiently long word contains a weak square. Keränen [1992] has shown that Pleasants' $a = 5$ can be replaced by $a = 4$. Historically, the concept "strongly squarefree" was first, and it suggested to us to define the notion of weak squares.

Every binary word of length at least 4 has a weak square because it has a square; so it is natural to ask for the longest word having only k weak squares. In the present paper we take a first step in this direction, by solving this problem for the case of weak circular squares.

We also remark that questions regarding squares in sequences arise in molecular biology, where they are known as *repeats*, or *tandem repeats*. In fact, the most frequent repeat in the human genome seems to be the word GT, with high *copy number* (the number of times GT is repeated). Only about 10% of the human genome codes for proteins; the rest, called by biologists *junk DNA*, contains the tandem repeats. Biologists are slowly modifying their view as to the appropriateness of the terminology "junk DNA". For example, Trifonov [1989] argues that the copy number influences the functions of DNA chains adjacent to the repeated word, such as their binding power and gene expression; it can even cause certain diseases if too high or too low; and it also influences the unwinding capability of the DNA helix. Algorithms for identifying repeats and databases of repeats in the human genome are maintained by Milosavljević [$\geq$ 1997].

Since the copy number at a given site changes from one individual to another, the copy number has also been used in DNA-*fingerprinting*. This application appears to have been originated by Alec Jeffreys' group in Leicester. See, e.g., Jeffreys, Wilson and Thein [1985] and Jeffreys, Turner and Debenham [1991]. Further elaborations on applications of DNA-fingerprinting to medicine and forensic medicine are given in Raskó and Downes [1995 (ch. 6, especially p. 156, and ch. 12, especially pp. 379–380)], where it is also stated that the human genome contains some 500,000 repeated words. (Keywords for human genome applications are VNTR (Variable Number Tandem Repeats) and mini- and microsatellite sequences for the basic subwords that are repeated.)

3   A Lemma

We may represent any word w of length m containing n 1s by the sequence

$$(a_0, a_1, \ldots, a_{n-1}; m), \tag{1}$$

where the j-th digit of w is 1 if and only if $j = a_i$ for some i ($a_i, j \in \{0, \ldots, m-1\}$). Thus the word $w = 100100011$ corresponds to the sequence $(0, 3, 7, 8; 9)$. Since we are concerned with *circular* squares, arithmetic on the a_is is done over the ring $\mathbb{Z}_m$ of integers modulo m, i.e., it is to be interpreted modulo m, with the result in the interval $[0, m-1]$, while arithmetic on the subscripts of the a_is is over $\mathbb{Z}_n$, i.e., it is to be interpreted modulo n, with the result in the interval

$[0, n-1]$. The inequalities below involving the a_is are over $\mathbb{Z}_m$, for which we use $\prec, \preceq, \succ, \succeq$. Arithmetic on the indexes is over $\mathbb{Z}_n$.

As a warm-up, we consider the ways in which a weak circular square with 1s at the *specified locations* $a_{i+1}, \ldots, a_{i+2k}$ $(2k \leq n)$ and in the *specified order* does not exist in a given word w of the form (1). First, it does not exist when the left-most set of k 1s is separated by more 0s than lie in between and around the k 1s to their right. Example: $w = 0100001010101$, where $m = 13$, $n = 5$, $(a_0, a_1, a_2, a_3, a_4) = (1, 6, 8, 10, 12)$. Let $k = 2$, $i = 4$. Then $i+1 = 0$, and there is no weak circular square consisting of a_0, a_1, a_2, a_3. However, w clearly contains a weak circular square with four 1s, e.g., $0101\,0101$, which consists of a *different* set of a_is. The same situation occurs when "left" and "right" are interchanged. Secondly, it does not exist when there is no 0 to the left of the left-most 1 nor to the right of the right-most 1, and when the total number of 0s between them is odd. Example: $w=010011010$ with $m = 9$, $n = 4$, $(a_0, a_1, a_2, a_3) = (1, 4, 5, 7)$. Let $k = 2$, $i = 1$. There is no weak circular square consisting of a_2, a_3, a_0, a_1 in this order. Again we see, however, that w does contain a weak circular square, e.g., $0110\,1001$, which consists of the *same* set of a_i but in a different order. We can in fact show that the conditions

(a) $a_{i+k} - a_{i+1} + 1 \preceq a_{i+2k+1} - a_{i+k} - 1$,
(b) $a_{i+k+1} - a_i - 1 \succeq a_{i+2k} - a_{i+k+1} + 1$,
(c) if $a_{i+1} - a_i = 1$, $a_{i+2k+1} - a_{i+2k} = 1$ then $a_{i+2k} - a_{i+1} - 2k + 1$ is even,

are necessary and sufficient for the existence of a weak circular square with $2k$ 1s at *specified* locations and in a *given* order, but we won't need this fact in the sequel.

Lemma 1. *Given a word w of the form (1), where k is a fixed positive integer in $\mathbb{Z}_n$. Then the inequality*

$$a_{i+k} - a_i \succ a_{i+2k} - a_{i+k} \qquad (2)$$

cannot hold for all i in $\mathbb{Z}_n$. Similarly, (1) with the inequality sign reversed, namely

$$a_{i+k} - a_i \prec a_{i+2k} - a_{i+k}, \qquad (3)$$

cannot hold for all i in $\mathbb{Z}_n$.

Proof. Suppose that (2) holds for all i. Iterating this inequality with i replaced by $i + k$, we get the infinite chain,

$$a_{i+k} - a_i \succ a_{i+2k} - a_{i+k} \succ a_{i+3k} - a_{i+2k} \succ \ldots.$$

In any block of length $n+1$ of this chain there must be two equal terms, which is a contradiction. This proves the first statement of the lemma. The second part is proved in the same way with the inequalities reversed. $\qquad \square$

For the noncircular case Lemma 1 fails — the chain we used in the proof is no longer finite.

4 Two Theorems

We use Lemma 1 to prove our first theorem.

Theorem 1. *Let $k \geq 1$ and $n \geq 2k$. Then there exists some $i \in \mathbb{Z}_n$ such that the word (1) contains a weak circular square using the $2k$ 1s at positions $a_{i+1}, \ldots, a_{i+2k}$ in this order.*

Proof. If there is some i such that (2) becomes an equality, i.e., $a_{i+k} - a_i = a_{i+2k} - a_{i+k}$, we are done, since the left-hand side is the number of bits in the semi-closed interval $I_1 = (a_i, a_{i+k}]$, and the right-hand side is the number of bits in $I_2 = (a_{i+k}, a_{i+2k}]$, so we have the weak circular square consisting of the $2k$ 1s at $a_{i+1}, \ldots, a_{i+2k}$, with k of them in I_1 and the other k in I_2. So we assume that there is no such i. Then for every i, either (2) holds or (3) holds. By Lemma 1 we may thus assume, without loss of generality, that there is some i for which (2) holds, but when we replace i by $i + 1$, then the reverse inequality, namely,

$$a_{i+k+1} - a_{i+1} \prec a_{i+2k+1} - a_{i+k+1} \qquad (4)$$

holds. Let $I_3 = [a_{i+1}, a_{i+k+1})$ and $I_4 = [a_{i+k+1}, a_{i+2k+1})$. (See Figure 1.)

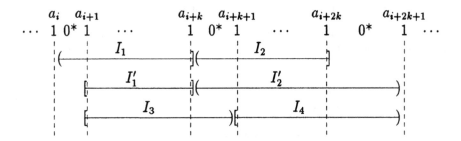

Figure 1.

By (2), $|I_1| > |I_2|$. We can slide the left (open) boundary of I_1 to the right, bit by bit, up to the 0 to the left of a_{i+1}, thus reducing the size of I_1, without changing the number of 1s it contains. If at any stage equality holds, we are done. Similarly, we may slide the right (closed) boundary of I_2 to the right, without changing the number of 1s, thus increasing the size of I_2. If there is never equality, we have even $|I_1'| > |I_2'|$, where $I_1' = [a_{i+1}, a_{i+k}]$ is closed, and $I_2' = (a_{i+k}, a_{i+2k+1})$ is open. Since $|I_3| \geq |I_1'|$ and $|I_2'| \geq |I_4|$, we therefore have $|I_3| > |I_4|$, which contradicts (4). Thus equality must hold at some stage, so w contains a weak circular square. $\qquad \square$

Theorem 2. *For any $k \geq 0$, a longest word containing only k inequivalent weak circular squares has length $4k + 2$ and has the form $(01)^{2k+1}$ or its complement.*

Proof. It is clear that the word $(01)^{2k+1}$ and its complement each have length $4k + 2$ and contain precisely k weak circular squares. We must show that there does not exist a longer word containing only k weak circular squares, and that $(01)^{2k+1}$ and its complement are the only words of length $4k + 2$ containing k weak circular squares.

Given any word of length at least $4k + 3$, without loss of generality suppose it contains at least as many 1s as 0s. Then it contains at least $2k + 2$ 1s. By Theorem 1 it then contains weak circular squares with $2i$ 1s for each value of i from 1 to $k + 1$. Thus it contains at least $k + 1$ weak circular squares. It follows that the longest word containing only k weak circular squares has length $4k + 2$.

Now suppose that we have a word of length $4k + 2$ containing k weak circular squares. Again we may assume that it contains at least as many 1s as 0s, and by an argument similar to the above conclude that it has exactly $2k + 1$ 1s, $2k + 1$ 0s and weak circular squares with $2i$ 1s for each value of i from 1 to k. If the word contains 2 adjacent 0s then it will also contain the weak square 00, so this cannot occur. Interchanging 0s and 1s does not alter the number of weak squares, so we conclude that the word does not contain adjacent 1s or adjacent 0s. It must therefore be $(01)^{2k+1}$ or its complement as required. $\square$

References

1. J. Berstel [1992], Axel Thue's work on repetitions in words, in: *Séries Formelles et Combinatoire Algébrique* (P. Leroux and C. Reutenauer, eds.), Publ. du LaCIM, Vol. 11, Université de Québec à Montréal, Canada, pp. 65-80.

2. J. Berstel [1995], Axel Thue's papers on repetition in words: an English translation, Publ. du LACIM, Vol. 20, Université de Québec à Montréal, Canada.

3. S. Burris and E. Nelson [1971/72], Embedding the dual of π_∞ in the lattice of equational classes of semigroups, *Algebra Universalis* 1, 248–153.

4. J.D. Currie [1993], Open problems in pattern avoidance, *Amer. Math. Monthly* 100, 790–793.

5. A. del Junco [1977], A transformation with simple spectrum which is not rank one, *Canad. J. Math.* 29, 655–663.

6. A. Ehrenfeucht and G. Rozenberg [1983], On the separating power of EOL systems, *RAIRO Inform. Théor.* 17, 13–22.

7. R. Entringer, D. Jackson and J. Schatz [1974], On nonrepetitive sequences, *J. Combin. Theory* (Ser. A) 16, 159–164.

8. P. Erdős [1961], Some unsolved problems, *Magyar Tud. Akad. Mat. Kutato. Int. Kozl.* 6, 221–254.

9. A.S. Fraenkel and R.J. Simpson [1995], How many squares must a binary sequence contain? *Electronic J. Combinatorics* 2 (1995) R2, 9pp.
   http://ejc.math.gatech.edu:8080/Journal/journalhome.html

10. D. Hawkins and W.E. Mientka [1956], On sequences which contain no repetitions, *Math. Student* 24, 185–187.

11. A.J. Jeffreys, M. Turner and P. Debenham [1991], The efficiency of multilocus DNA fingerprint probes for individualization and establishment of family relationships, determined from extensive casework, *Am. J. Hum. Genet.* 48, 824–840.

12. A.J. Jeffreys, V. Wilson and S.L. Thein [1985], Hypervariable 'minisatellite' regions in human DNA, *Nature* **314**, 67–73.

13. A.J. Jeffreys, V. Wilson and S.L. Thein [1985], Individual-specific 'fingerprints' of human DNA, *Nature* **316**, 76–79.

14. V. Keränen [1992], Abelian squares are avoidable on 4 letters, Automata, Languages and Programming: Lecture Notes in Computer Science 623, 41–52, Springer Verlag.

15. J.A. Leech [1957], A problem on strings of beads, *Math. Gaz.* **41**, 277–278.

16. A. Milosavljević [≥ 1997], Repeat Analysis, Ch. 13, Sect. 4, Imperial Cancer Research Fund Handbook of Genome Analysis, Blackwell Scientific Publications, in press.

17. M. Morse and G.A. Hedlund [1944], Unending chess, symbolic dynamics and a problem in semigroups, *Duke Math. J.* **11**, 1–7.

18. P.S. Novikov and S.I. Adjan [1968], Infinite periodic groups I, II, III, *Izv. Akad. Nauk. SSSR Ser. Mat.* **32**, 212–244; 251–524; 709–731.

19. P.A.B. Pleasants [1970], Non-repetitive sequences, *Proc. Cambridge Phil. Soc.* **68**, 267–274.

20. I. Raskó and C.S. Downes [1995], *Genes in Medicine: Molecular Biology and Human Genetic Disorders*, Chapman & Hall, London.

21. A. Thue [1912], Über die gegenseitige Lage gleicher Teile gewisser Zeichenreihen, *Norske Vid. Selsk. Skr., I. Mat. Nat. Kl. Christiania* **I**, 1–67.

22. E.N. Trifonov [1989], The multiple codes of nucleotide sequences, *Bull. Math. Biology* **51**, 417-432.

An Easy Case of Sorting by Reversals

Nicholas Tran

Department of Computer and Information Science
University of Pennsylvania
Philadelphia, PA 19104
nick@central.cis.upenn.edu

Abstract. We show that sorting by reversals can be performed in polynomial time when the number of breakpoints is twice the distance. This result answers an open question in [KS95].

1 Introduction

Biologists have discovered that genome rearrangements is a common mode of molecular evolution: the chromosomes of two related species often have the same set of genes arranged in different orders [PH88]. Genome rearrangements are brought about by four basic mechanisms: the reversal in order of a chromosome portion, (inversion), the relocation of chromosome portions (transposition), the exchange of prefixes and suffixes of two chromosomes (translocation), and the copying of a chromosome portion (duplication). An important question in biology is concerned with the minimum number of these operations (called distance) required to transform one genome to another; this distance is believed to be a useful measure of the evolutionary distance between two species.

Inversions are the most common among the four types of operations, especially in organisms with one chromosome. Kececioglu and Sankoff were the first to frame and study the problem of finding the reversal distance in terms of permutations. They gave an approximation algorithm with a performance ratio of 2 and conjectured that in general the problem is NP-hard [KS95]. Recently, this conjecture was confirmed by Caprara [Cap97]. However, the reversal distance problem becomes easy if additional assumptions are made. For example, building on work by Bafna and Pevzner [BP93], Hannenhalli and Pevzner have shown that the following restrictions of the problem can be solved in polynomial time: i) if the permutations have a "small" number of blocks (of consecutive elements) of size one, and ii) if the directions of the genes are known, i.e. the permutations are signed [HP96, HP95]. Their algorithm for the second case has recently been improved and simplified by Berman and Hannenhalli, and by Kaplan, Shamir, and Tarjan [BH96, KST97].

Another restriction of this problem involves permutations whose reversal distances are twice the number of "breakpoints", which are boundaries between blocks of consecutive elements. It was conjectured in [PW95, KS95] that even this special case is NP-hard. In this note, we show that in fact it can be solved in polynomial time by giving a graph-theoretic characterization of these permutations.

2 Definitions

A *permutation* $\pi = (\pi_1 \pi_2 \ldots \pi_n)$ is a 1-1 function $\pi : [0, n+1] \mapsto [0, n+1]$, where $\pi(0) = 0, \pi(n+1) = n+1$, and $\pi(i) = \pi_i$ for $1 \leq i \leq n$. A *reversal of interval* $[i, j]$ is the permutation

$$\rho_{ij} = (1\ 2\ \ldots\ i\ j\ j-1\ \ldots\ i+2\ i+1\ j+1\ j+2\ \ldots\ n).$$

Given permutations π and σ, the *reversal distance between π and σ* is the length of a shortest sequence of reversals $\rho_1, \rho_2, \ldots, \rho_k$ such that $\pi \cdot \rho_1 \cdot \rho_2 \cdots \rho_k = \sigma$. (Note that this definition is robust since the reversals generate the permutation group S_n.) It is easy to see that this distance is at most $n - 1$ [WEHM82]. *Sorting by reversals* is the problem of finding the reversal distance $d(\pi)$ between a permutation π and the identity permutation $\imath$.

Fix a permutation $\pi \in S_n$. For $0 \leq i \leq n$, we call (π_i, π_{i+1}) an *adjacency* of π if $\pi_i \sim \pi_{i+1}$ ($i \sim j$ means $|i - j| = 1$); otherwise, (π_i, π_{i+1}) is called a *breakpoint* of π. Let $bp(\pi)$ denote the number of breakpoints of π; note that $bp(\pi) \leq n+1$, and $bp(\imath) = 0$. Two breakpoints of π (π_i, π_{i+1}) and (π_j, π_{j+1}) define an *active interval* $[i, j]$ if $\pi_i \sim \pi_j$ and $\pi_{i+1} \sim \pi_{j+1}$; similarly they define a *passive interval* $]i, j[$ if $\pi_i \sim \pi_{j+1}$ and $\pi_{i+1} \sim \pi_j$.

Note that reversals removing two breakpoints are the only ones that can appear in any sequence of reversals reducing a permutation π to $\imath$ in $bp(\pi)/2$ steps. This observation allows us to restrict our attention to active and passive intervals, which are much simpler notions than *proper* intervals used by Pevzner et al. in defining oriented/non-oriented cycles.

Let B_π be the graph whose vertices are breakpoints of π, and whose edges connect those breakpoints that form active or passive intervals. If B_π has a perfect matching M, let I_M be the graph whose vertices are the intervals defined by the edges of M, and whose edges connect intersecting intervals. Two intervals $[i, j]$ and $[k, l]$ *intersects* each other if $i < k < j < l$ or $k < i < l < j$.

Figure 1 gives an example of these definitions for $\pi = (0\ 3\ 7\ 5\ 1\ 4\ 2\ 6\ 8)$. π has one adjacency $(8,9)$ and eight breakpoints $(0,3), (3,7), (7,5), (5,1), (1,4), (4,2), (2,6),$ and $(6,8)$. π has three active intervals, drawn with unbroken lines, and two passive intervals, drawn with dotted lines. The vertices of the graph B_π are the breakpoints of π, and the edges connect pairs of breakpoints that form an interval. M is a perfect matching of B_π. The vertices of the graph I_M are the intervals of π; for example, the interval connecting the two breakpoints $(0,3)$ and $(1,4)$ corresponds to the reversal of interval $[0, 4]$ and is labeled as $\rho(0, 4)$. The edges of I_M connect pairs of intersecting intervals.

3 Main Result

Suppose π is a permutation that satisfies $bp(\pi) = 2d(\pi)$, and $\rho_1, \rho_2, \ldots, \rho_{d(\pi)}$ be a solution of the reversal distance problem for π. We already observed that each ρ_i must remove a pair of breakpoints. The following lemma says that these pairs of breakpoints must already be present in π.

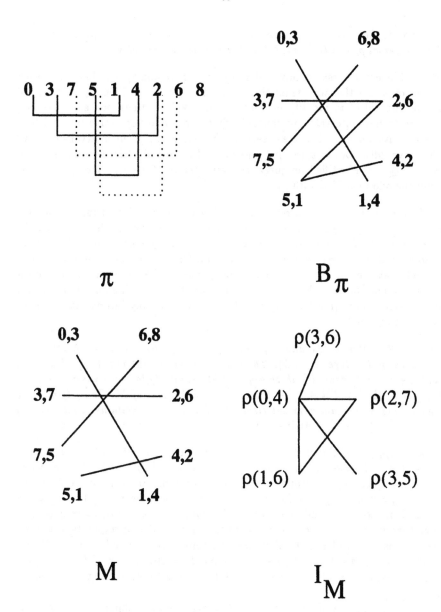

Fig. 1. A permutation π and its associated graphs

Lemma 1. *Let $\pi \in S_n$ satisfy $bp(\pi) = 2d(\pi)$, and suppose $\pi \cdot \rho_1 \cdot \rho_2 \cdots \rho_{d(\pi)} = \imath$. Each reversal ρ_i can be identified with a unique interval of π.*

Proof. Since a reversal removes at most two breakpoints, it follows that each ρ_i removes exactly two breakpoints from $\pi \cdot \rho_1 \cdots \rho_{i-1}$. Thus ρ_1 reverses an active interval of π; identify ρ_1 with this interval. Furthermore, since ρ_1 does not create new intervals and can only change a remaining active interval to a passive interval and vice-versa, each interval of $\pi \cdot \rho_1$ is an interval of π. We also have $2d(\pi \cdot \rho_1) = bp(\pi \cdot \rho_1)$ and hence by the induction hypothesis, each $\rho_2, \cdots, \rho_{d(\pi)}$ is identified uniquely with an interval of $\pi \cdot \rho_1$, which is different from the one identified with ρ_1.

Lemma 1 says that each reversal in a shortest sequence reducing a permutation π to $\imath$ in $bp(\pi)/2$ steps can be identified with an edge in the graph B_π. In fact, we will now show that each such shortest sequence corresponds to a perfect matching of B_π. First we need to prove that the intervals corresponding to a perfect matching of B_π are pairwise "independent", i.e. reversing an active interval does not destroy the other intervals. Note that reversing an active interval $[i, j]$ destroys only $[i + 1, j + 1]$, $[i - 1, j - 1]$ besides any interval that has an endpoint at i or j.

Lemma 2. *Let $\pi \in S_n$ and suppose B_π has a perfect matching M that has no edges of the type $]i, i + 2[$. Then for each active interval $[i, j]$ of π, $[i, j]$ and $[i + 1, j + 1]$ cannot both be edges of M; similarly for each passive interval $]k, l[$ of π, $]k, l[$ and $]k + 1, l - 1[$ cannot be both edges of M. Thus, if $\pi_i \sim \pi_j$ and π_i and π_j are not adjacent elements of π, then M contains exactly one of $[i, j],]i, j - 1[, [i - 1, j - 1],]i - 1, j[$.*

Proof. Suppose to the contrary that M contains a pair of intervals $[i, j], [i + 1, j + 1]$ or $]k, l[,]k + 1, l - 1[$. Associate with each pair $[i, j], [i + 1, j + 1]$ the value $v_{i,j} = max(\pi_{i+1}, \pi_{j+1})$; similarly, associate with each pair $]k, l[,]k + 1, l - 1[$ the value $v_{k,l} = max(\pi_{k+1}, \pi_l)$. Let π_b be the maximum value among the $v_{i,j}$'s and $v_{k,l}$'s. Since π_b is the endpoint of some interval, $\pi_b \leq n$. Thus, there is some c such that $\pi_c = \pi_b + 1$. If (π_{c-1}, π_c) and (π_c, π_{c+1}) are two breakpoints of π, then since M is a perfect matching, it must contain another pair $[c - 1, d - 1]$ and $[c, d]$, or $]c - 1, d + 1[$ and $]c, d - 1[$ for some d (where $\pi_d = \pi_c + 1$). The value $v_{c-1,d-1}$ associated with this pair is $\pi_c + 1 = \pi_b + 2$, contradicting our choice of π_b.

Else exactly one of (π_{c-1}, π_c) and (π_c, π_{c+1}) is a breakpoint of π; without loss of generality, say (π_{c-1}, π_c). M is a perfect matching, so it must contain an interval starting from this breakpoint. But the only choices are $]c - 1, c + 1[$, which by assumption does not belong to M, and those intervals with a breakpoint involving π_b. But we already have two intervals in M of this type (the pair of intervals whose value is π_b), so M cannot be a perfect matching, a contradiction.

Thus, it follows that if π_i and π_j are two adjacent integers that are not adjacent members of π, then exactly one of $[i, j],]i, j - 1[, [i - 1, j - 1],]i - 1, j[$ is an edge of any perfect matching M of B_π.

Figure 2 illustrates the two cases considered in the proof of Lemma 2.

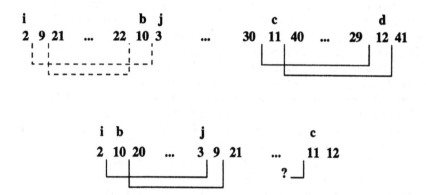

Fig. 2. $[i, j]$, $[i + 1, j + 1]$ cannot both be edges of a perfect matching of B_π

We are now ready to give a graph-theoretic characterization of permutations which satisfy $2d(\pi) = bp(\pi)$.

Theorem 3. *Let $\pi \in S_n$. Then $2d(\pi) = bp(\pi)$ iff there exists a perfect matching M of B_π such that each connected component of the graph I_M includes at least one active interval of π.*

Proof. Let $\rho_1, \ldots, \rho_{d(\pi)}$ be a shortest sequence of reversals reducing π to ι. Then by Lemma 1, each reversal can be identified with a unique interval of π. Representing each reversal as an edge of B_π we obtain a subgraph M of $d(\pi)$ edges. Furthermore, no two edges share a vertex since a breakpoint cannot be removed twice. Hence the subgraph M is a perfect matching of B_π. Finally, note that a reversal can affect only reversals in its connected component of I_M. Hence, the first reversal of each connected component reverses an active interval of π.

Conversely, suppose B_π has a perfect matching M such that each connected component of the graph I_M includes at least one active interval of π. In particular, M has no intervals of the type $]i, i + 2[$ (since such passive intervals are components by themselves), i.e. the condition of Lemma 2 is met. We show by induction on $bp(\pi)$ (which must be even since B_π has a perfect matching) that $2d(\pi) = bp(\pi)$.

When $bp(\pi) = 2$, we have $d(\pi) = 1$. Suppose the claim is true for $n \geq 2$, and let π be a permutation such that $bp(\pi) = n + 2$ and π satisfies the condition of this theorem. Let M be the matching of B_π. Select an active interval $[i, j]$ among the edges of M such that the permutation $\sigma = \pi \cdot \rho_{ij}$ obtained by reversing the interval $[i, j]$ of π has the most active intervals. If we can show that σ also satisfies the condition of this theorem then by the induction hypothesis $2d(\sigma) = bp(\sigma)$ and hence $2d(\pi) \leq 2(d(\sigma) + 1) = bp(\sigma) + 2 = bp(\pi)$.

First by Lemma 2 the matching M minus the edge $[i, j]$ is a perfect matching of B_σ, since the reversal $[i, j]$ can only destroy either reversals that share one of its breakpoints, or $[i-1, j-1]$ or $[i+1, j+1]$. Call this matching N. It remains to show each connected component of the graph I_N has an active interval. Each such connected component of I_N is either a connected component of I_M that does not include $[i, j]$ (and thus has an active interval unaffected by the reversal of $[i, j]$) or a fragment of the connected component C_{ij} of I_M that includes $[i, j]$. A connected component of the second type must have an interval $[k, l]$ or $]k, l[$ intersecting $[i, j]$ in I_M. If this interval is passive in I_M, it becomes active in I_N and we are done. So assume the interval is an active one. If in I_M this interval $[k, l]$ intersects with an active interval which does not intersect $[i, j]$, or if in I_M there is a passive interval intersecting $[i, j]$ but not $[k, l]$ then in I_N $]k, l[$ intersects with an active interval, and we are done.

Thus, suppose in I_M i) the interval $[k, l]$ is active and intersects $[i, j]$, ii) each active interval intersecting $[k, l]$ also intersects $[i, j]$, and iii) each passive interval intersecting $[i, j]$ also intersects $[k, l]$. In other words, reversing $[k, l]$ destroys no more active intervals and creates no less active intervals than reversing $[i, j]$. Since we pick $[i, j]$ to maximize the number of active intervals after reversing it, it follows that an interval (active or passive) intersects $[i, j]$ if and only if it intersects $[k, l]$. Without loss of generality, assume $i < k < j < l$. Let $v = \pi_r$ be the largest integer among $\pi_{i+1}, \pi_{i+2}, \ldots, \pi_k$, and $\pi_{j+1}, \pi_{j+2}, \ldots, \pi_l$. Clearly $v \leq n$, and so $v + 1 = \pi_s$ for some s. By Lemma 2, M includes exactly one of $[r, s],]r, s-1[, [r-1, s-1],]r-1, s[$. This interval cannot intersect both $[i, j]$ and $[k, l]$, contradicting the assumption at the beginning of this paragraph.

Thus we conclude every connected component of I_N has an active interval, and the theorem follows.

It follows from Theorem 3 that the problem of sorting by reversals can be solved efficiently for the special cases when the number of breakpoints is twice the distance.

Theorem 4. *Deciding whether $bp(\pi) = 2d(\pi)$ for any permutation $\pi \in S_n$ (and sorting it by reversals) is in P.*

Proof. Given π, we construct the graph B_π and assign to each active interval the weight $+1$ and each passive interval the weight -1. Then find a perfect matching M that has maximum weight. If M satisfies the condition of Theorem 3 then $2d(\pi) = bp(\pi)$. Suppose I_M has a connected component C consisting of only passive intervals. Let $]i, x[$ and $]y, j[$ be the leftmost and rightmost intervals of C, respectively. It is clear that every breakpoint between i and j is an endpoint of some interval in C: otherwise, select such a breakpoint (π_z, π_{z+1}) so that $max(\pi_z, \pi_{z+1})$ is maximum among all breakpoints that are not endpoints of some interval in C. Then C must contain an interval $]z, z'[$ and $z' > j$, or $]z', z[$ and $z' < i$, contradicting our choice of i and j.

By tracing the endpoints of the intervals in C, we conclude that $\pi_i, \pi_{i+1}, \ldots, \pi_j, \pi_{j+1}$ are all the integers bounded by π_i and π_{j+1}. As a result, no intervals of π (even those that are not in M) has an endpoint between i and j and the other endpoint outside this range.

Thus if B_π has another perfect matching M' that satisfies the condition of Theorem 2, then it must have a connected component C' whose intervals are pairs of breakpoints between i and j. Furthermore, C' has at least one active interval. We can construct from M and M' a new matching N by replacing the connected component C of M with C' of M'. But the weight of N is greater than that of M, a contradiction. Hence $2d(\pi) \neq bp(\pi)$.

The graph B_π can be constructed from π (and π^{-1}) in time $O(n)$. In general, a maximum-weighted matching of a graph G of v vertices and e edges can be obtained in time $O(velogv)$ [GMG82]. Since the number of vertices and edges of B_π are both $O(n)$, the time complexity of this algorithm is $O(n^2 \log n)$.

Once we have the matching, π can be sorted in $bp(\pi)/2$ reversals by always reversing an active interval that leaves the most number of active intervals (see the proof of Theorem 3).

References

[BH96]   P. BERMAN AND S. HANNENHALLI, *Fast sorting by reversal*, in Proc. 7th Combinatorial Pattern Matching, 1996, pp. 168–185.

[BP93]   V. BAFNA AND P. PEVZNER, *Genome rearrangements and sorting by reversals*, in Proc. 34th FOCS, IEEE, 1993, pp. 148–157.

[Cap97]   A. CAPRARA, *Sorting by reversals is difficult*, in Proc. 1st International Conference on Computational Molecular Biology, 1997.

[GMG82]   Z. GALIL, S. MICALI, AND H. GABOW, *Maximal weighted matching on general graphs*, in Proc. 23rd FOCS, IEEE, 1982, pp. 255–261.

[HP95]   S. HANNENHALLI AND P. PEVZNER, *Transforming cabbage into turnip (polynomial algorithm for sorting signed permutations by reversals)*, in Proc. 27th STOC, ACM, 1995, pp. 178–189.

[HP96]   ———, *To cut ... or not to cut (applications of comparative physical maps in molecular evolution*, in Proc. 7th SODA, ACM-SIAM, 1996, pp. 304–313.

[KS95]   J. KECECIOGLU AND D. SANKOFF, *Exact and approximation algorithms for sorting by reversals, with application to genome rearrangement*, Algorithmica, 13 (1995), pp. 180–210.

[KST97]   H. KAPLAN, R. SHAMIR, AND R. TARJAN, *Faster and simpler algorithm for sorting signed permutations by reversals*, in Proc. 8th SODA, ACM, 1997, pp. 344–351.

[PH88]   J. D. PALMER AND L. A. HERBON, *Plant mitochondrial DNA evolves rapidly in structure, but slowly in sequence*, J. of Mol. Evol., 27 (1988), pp. 87–97.

[PW95]   P. PEVZNER AND M. WATERMAN, *Open combinatorial problems in computational molecular biology*, in Proceedings of the 3rd Israel Symposium on the Theory of Computing and Systems, 1995, pp. 148–173.

[WEHM82]   G. A. WATTERSON, W. J. EWENS, T. E. HALL, AND A. MORGAN, *The chromosome inversion problem*, Journal of Theoretical Biology, 99 (1982), pp. 1–7.

External Inverse Pattern Matching

Leszek Gąsieniec[1*] and Piotr Indyk[2**] and Piotr Krysta[3***]

[1] Max-Planck Institut für Informatik, Im Stadtwald, Saarbrücken D-66123, Germany.
[2] Computer Science Department, Stanford University, Gates Building, CA–94305, USA.
[3] Institute of Computer Science, University of Wrocław, Przesmyckiego 20, PL–51151, Wrocław, Poland.

Abstract. In this paper we consider the *external inverse pattern matching* problem. Given a text $\mathcal{T}$ of length n over an ordered alphabet Σ and a number $m \leq n$, the goal is to find a pattern $\tilde{\mathcal{P}}_{MAX} \in \Sigma^m$ which is not a subword of $\mathcal{T}$ and which maximizes the sum of Hamming distances between $\tilde{\mathcal{P}}_{MAX}$ and all subwords of $\mathcal{T}$ of length m. We present an optimal $O(n \log \sigma)$-time (where $\sigma = |\Sigma|$) algorithm for the external inverse pattern matching problem. This substantially improves the $O(nm \log \sigma)$-time algorithm given in [2]. Moreover we discuss briefly fast parallel implementation of our algorithm on the CREW PRAM model.

1 Introduction

Given a string $\mathcal{T}$ (called later a text) of length n over an alphabet Σ, the *inverse pattern matching problem* is to find a word $\mathcal{P}_{MIN} \in \Sigma^m$ (or $\mathcal{P}_{MAX} \in \Sigma^m$) which minimizes (maximizes) the sum of Hamming distances [9] between $\mathcal{P}_{MIN}$ ($\mathcal{P}_{MAX}$) and all subwords of length m in the text $\mathcal{T}$. Two variations of this problem are also considered, when the optimal word is supposed to occur in the text $\mathcal{T}$ (the *internal* inverse pattern matching) or oppositely when its occurrence in $\mathcal{T}$ is forbidden (the *external* inverse pattern matching). It is assumed that in case of internal inverse pattern matching the required internal pattern $\tilde{\mathcal{P}}_{MIN}$ must minimize the sum of distances, whereas in the external case optimal external pattern $\tilde{\mathcal{P}}_{MAX}$ maximizes the sum. As reported in [2] the inverse pattern matching problems appear naturally and find applications in several fields like: information retrieval, data compression, computer security and molecular biology.

It was shown by Amir, Apostolico and Lewenstein in [2] that the inverse pattern matching problem can be solved in time $O(n \log \sigma)$ when no additional restrictions on $\mathcal{P}_{MAX}$ (or $\mathcal{P}_{MIN}$) are assumed. However, the *internal* and *external* inverse pattern matching problems appeared to be significantly harder. Amir

---

* E-mail: leszek@mpi-sb.mpg.de

** E-mail: indyk@cs.stanford.edu

*** Research of this author was partially supported by KBN grant 2 P301 034 07. E-mail: pkrysta@ii.uni.wroc.pl

et al. presented two algorithms for the internal case: a simple one, with the running time $O(nm \log \sigma)$ and the more sophisticated one (using convolutions [5]) running in time $O(n\sqrt{m} \log^2 m)$. They also pointed out a close correspondence between the internal inverse pattern matching and *all mismatches problem* (see [1]). Any improvement to the latter question is a long standing open problem, thus it seems to be quite unlikely to get a faster algorithm for the internal inverse pattern matching.

For the external case, obtaining any polynomial time algorithm was a nontrivial task, as the space of possible solutions is exponential. To overcome this problem, Amir *et al.* introduced the notion of *m-stems* for the text $\mathcal{T}$. The *m*-stems are defined as all possible words of length at most m not belonging to $\mathcal{T}$ but whose all proper prefixes form subwords of $\mathcal{T}$. It was shown in [2] that the optimal external pattern $\tilde{P}_{MAX}$ can be composed of some *m*-stem of $\mathcal{T}$ extended by a proper size suffix of the maximal word P_{MAX}. The application of the *m*-stem approach enabled Amir *et al.* to achieve a $O(nm \log \sigma)$-time solution by testing all text subwords of length at most m.

The apparent similarity to the internal case, and the high complexity of the algorithm of [2] suggest hardness of the external inverse pattern matching problem. In this paper we show the contrary. We present first *optimal* $O(n \log \sigma)$-time algorithm for the external inverse pattern matching problem indicating that the internal case is the hardest problem in the inverse pattern matching family. The optimality of our algorithm comes from the complexity of element distinctness problem, which can be reduced to the external inverse pattern matching. The new efficient solution is a consequence of deeper analysis of relation between the maximal words P_{MAX}, $\tilde{P}_{MAX}$ and the text $\mathcal{T}$. Our main algorithm uses efficient algorithmic techniques like: compact suffix trees [14], range minimum queries [7] and lowest common ancestor queries in trees [10] supported by an on-line computations of symbol weights (defined later).

The rest of the paper is organized as follows. In section 2 we introduce notation and basic techniques used in our algorithm. Section 3 contains the main algorithm with complexity analysis and proof of correctness. In this section we also discuss a parallel implementation of our algorithm. Section 4 contains the final remarks and states some open problems in the related areas.

2   Preliminaries

Given an ordered *alphabet* Σ containing σ symbols, i.e. $|\Sigma| = \sigma$. Any sequence of concatenated symbols from Σ is called a *word* or a *string*. We use symbol $\cdot$ to denote operation of concatenation, but the symbol is omitted in cases where the use of concatenation is natural. We use a notation $w[i]$ for the ith symbol of the word w, $w[i..j]$ for the substring of w which starts at position i and ends at j, ^-w stands for the string w without its first symbol, while symbol ε stands for the empty string. For example, let $w \in \Sigma^*$ be a string of length n, i.e. $|w| = n$. Then $w = w[1..n]$, $^-w = w[2..n]$, $w[i, i] = w[i]$ and $w[i, j] = \varepsilon$ when $i > j$. Any subword of w of the form $w[1..i]$, for all $i \in \{1, \ldots, n\}$, is called a *prefix* of w and a subword of the form $w[j..n]$, for all $j \in \{1, \ldots, n\}$, is called a *suffix* of w. We use notation $u \in w$ $(u \notin w)$ when u is (not) a subword of string w. In case $u \notin w$ we say that the word u is *external string* for w. A search tree for a given set of words $S \subseteq \Sigma^*$ whose edges are labeled by symbols drawn from the alphabet Σ is called a *trie* [6] for S. Any sequence $v = v_1, \ldots, v_k$ of neighboring nodes (by parent-child relation) in a tree, such that all v_is are pairwise disjoint, is called a *path*. Each word $w \in S$ is represented in the trie as a path from the root to some leaf. Recall that a trie is a prefix tree, i.e. two words have a common path from the root as long as they have the same prefix. A path $v = v_1, \ldots, v_k$, whose all internal nodes but last have degree equal to 1, is called a *chain*.

2.1 Problem Definition

Let $\mathcal{T}$ be a text over Σ such that $|\mathcal{T}| = n$, and $\mathcal{P}$ be a string over Σ such that $|\mathcal{P}| = m \leq n$. The *Hamming distance* [9] between the word $\mathcal{P}$ and a subword of $\mathcal{T}$ starting at position i is defined as follows:

$$H(\mathcal{P}, \mathcal{T}[i..i + |\mathcal{P}| - 1]) = \sum_{j=1}^{|\mathcal{P}|} h(\mathcal{P}[j], \mathcal{T}[i + j - 1]),$$

where for any two symbols $a, b \in \Sigma$

$$h(a, b) = \begin{cases} 0, & a = b \\ 1, & a \neq b. \end{cases}$$

In other words the Hamming distance $H(\mathcal{P}, \mathcal{T}[i..i + |\mathcal{P}| - 1])$ gives the number of mismatches between symbols of the aligned words $\mathcal{T}[i..i + |\mathcal{P}| - 1]$ and $\mathcal{P}$. In this paper we are primarily interested in the *total Hamming distance* between the word $\mathcal{P}$ and all its alignments in the text $\mathcal{T}$, which is defined as:

$$\mathcal{H}(\mathcal{P}, \mathcal{T}) = \sum_{i=1}^{|\mathcal{T}| - |\mathcal{P}| + 1} H(\mathcal{P}, \mathcal{T}[i..i + |\mathcal{P}| - 1]). \tag{1}$$

Now we are ready to introduce the entire problem:

Problem: *External Inverse Pattern Matching.*
Given a text string $\mathcal{T} \in \Sigma^n$, where $|\Sigma| = \sigma$, and a positive integer m, s.t. $m \leq n$. The entire problem is to find a pattern $\tilde{\mathcal{P}}_{MAX} \in \Sigma^m$, s.t. $\tilde{\mathcal{P}}_{MAX} \notin \mathcal{T}$ and $\mathcal{H}(\tilde{\mathcal{P}}_{MAX}, \mathcal{T}) \geq \mathcal{H}(\mathcal{P}, \mathcal{T})$, for all strings $\mathcal{P} \in \Sigma^m$, which do not belong to $\mathcal{T}$.

Notice that if text $\mathcal{T}$ contains all possible strings of length m then the external inverse pattern matching has no solution.

According to the definition of entire problem the ith symbol of the desired pattern $\tilde{\mathcal{P}}_{MAX}$ can be aligned only with positions from i to $n - m + i$ in the text $\mathcal{T}$, since we are interested only in full alignments of the pattern $\tilde{\mathcal{P}}_{MAX}$ in $\mathcal{T}$. The latter observation defines naturally m different ranges in the text $\mathcal{T}$, s.t. the

i^{th} range is associated with the ith position in pattern $\tilde{P}_{MAX}$. We call these m consecutive ranges *windows* $Win_1, \ldots, Win_m$. So simply $Win_i = \mathcal{T}[i..n-m+i]$. Let ϕ_i be a *frequency* function defined in the ith window Win_i, i.e. $\phi_i(a)$ equals to the number of all occurrences of symbol a in Win_i, for all $a \in \Sigma$ and $i \in \{1, \ldots, m\}$. The weight w_i of symbols in the ith window is defined as follows:

$$w_i(a) = (n - m + 1) - \phi_i(a), \quad \text{for all} \quad i \in \{1, \ldots, m\} \text{ and } a \in \Sigma.$$

In terms of the weights the entire problem can be viewed as looking for a string $\mathcal{P} \notin \mathcal{T}$ of length m, which maximizes the sum:

$$\sum_{i=1}^{m} w_i(\mathcal{P}[i]). \tag{2}$$

Notice that the sum (2) is maximized when the ith position in the pattern $\mathcal{P}$ is occupied by the least frequent, or equivalently the heaviest symbol in the window Win_i, which corresponds to the definition of the maximal word $\mathcal{P}_{MAX}$ (the maximal solution of the general inverse pattern matching). However in the external inverse pattern matching optimal pattern $\tilde{P}_{MAX}$ maximizes the sum (2) among all external strings for the text $\mathcal{T}$.

Through the rest of the paper we will use the weighted version of the external inverse pattern matching. Moreover, before the main algorithm starts, we transform the input string to one which consists of numbers from the range 1 to σ, s.t. every symbol from the alphabet Σ is substituted by a unique number. Thus from now on we assume that symbols can be treated as small numbers. Since the alphabet Σ is ordered, the transformation can be simply performed in time $O(n \log \sigma)$, which does not violate the complexity of the entire algorithm.

2.2 Basic Techniques

In the following section we recall some basic techniques used in our algorithm.
Compact suffix tree and compact trie. A *suffix tree* [14] of a word $w \in \Sigma^*$ is a trie which represents all suffixes of w. Notice that in the worst case the size of the suffix tree can be quadratic in the size of the input string. However, since the suffix tree has exactly $|w| = n$ leaves (corresponding to all suffixes) it can be stored in linear space as follows. Every chain in the suffix tree is represented by a pair of integers (i, j) which refers to the subword $w[i..j]$. There are exactly n leaves in the suffix tree, thus the number of internal nodes of degree greater than one and the number of chains are both not greater than n. The linear representation of a suffix tree is called *compact suffix tree* and it is a known fact [14] that for a word w, such that $|w| = n$, it can be constructed in time $O(n \log \sigma)$. In this paper we consider tries with compact description of chains. Recall that a chain is a path $v = v_1, \ldots, v_k$ whose all nodes but last have degree 1. For our purposes there are stored subwords of only one text in the trie, which means that all the chains in the trie represent substrings of the same text. All the chains are exchanged by edges labeled by pairs of indices describing a position

of the corresponding subword in the text. It is important that our definition of the chain implies that each node of the trie of degree ≥ 2 has all outgoing edges of length 1. This means that these edges are labeled by single symbols.

Range minimum search. Given a vector $V = V[1..n]$ of n numbers. A *range query* for a pair (i, j), where $1 \leq i \leq j \leq n$, is a question about minimum among all numbers in the range $V[i..j]$. The main goal in *range minimum search problem* is to preprocess efficiently vector V, such that the range queries can be answered as fast as possible. Gabow et al. [7] gave a linear time preprocessing algorithm for the range minima that results in constant-time query retrieval.

Lowest common ancestor in a tree. Let T be a rooted tree with a root r. For any node $x \in T$ let *branch(x)* denote a path from the node x to the root r, and *depth(x)* denote a distance (length of the path) from the node x to the root r. Given two nodes $v, w \in T$. Node v is an *ancestor* of node w iff $v \in branch(w)$. For example the root r is an ancestor of all nodes in T. A *lowest common ancestor* for any pair of nodes $v, w \in T$ is a node $u \in T$ with the greatest possible *depth(u)*, such that $u \in branch(v)$ and $u \in branch(w)$. In [10] there was shown that after a linear preprocessing of a tree T all lowest common ancestor queries can be answered in constant time. Moreover [13] introduced a simpler algorithm with the same sequential time bounds and its parallel counterpart with linear $O(\log n)$-time preprocessing and constant time queries.

3   External Inverse Pattern Matching Algorithm

We start this section recalling known facts about the external inverse pattern matching introduced by Amir *et al.* in [2]. The following notion of m-stems plays a crucial role in Amir *et al.* approach, as well as in our algorithm.

Definition 1. Any string $\mathcal{R} = \mathcal{R}[1..l]$ over the alphabet Σ, for $l \in \{2, \ldots, m\}$, is called an *m-stem* for the text $\mathcal{T} = \mathcal{T}[1..n]$ iff the whole word $\mathcal{R} \notin \mathcal{T}[1..n-m+l]$, but $\mathcal{R}[1..l-1] \in \mathcal{T}[1..n-m+l-1]$.

Assume that we have already computed the optimal maximal word $\mathcal{P}_{MAX}$ using techniques from [2]. Recall that the cell $\mathcal{P}_{MAX}[i]$ contains the heaviest symbol in the window Win_i. Roughly speaking construction of the word $\mathcal{P}_{MAX}$ can be done as follows. First one has to compute the frequencies of all symbols in the window Win_1, and this can be done in linear time. Since the difference between symbol frequencies in any two neighboring windows is small (only one symbol comes in and only one comes out when we change the window), we can compute the heaviest symbols in the consecutive windows $Win_2, \ldots, Win_m$ allowing constant time for each window. Additionally we compute two arrays $\mathcal{F}_1[1..m]$ and $\mathcal{F}_2[1..m]$, where $\mathcal{F}_1[i]$ contains the second heaviest symbol in the window Win_i, and $\mathcal{F}_2[i]$ contains the difference $w_i(\mathcal{P}_{MAX}[i]) - w_i(\mathcal{F}_1[i])$. The arrays are called *tables of flips* for the pattern $\mathcal{P}_{MAX}$. More detailed description of a data structure, which gives the weights of symbols in the consecutive windows, is given in section 3.1.

If $\mathcal{P}_{MAX} \notin \mathcal{T}$ then we take $\mathcal{P}_{MAX}$ as desired pattern $\tilde{\mathcal{P}}_{MAX}$ and the external inverse pattern matching is solved. Otherwise if $\mathcal{P}_{MAX}$ is a subword of $\mathcal{T}$, then the following fact holds:

Fact 3.1 [2] *If $\tilde{\mathcal{P}}_{MAX}$ is the solution of the external inverse pattern matching problem, then $\tilde{\mathcal{P}}_{MAX} = \alpha \cdot \beta$, where α is some m-stem for the text $\mathcal{T}$ and $\beta = \mathcal{P}_{MAX}[|\alpha|+1..m]$.*

According to the Fact 3.1 searching for the optimal pattern $\tilde{\mathcal{P}}_{MAX}$ has been reduced to testing (weights) of all $O(nm\sigma)$ possible words of the form $\alpha \cdot \beta$. In fact Amir et al. in [2] reduced the number of words for testing to $O(nm)$, skipping non-reasonable solutions. More precisely they build a trie for all the substrings of $\mathcal{T}$ of length m and they traverse it (node by node) in BFS order, testing a maximal external string leaving the trie at a current node. Since the size of the trie is $O(nm)$, their approach gives an algorithm with running time $O(mn \log \sigma)$. In this paper we show how to search the nodes of the trie more efficiently.

Let v be a node in the trie on depth k. Let $C(v) = \{c_1, \ldots, c_l\}$ be set of children of the node v and $X(v) = \{x_1, \ldots, x_l\}$ be set of symbols on first positions of edges $e_1, \ldots, e_l$ connecting the node v to its children respectively. Let s be a string represented by a path from the root of the trie to v, see Figure 1. Moreover let y be the heaviest symbol in the window Win_{k+1} which is not in

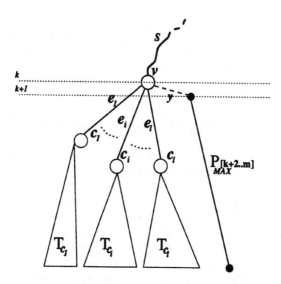

Fig. 1. Heaviest external string leaving the trie at the node v.

$X(v)$, i.e. $w_{k+1}(y) \geq w_{k+1}(z)$, for all $z \in \Sigma \setminus X(v)$. The following lemma shows the benefit of m-stem approach.

Lemma 2. *The string $s \cdot y \cdot \mathcal{P}_{MAX}[k + 2..m]$ is the heaviest possible external string leaving the trie at the node v.*

Proof. Since the weight of the string s is fixed and the suffix $\mathcal{P}_{MAX}[k + 2..m]$ is the heaviest possible extension (see the definition of $\mathcal{P}_{MAX}$), thus the string $s \cdot y \cdot \mathcal{P}_{MAX}[k + 2..m]$ is the heaviest possible external string leaving trie at the node v.

Let $\bar{X}(v) = \{\bar{x}_1, \ldots, \bar{x}_{l'}\} \subset X(v)$ be a set of symbols on first positions in the edges $\bar{e}_1, \ldots, \bar{e}_{l'}$ ($\bar{e}_p = e_q$ iff $\bar{x}_p = x_q$), such that $w_{k+1}(\bar{x}_i) \leq w_{k+1}(y)$ for all $i \in \{1, \ldots, l'\}$.

Lemma 3. *Maximal external string $s \cdot y \cdot \mathcal{P}_{MAX}[k + 2..m]$ leaving the trie at the node v is heavier than all strings of the form $s \cdot \bar{x}_i \cdot \mathcal{P}_{MAX}[k + 2..m]$ for all $i \in \{1, \ldots, l'\}$, and there is no need to test external strings going out of the trie at and below the edges $\bar{e}_1, \ldots, \bar{e}_{l'}$.*

Proof. Notice that all external strings passing through the node v have the same prefix s. Since the suffix $y \cdot \mathcal{P}_{MAX}[k + 2..m]$ is heavier than all possible suffixes starting from symbols of the set $\bar{X}(v)$, the results follows.

Lemma 3 has interesting consequences if the maximal external symbol $y = \mathcal{P}_{MAX}[k + 1]$.

Corollary 4. *If the maximal external symbol $y = \mathcal{P}_{MAX}[k + 1]$, then the string $s \cdot \mathcal{P}_{MAX}[k + 1..m]$ is the heaviest possible external string passing through the node v, and there is no need to traverse the trie below the node v.*

The advantage of the Corollary 4 becomes more clear when it is used in the context of the nodes of some chain in the trie. The following lemma plays a crucial role in our efficient searching of chains in the trie of the text subwords.

Lemma 5. *Let $u = u[1..r]$ be a string which is represented by the only chain ρ_u going out from a node v (v has degree one). If $u[1..r] \neq \mathcal{P}_{MAX}[k+1..k+r]$ then:*

A. *the string $s \cdot \mathcal{P}_{MAX}[k + 1..m]$ is the heaviest possible external string passing through the node v and there is no need to search the trie below v, otherwise*

B. *let j be the position in the word $u[1..r]$ for which the corresponding difference in the flip table $\mathcal{F}_2[k + j]$ is the smallest in range $k + 1, \ldots, k + r$, then the word $s \cdot \mathcal{P}_{MAX}[k + 1..k + j - 1] \cdot \mathcal{F}_1[k + j] \cdot \mathcal{P}_{MAX}[k + j + 1..m]$ is the heaviest possible external string among all external strings leaving the trie at nodes of the chain ρ_u. In this case a part of the trie below the chain ρ_u is a subject for further search.*

Proof. ad A. The string $s \cdot \mathcal{P}_{MAX}[k + 1..m]$ is an external string for $\mathcal{T}$ and it is the heaviest possible external string which passes through the node v in the trie.

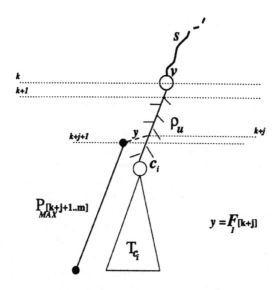

Fig. 2. Heaviest external string leaving the trie at the chain ρ_u.

ad B. It is enough to change only one symbol in the word u to create an external
string which leaves the trie at some node of the chain ρ_u, see Figure 2 and
Fact 3.1. According to the definition of the tables of flips, the position j in
u gives the minimal lose of weight among all possible swaps of one symbol
in u. Since it is still possible that the maximal external string leaves the
trie below the chain ρ_u, the part of the trie hanged below ρ_u is a subject of
further search.

Now we are ready to present our main algorithm.

3.1 Algorithm

The algorithm consists of two stages. The first one, called *preprocessing*, contains
a construction and initialization of all data structures used later during the
actual search. The second stage, called *searching phase*, consists of an actual
construction of the desired optimal external solution $\tilde{\mathcal{P}}_{MAX}$.

Preprocessing First of all, we find the maximal pattern $\mathcal{P}_{MAX}$ using techniques from [2]. If $\mathcal{P}_{MAX}$ is an external string for the text $\mathcal{T}$ (which can be
checked by any string matching algorithm, e.g. see [12]) then we are done, otherwise instead of the full trie of text subwords we build a compact trie $T_{\mathcal{T}}$. It is
reconstructed from a compact suffix tree for the text $\mathcal{T}$ by cutting all deep paths
(from the root to leaves) on depth m, and skipping all shallow paths (shorter
than m). At every node v of $T_{\mathcal{T}}$ we keep information about the string s (subword
of the text $\mathcal{T}$) which is represented by the path from root of the trie to the node

v. Additionally we build a common suffix tree T^* for the text T and the maximal word $\mathcal{P}_{MAX}$ (i.e. a suffix tree for the word $T\$\mathcal{P}_{MAX}$) and we preprocess it for LCA queries. Construction of all the trees can be done in time $O(n \log \sigma)$ as well as the preprocessing for LCA queries.

An on-line computation of the weights of symbols in the consecutive windows plays a crucial role in the preprocessing and the searching phase. According to the need of the algorithm a data structure which represents the weights of symbols must keep also the current order between the weighted symbols. The data structure is represented by an array $M = M[1..n]$, s.t. the ith cell of the array contains a pointer to a (double-linked) *horizontal* list of all symbols having weight i. Moreover all non-empty cells of the array are connected into a (double-linked) *vertical* list. The non-empty cell in the array M with the largest index (which contains a list of the heaviest symbols) is accessible directly by a variable max. Symbols in the lists are also accessible directly by the *symbol index*. The construction (initialization) of the data structure, which corresponds to Win_1, can be simply done in linear $O(n)$ time, since we assumed that symbols from the alphabet Σ are substituted by unique numbers from the range $1, \ldots, \sigma$. The data structure supports the following three operations.

The *first* operation gives the weight of any symbol $a \in \Sigma$ in the current window. The weight of the symbol a corresponds to the position of a horizontal list containing a in the array M. Since the symbols in the horizontal lists are accessible by the symbol index thus this operation works in constant time.

The *second* operation is needed when the algorithm changes a window from Win_i to Win_{i+1}, for all $i = 1, \ldots, n-m$. When the window is changed, only two symbols change slightly their weights, i.e. $T[i]$ comes out and the symbol $T[n-m+i]$ comes into the window. We find the weight $w_i(T[i])$ using the symbol index, then we exclude the symbol $T[i]$ from the list linked at $M[w_i(T[i])]$, and then the symbol $T[i]$ is inserted at the beginning of the list linked at $M[w_i(T[i])-1]$. In the meantime the pointers to the neighbors of $M[w_i(T[i])]$ and $M[w_i(T[i])-1]$ in the vertical list are modified if necessary. Finally the symbol index is decreased by one at the position $T[i]$. Similar operation is performed when the symbol $T[n-m+i]$ increases by one its level in the array M. Thus the whole step can be implemented in constant time.

The *third* operation is performed when we look for the heaviest symbol in a window Win_i, not belonging to the given set of symbols $X(v) = \{x_1, \ldots, x_l\} \subset \Sigma$. After a sequence of l deletions in the horizontal lists and at most l deletions in the vertical list, the desired symbol is accessible at $M[max]$. Finally the current structure of symbol weights is restored by the reverse sequence of insertions to the horizontal lists and the vertical list. And the whole step can be implemented in time $O(l)$.

Using the on-line computation of weights we compute in time $O(m)$ the flip tables: $\mathcal{F}_1[1..m]$ which contains the second heaviest symbols in the consecutive windows and $\mathcal{F}_2[1..m]$ whose i^{th} cell contains the difference $w_i(\mathcal{P}_{MAX}[i]) - w_i(\mathcal{F}_1[i])$. The second table $\mathcal{F}_2$ is preprocessed in linear time for the minimum range queries. Finally we compute the weights of all suffixes of the maximal pattern $\mathcal{P}_{MAX}$ and we store them in a table $\mathcal{S}[1..m]$ also in time $O(m)$.

Searching phase The searching phase consists of two rounds. During the first search of T_T for every node v we compute the heaviest external string, called a *candidate*, which leaves the trie at a node v or at a chain which is placed under the node v. Finally, if there is any candidate, the trie is searched again to find the maximal external pattern $\tilde{P}_{MAX}$. Otherwise the entire problem has no solution.

During the first round the algorithm traverses the tree T_T in the BFS-like order, s.t. children are inserted into a waiting list according to their depth in the tree. Assume that the algorithm just took from the waiting list a node v of depth k in T_T. It is assumed recursively that the weight of a string s, which is represented by a path from the root of T_T to the node v, has been already computed and the on-line weight data structure is currently set to answer queries in the window Win_{k+1}.

If the node v is of degree ≥ 2, all edges coming out of v are labeled by single symbols (definition of the compact trie, see section 2.2). We find the maximal external symbol y using the on-line weight data structure. The weight of the word $s \cdot y \cdot P_{MAX}[k+2..m]$ is clearly composed of weights of: the string s (stored at the node v), the symbol y (described by weight function in the current window) and the suffix of pattern P_{MAX} (stored in the table S). The weight of the word $s \cdot y \cdot P_{MAX}[k+2..m]$ is stored at the node v. According to Lemma 3 all *light* edges (and corresponding subtrees hanged under them) with symbols lighter than y can be ignored. For the rest of edges we update at their ending nodes information about the weight of a string which is represented by the path coming from the root, to fulfill the recursive assumption. The weight of the string is composed of the weight of the string s (stored at v) and the weight of a symbol placed on the edge (given by the weight function). All nodes below *heavy* edges are inserted into the waiting list on level $k+2$.

If the node v is a first node (its degree is 1) of a chain ρ_u, we check if a string $u = u[1..r]$ represented by the chain symbols, is a subword of the pattern P_{MAX}, i.e. if $u[1..r] = P_{MAX}[k+1..k+r]$. This can be done by asking for a lowest common ancestor of the proper suffix of T (string s and its extension in T) and the pattern suffix $P_{MAX}[k+1..m]$. If the lowest common ancestor for both suffixes is placed on a level $< r$ in T^*, then we know that $u[1..r] \neq P_{MAX}[k+1..k+r]$ and according to part A of Lemma 5 we have only one candidate $s \cdot P_{MAX}[k+1..m]$, and we do not search the trie below the chain ρ_u. Otherwise, when the equality holds, we recover the candidate from the flip tables $\mathcal{F}_1$ and $\mathcal{F}_2$. First we ask a minimum range query in $\mathcal{F}_2[k+1..k+r]$, getting index of a position j whose change gives the smallest lose of weight, and getting the candidate $s \cdot P_{MAX}[k+1..j-1] \cdot \mathcal{F}_1[k+j] \cdot P_{MAX}[k+j+1..m]$. The information about the weight of a string represented by path coming from the root to the node under the chain ρ_u is updated with a help of the table S.

In both cases the time at node v is proportional to degree of the node v, thus searching of the whole trie T_T can be done in time proportional to the size of the trie, i.e. in time $O(n)$. At last the trie T_T is searched again to find the maximal weight, which is the weight of the maximal external pattern $\tilde{P}_{MAX}$.

Theorem 6. *The external inverse pattern matching problem can be solved in optimal time $O(n \log \sigma)$.* □

3.2 Parallel Approach

In this section we discuss shortly a parallel implementation of our external inverse pattern matching algorithm on the CREW PRAM model. Most of the steps in our algorithm can be easily parallelized when we allow for superlinear work and space.

Theorem 7. *The external inverse pattern matching algorithm can be implemented in time $O(\log n)$ and work $O(n \log n + m\sigma \log \sigma)$ on the CREW PRAM.*

Proof. Both trees T_T and T^* can be computed in $O(\log n)$-time and $O(n \log n)$ work when subquadratic space is available, see [3]. Moreover the tree T^* can be preprocessed in time $O(\log n)$ and linear work for LCA queries, see [13]. Since we can not use on-line computation of weights in all windows at the same time we have to compute the whole table of weights which is of size $O(m\sigma)$. The table of weights can be easily computed in time $O(\log m)$ and work $O(m\sigma)$ but since we still need to keep order between weights of symbols we have to sort all m columns, by parallel merge-sort [4], which gives total work $O(m\sigma \log \sigma)$. When the table is ready, we compute the pattern $\mathcal{P}_{MAX}$, flip tables $\mathcal{F}_1$, $\mathcal{F}_2$ and the table S in time $O(\log m)$ and linear work. Then the table $\mathcal{F}_2$ is preprocessed for minimum range queries in logarithmic time and work $O(m \log m)$ (computing minimum in every block of size 2^i, for $i = 1, \ldots, m$). When all data structures are ready we start the searching algorithm. We assume that at every node v of the trie T_T there is a linear number of processors according to the degree of v. We compute in constant time the weights of all *feasible edges*, i.e. the edges of length 1 placed under nodes of degree ≥ 2 (with help of the table of weights) and the edges representing chains which are subwords of $\mathcal{P}_{MAX}$ (using LCA queries and table S). If the path from root of the trie to the node v is composed only of feasible edges, then the node v is called *feasible node*. We use any *Euler tour technique*, see e.g. [8], to compute all feasible nodes and the weights of strings which are represented by paths from the root. Computation of the feasible nodes is done in time $O(\log n)$ and linear work. Now if a feasible node v (placed on depth k and under a string s) is the first node of a chain, the weight of a candidate is composed from the weight of the string s, table S and flip tables. If the node v has degree $l \geq 2$, then $O(l)$ processors associated with the node find in logarithmic time the heaviest symbol y in column $k + 1$ after deletion of l symbols placed in edges coming out of v. This can be done by testing only $l + 1$ heaviest symbols in column $k + 1$. In this case the weight of the candidate is composed from the weight of the string s, symbol y and table $S[1..m]$. When the weights of candidates are ready, we apply any tree contraction algorithm, see e.g. [8], looking for the maximum in the tree representing optimal pattern $\tilde{\mathcal{P}}_{MAX}$.

4 Conclusion

We have presented first optimal $O(n \log \sigma)$-time algorithm for the sequential external inverse pattern matching, showing that the internal case is the hardest problem in the inverse pattern matching family. The interesting task for further research is to improve the bounds of the external inverse pattern matching in the parallel case. Notice that if the product of m and σ is small, i.e. $m\sigma = O(n)$, our parallel implementation is fast and efficient. But if we want to keep a linear complexity for all feasible values of n, m and σ, there is a bottleneck hidden in the computation of the weights of symbols in every window Win_i. Finally, the existence of a fast algorithm for the internal case remains the most important (though seemingly hard) open problem.

References

1. K. Abrahamson, Generalized String Matching, *SIAM Journal on Computing*, 16(6):1039-1051, 1987.
2. Amihood Amir, Alberto Apostolico and Moshe Lewenstein, Inverse Pattern Matching, Manuscript, to appear in *Journal of Algorithms*.
3. A. Apostolico, C. Ilioppoulos, G.M. Landau, B. Schieber and U. Vishkin, Parallel construction of a suffix tree with applications, *Algorithmica*, 3:347-365, 1988.
4. R. Cole, Parallel merge sort, *SIAM, J. Computing*, 4(1988), pp 770-785.
5. M.J. Fischer and M.S. Paterson, String matching and other products, *Complexity of Computation*, R.M. Karp (editor), SIAM-AMS Proceedings, 7:113-125, 1974.
6. E. Fredkin, Trie Memory, *Communications of the ACM*, 3:490-499, 1962.
7. H.N. Gabow, J.L. Bentley and R.E. Tarjan, Scaling and related techniques for geometry problems, In Proceedings of *16th ACM Symposium on Theory of Computing (STOC)*, pp. 135-143, 1984.
8. A. Gibbons and W. Rytter, Efficient Parallel Algorithms, *Cambridge University Press*, 1988.
9. R.W. Hamming, Error detecting and error correcting codes, *Bell. Sys. Tech. Journal*, 26(2):147-160, 1950.
10. D. Harel and R.E. Tarjan, Fast algorithms for finding nearest common ancestors, *SIAM Journal on Computing*, 13:338-355, 1984.
11. H. Karloff, Fast algorithms for approximately counting mismatches, *Information Processing Letters*, 48(2):53-60, 1993.
12. D.E. Knuth, J.H. Morris, and V.B. Pratt, Fast pattern matching in strings, *SIAM J. Comput.* 6 (1977), 323-350.
13. B. Schieber and U. Vishkin, On finding lower common ancestors: simplification and parallelization, In Proceedings of *3rd Aegean Workshop on VLSI Algorithms and Architecture*, LNCS 319:111-123, 1988.
14. P. Weiner, Linear pattern matching algorithms, In Proceedings of *14th IEEE Symposium on Foundations of Computer Science (FOCS)*, pp. 1-11, 1973.

Distributed Generation of Suffix Arrays

Gonzalo Navarro [1] *
João Paulo Kitajima [2] **
Berthier A. Ribeiro-Neto [2] ***
Nivio Ziviani [2] †

[1] Dept. of Computer Science, University of Chile, Chile.
[2] Dept. of Computer Science, Federal University of Minas Gerais, Brazil.

Abstract. An algorithm for the distributed computation of suffix arrays for large texts is presented. The parallelism model is that of a set of sequential tasks which execute in parallel and exchange messages among them. The underlying architecture is that of a high bandwidth network of processors. Our algorithm builds the suffix array by quickly assigning an independent sub-problem to each processor and completing the process with a final local sorting. We demonstrate that the algorithm has time complexity of $O(b \log n)$ computation and $O(b)$ communication in the average case, where b corresponds to the local text size on each processor (i.e., text size n divided by r, the number of processors). This is faster than the best known sequential algorithm and improves over previous parallel algorithms to build suffix arrays, both in time complexity and scaling factor.

1 Introduction and Motivation

We present a new algorithm for distributed parallel generation of large suffix arrays in the context of a high bandwidth network of processors. The motivation is three-fold. First, the high cost of the best known sequential algorithm for suffix array generation leads naturally to the exploration of parallel algorithms for solving the problem. Second, the use of a set of processors (connected by a fast switch like ATM, for example) as a parallel machine is an attractive alternative nowadays [1]. Third, the final index can be left distributed to reduce the query time overhead. The distributed algorithm we propose is based on a parallel generalized quicksort presented in [7, 15]. The algorithm is an alternative to a previous mergesort-based distributed algorithm [10, 16] and to a pure quicksort-based algorithm [18]. We show that the here proposed algorithm is faster and, more important, that it scales up well while the mergesort-based algorithm does not.

The problem of generating suffix arrays is equivalent to sorting a set of unbounded-length and overlapping strings. Because of those unique features, and because our parallelism model is not a classical one, the problem cannot be solved directly with a classical parallel sorting algorithm. For the PRAM model, there are several studies on parallel sorting. For instance, Jájá et al. [8] describe two optimal-work parallel algorithms for sorting a list of strings over an arbitrary alphabet. Apostolico et al. [2] build the suffix tree of a text of n characters using n processors in $O(\log n)$ time, in the CRCW PRAM model. Retrieval of strings in both cases is performed directly. In a suffix array, strings are pointed to and the pointers are the ones which are sorted. If a distributed memory is used, such indirection makes the sorting problem more complex and requires a more careful algorithm design.

The parallelism model we adopt is that of parallel machines with distributed memory. In such context, different approaches for sorting can be employed. For instance, Quinn [15] presents a quicksort for a hypercube architecture. That algorithm does not take into account the variable size and overlapping in the elements of our problem. Further, the behavior of the communication network in Quinn's work is different (processors are not equidistant one from each other) from the one we adopt here.

* This author has been partially supported by Fondecyt grant 1-950622 (Chile).
** This author has been partially supported by CNPq Project 300815/94-8.
*** This author has been partially supported by CNPq Project 300188/95-1.
† This author has been partially supported by CNPq Project 520916/94-8 and Project RITOS/CYTED.

1.1 Suffix Arrays

The advent of powerful processors and cheap storage has allowed the consideration of alternative models for information retrieval other than the traditional one of a collection of documents indexed by keywords. One such a model which is gaining popularity is the *full text* model. In this model documents are represented by either their complete full text or extended abstracts. The user expresses his information need via words, phrases or patterns to be matched for and the information system retrieves those documents containing the user specified strings. While the cost of searching the full text is usually high, the model is powerful, requires no structure in the text, and is conceptually simple [5].

To reduce the cost of searching a full text, specialized indexing structures are adopted. The most popular of these are *inverted lists*. Inverted lists are useful because their search strategy is based on the vocabulary (the set of distinct words in the text) which is usually much smaller than the text and thus, fits in main memory. For each word, the list of all its occurrences (positions) in the text is stored. Those lists are large and take space which is close to the text size.

Suffix arrays [13] or PAT *arrays* [4, 5] are more sophisticated indexing structures which also take space close to the text size. Their main drawback is their costly construction and maintenance procedures (i.e., creating and updating a suffix array). However, suffix arrays are superior to inverted lists for searching phrases or complex queries such as regular expressions [5, 13].

In this model, the entire text is viewed as one very long string. In this string, each position k is associated to a semi-infinite string or *suffix*, which initiates at position k in the text and extends to the right as far as needed to make it unique. Retrieving the "occurrences" of the user-provided patterns is equivalent to finding the positions of the suffixes that start with the given pattern.

A *suffix array* is a linear structure composed of pointers (here called *index pointers*) to every suffix in the text (since the user normally bases his queries upon words and phrases, it is customary to index only word beginnings). These index pointers are sorted according to a *lexicographical ordering* of their respective suffixes and each index pointer can be viewed simply as the offset (counted from the beginning of the text) of its corresponding suffix in the text. Figure 1 illustrates the suffix array for a text example with nine text positions.

To find the user patterns, binary search is performed on the array at $O(\log n)$ cost (where n is the text size). The construction of a suffix array is simply an *indirect sort* of the index pointers. The difficult part is to do this sorting efficiently when large texts are involved (i.e., texts of gigabytes). Large texts do not fit in main memory and an external sort procedure has to be used. The best known sequential procedure for generating large suffix arrays takes time $O(n^2 \log n / m)$ where n is the text size and m is the size of the main memory [5].

Suffix arrays come from the idea of building a digital search tree on all the suffixes of a text. Such

This text is an example of a textual database

Fig. 1. A suffix array.

search tree allows one to find all the occurrences of a pattern of length m in $O(m)$ time. To reduce the high space requirements, a Patricia tree can be used [14], which compresses unary paths to achieve $O(n)$ storage cost. A Patricia tree built over all suffixes of the text is called a suffix tree [12]. Suffix trees take time $O(n)$ to build [20]. However, this construction is only practical if the tree fits in main memory. Suffix arrays further reduce the space requirements by storing only the leaves of suffix trees. Recently, an intermediate structure between suffix trees and suffix arrays has been proposed [9].

1.2 Distributed Parallel Computers

Parallel machines with distributed memory (multicomputers or message passing parallel computers) are a good cost-performance tradeoff. The emergent fast switching technology has allowed the dissemination of high-speed networks of processors at relatively low cost. The underlying high-speed network could be, for instance, an ATM network running at a guaranteed rate of hundreds of megabits per second. In an ATM network, all processors are connected to a central ATM switch which runs internally at a rate much higher than the external rate. Any pair of processing nodes can communicate at the guaranteed rate without contention. Further, the communication between machines A and B does not interfere with the communication between machines C and D and broadcasting can be done efficiently. Other possible implementations are the IBM SP machine or a Myrinet cluster.

Our idea is to use the aggregate distributed memory of the parallel machine to hold the text. Accessing this aggregate memory requires frequent accesses to remote data (across the network) which take time similar to the time to get data from a local disk at transfer rate [10, 16]. Despite this relatively high remote data access time, use of the distributed aggregate memory to hold the text gives us two critical advantages. First, the aggregate memory allows random access to the data at uniform cost, which we do not have with local disks. Second, we can split our problem in smaller parts and work on them in parallel.

The algorithm we propose is suitable for an environment in which the indexing task is parallelized but the final index is stored at a single processor for sequential query processing. However, the final index may be left distributed along the participant machines.

In a distributed environment, the index can be distributed in two different ways. In the first one, each processor builds a local separate index relative to its local text only. The main drawback of this approach is that each query must be broadcast to every processor and the partial results must be later merged. Despite the high parallelism among processors, this strategy reduces concurrency because queries have to be processed sequentially (i.e., one after the other). In the second and more challenging scheme, a global index is computed and then partitioned among the processors, such that each processor holds a lexicographical interval of the index (e.g. a range of words in dictionary order). In this case, a query is normally directed to a few processors. Despite the low parallelism, concurrency is increased at query time and the system throughput (i.e., number of queries processed in a unit of time) tends to improve.

2 Preliminaries

Our parallelism model is that of a parallel machine with distributed memory. Assume that we have a number r of processors, each one storing b text positions, composing a total distributed text of size $n = rb$. Our final suffix array will also be distributed, and a query is solved with only $O(\log n)$ remote accesses. We assume that the parallelism is coarse-grained, with a few processors, each one with a large main memory. Typical values are r in the tenths or hundreds and b in the millions.

The fact that sorting is indirect poses the following problem when working with distributed memory. A processor which receives a suffix array cell (sent by another processor) is not able to directly compare this cell because it has no local access to the suffix pointed to by the cell (such suffix is stored in the original processor). Performing a communication to get (part of) this suffix from the original processor each time a comparison is to be done is very expensive. To deal with this problem, we use a technique called *pruned suffixes* which works as follows. Each time a suffix array cell is sent to a processor, the first ℓ characters of the corresponding suffix (which we call a *pruned suffix*) are also sent together. This allows the remote processor to perform comparisons locally if they can be decided looking at the first ℓ characters only. Otherwise, the remote processor requests more characters to the processor owning the text suffix cell [5]. We try to select ℓ large enough to ensure that most comparisons can be decided without extra communication and small enough to avoid very expensive exchanges and high memory requirements. In Section 6.2 we find experimentally good values for ℓ.

---

[5] As we will see, in some cases this is not necessary and one might assume that the suffixes are equal if the comparison cannot be locally decided.

Before entering into the algorithm itself we put in clear what we understand by a "worst-on-average-text" (WAT) case analysis. If we consider a pathological text such as "a a a a a a ...", the classical suffix array building algorithm will not be able to handle it well. This is because each comparison among two positions in the text will need to reach the end of the text to be decided, thus costing $O(n)$. Since we find such worst-case analysis unrealistic and probably useless, our analysis deal with *average* random or natural language text. In such text the comparisons among random positions take $O(1)$ time (because the probability of having to look at more than i characters is $1/\sigma^i$ for some $\sigma > 1$). Also, the number of index points (e.g., words) at each processor (and hence the size of its suffix array) is roughly the same. A WAT-case analysis is therefore a worst-case analysis on *average* text. We perform WAT-case and average-case analysis.

3   The Proposed Algorithm

The central idea of the algorithm is as follows. Consider the global sorted suffix array which results of the sorting task. If we cut this array in b similarly-sized portions (which we call *slices*), we can think that each processor holds exactly one such slice at the end. Thus, the idea is to quickly deliver to each processor the index pointers corresponding to its slice.

We recall the definition of a *percentile*. An α-percentile is the value at position αn in the global sorted suffix array. For example, the $(1/r)$-percentile is the element at position b. Our algorithm partitions the data to be worked on by each processor by finding the percentiles $1/r, 2/r, ... (r-1)/r$. An alternative definition for slice is: the portion of the global suffix array between two consecutive (i/r)-percentiles.

The algorithm proceeds in four steps:

Step 1: Every processor builds internally its local suffix array.

Step 2: The processors cooperate to find the r global (i/r)-percentiles. This defines the portion of each slice stored on each processor.

Step 3: The processors engage in a distribution process so that every processor gets the part of its slice stored on any other processor.

Step 4: Every processor completes internally the sorting of its slice.

The analysis is divided in two parts: CPU internal cost for the processors, which is indicated by a factor I, and communication cost, which is indicated by a factor C. CPU operations occur in parallel while communication operations may occur in parallel between distinct pairs of processors.

3.1   Internal Sorting

For this first step, each processor traverses its local text, finds the index points of interest (e.g., beginning of words), and builds an array with all the positions of those index points. The pointers must be shifted to reflect the offsets in the global text, not the local one. Once this is done, the array must be sorted by the suffix each position points to.

Since the text is local, the cost of this step is $O(b \log b)$I in the average and WAT case.

3.2   Finding the Percentiles

Once every processor has sorted its local suffix array, all the processors must collaborate to find the r global percentiles. We first use the median (0.5-percentile) to explain the technique.

It is well known that given two sorted arrays A_1 and A_2 of total size n, the median of $A_1 \cup A_2$ can be obtained in $O(\log n)$. The algorithm proceeds by binary searching on both arrays simultaneously. The search of the median is performed even without knowing the median.

We keep two positions i_1 and i_2, one for each array. The sum of the two positions is always n. If we could find i_1, i_2 such that $i_1 + i_2 = n$ and $A_1[i_1] = A_2[i_2]$, that would be the median, since that value would be in the middle of the sorted union of both arrays.

We first look at the middle of both arrays, i.e., $i_1 = i_2 = n/2$. If $A_1[i_1] < A_2[i_2]$, we conclude that $A_1[i_1] \leq median \leq A_2[i_2]$, and therefore binary search adds $n/4$ to i_1 and subtracts $n/4$ from i_2. The other case is symmetric. In $O(\log n)$ steps the median is found. We are ignoring boundary conditions in this exposition (for instance, it might be that there is no exact median in an array of size $2n$), since their effect in the algorithm is negligible.

Now imagine we have r arrays of size b and want the global median. We begin in the middle of all of them. The $\lfloor r/2 \rfloor$ smaller values must increment their position, while the $\lfloor r/2 \rfloor$ larger must decrement it. At the end of the multiple binary search, the median of all the r final values is the global median.

If we consider that every array is hold by one processor, we obtain that at each step, every processor must broadcast its current value, which costs $O(r)C$. Since $O(\log b)$ steps are carried out, the cost to extract the global median is $O(r \log b)C$.

The algorithm to find general α-percentiles is conceptually the same. As before, all arrays start in their middle positions. However, instead of selecting the median of the r values, we select the α-percentile. Therefore, $\lfloor \alpha r \rfloor$ processors increment their position and $\lfloor (1 - \alpha)r \rfloor$ decrement it. The rest proceeds the same as before. A binary search is performed at each array, and the percentile sought drives the number of processors increasing or decreasing their position. Note that in this case it is not true that at any moment the sum of all the positions equals αn. However it is easy to show that with this strategy that sum converges to the correct value after $\log_2(n(1 - 2\alpha))$ steps for $\alpha \leq 1/2$ (the case $\alpha > 1/2$ is symmetric). Therefore, the algorithm converges to the correct sum before the end of the binary search.

Since each percentile must be found separately, the total cost of this algorithm is $O(r^2 \log b)C$ in the average and WAT case (it is true that a percentile found can reduce the search area for the others, but the gain is marginal).

Observe that the processors cannot send the complete text suffixes when they broadcast their values, but only pruned suffixes. The first ℓ characters are compared and equality is assumed if the comparison cannot be decided. Therefore, additional characters are never requested. This involves some details to deal with. First, care must be exercised to ensure that, at each step, exactly $\lfloor \alpha r \rfloor$ processors move their position forward and $\lfloor (1 - \alpha)r \rfloor$ move backward, even in the case of repeated values. Second, when the value of the final (pruned) percentile is known, the processors must agree on a complete (not pruned) percentile to perform all internal partitions consistently. For example, they can put to the left the suffixes that, once pruned, are smaller or equal to the pruned percentile.

Therefore, the obtained slices can be slightly different in size because of the possible lack of precision when comparing pruned suffixes. These errors are negligible on normal text and the affected processor can easily absorb the few extra items (cf. Section 6.1). Since the number of percentiles broadcast along this process is small, a large ℓ value can be used to ensure a good partition.

3.3 Redistributing the Slices

Once every processor knows the r uniform percentiles, it knows the local slice in its suffix array that must be sent to every other processor. At this point they engage in a redistribution process to send to each other processor the corresponding local slice. This process must ensure that every pair of processors gets a chance to exchange their slices.

We describe an exchange mechanism which allows every processor to be paired with each other at some moment. When two processors are paired, they exchange the appropriate portions of their arrays. The exchange mechanism progresses in stages. In the first stage, we make sure that every processor is paired to its previous and next processor (assuming that processors 0 and $r - 1$ are neighbors). In the second stage we do the same for every pair of processors at "distance" two, and so on. By doing so, only $\lfloor r/2 \rfloor$ stages are needed. Figure 2 illustrates this exchange mechanism for the case of seven processors. The stages are the rows in the Figure. In the ith stage, processors at distance i are paired.

In general, the exchange mechanism works as follows. At stage i, we ensure that every processor p is paired with processors $p + i$ and $p - i$ (for simplicity, we speak *modulo* r in this passage). This can always be accomplished with three rounds of pairing. To show this, we distinguish groupings of pairs of processors that can communicate all in parallel (grayed in the Figure). In the first round of stage i, the groupings of pairs are $[(0, i), (1, i+1), ..., (i-1, 2i-1)]$, $[(2i, 3i), (2i+1, 3i+1), ..., (3i-1, 4i-1)]$, and so on. In the second round of stage i, the groupings of pairs are $[(i, 2i), (i+1, 2i+1), ..., (2i-1, 3i-1)]$, $[(3i, 4i), (3i+1, 4i+1), ..., (4i-1, 5i-1)]$, and so on. Notice that, in general, the number of processors

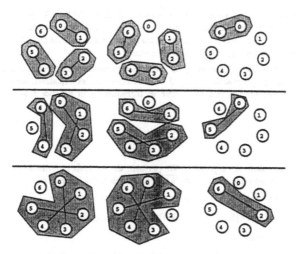

Fig. 2. A redistribution process with 7 processors.

in a grouping is $2i$. Whenever r is a multiple of $2i$, all pairings in stage i are accomplished with only two rounds. However, if r is not a multiple of $2i$, processors might be left unpaired in the first and second rounds. In this case, a third round is required, which pairs exactly the couples left out in the other two rounds.

To be more precise, let k and s be two integers such that $r = k(2i) + s$, where $k > 0$, $0 \le s < 2i$, and i is always a stage number. If r is not a multiple of $2i$ then $s > 0$. In this case, the number of unpaired processors is s when ($s < i$) and is $2i - s$ when $s > i$. Notice that, in the first round, the unpaired processors are near the highest-numbered processors, while in the second round they are among the lowest-numbered. Those processors unpaired in the first two rounds need precisely to be paired among them in an additional third round. Therefore, in at most 3 rounds we complete a stage.

Hence, we need a total of $3\lfloor r/2 \rfloor = O(r)$ exchange rounds. Since each round takes time proportional to the largest exchange in the round, we have a WAT case of $O(b)C$ cost per round, for a total WAT case of $O(n)C$. However, we prove in Appendix A that the cost of each round is on average $O(b/r)$, even taking into account that we wait for the slower exchange in the round. Therefore, the average cost of this step is $O(b)C$ (we verify this fact experimentally in Section 6.1). Recall that the processors need to exchange not only the elements of the suffix array, but also the first ℓ characters of each suffix pointed to by each element. This allows the target processor to complete the sorting without asking the suffix to the owner of the text in most cases.

3.4 Final Sorting

Once each processor obtained all the elements of its slice, the process is completed by an internal sorting. Observe, however, that the situation is not the same as in the initial sorting, because the elements point to remote text, and therefore reliance on the pruned suffixes transmitted together with the pointers is necessary. Another difference is that the elements are arranged in r sorted sequences (i.e., the slice sent by each processor).

Since the processor that sends a slice will send all the pruned suffixes in ascending order, most suffixes will share a common prefix with their neighbors. This can be used to reduce the amount of communication. This technique has been previously applied to compress suffix array indices [3], and works as follows: the first pruned suffix is sent complete. The next ones are coded in two parts: the length of the prefix shared with the previous pruned suffix; and the remaining characters. For example, to send "core", "court" and "custom", we sent "core", (2, "urt") and (1, "ustom"). In Section 6.2 we show that gains near 50% can be expected.

Since we have r sorted sequences, we use a heap to merge them at $O(b \log r)\mathbf{I}$ cost in the average and WAT case. This has an additional advantage: the r local slices received are accessed sequentially and therefore can be stored on disk with little penalty. This is important because the set of all suffix array cells plus their pruned suffixes may not fit in main memory.

Additional communication may be necessary to break ties between equal pruned suffixes. More text may be retrieved from the processors owning the texts. However, as explained, we use long enough pruned suffixes to guarantee that this will occur so infrequently in practice that its effect can be neglected (cf. Section 6.2).

4 Analysis

We compute the global cost of the algorithm. We perform the analysis in terms of r and b, as well as a simplification that is valid whenever $r = o(\sqrt{b/\log b})$, which holds in practice.

Summing up the costs of the algorithm shows an average case of

$$O(b \log n)\mathbf{I} + O(r^2 \log b + b)\mathbf{C} \;=\; O(b \log n)\mathbf{I} + O(b)\mathbf{C}$$

while in the WAT case we have

$$O(b \log n)\mathbf{I} + O(r^2 \log b + n)\mathbf{C} \;=\; O(b \log n)\mathbf{I} + O(n)\mathbf{C}$$

The CPU time improves over the sequential algorithm [5], which is $O(n^2 \log n/b)$ time (assuming $m = b$), by a factor of $\Theta(r^2)$ (this is because the sequential algorithm is not optimal but tries to minimize seek time). The improvement over an optimal sequential algorithm of cost $O(n \log n)$ is $\Theta(r)$, which is optimal.

To analyze the scalability of the algorithm, we consider how the cost is increased if we double the text size and the number of processors, i.e.,

$$\frac{C(2n, 2r)}{C(n, r)} = \frac{b \log(2n)}{b \log n}\,\mathbf{I} + \frac{4r^2 \log b + b}{r^2 \log b + b}\,\mathbf{C} = 1 + O\left(\frac{1}{\log n}\right)\mathbf{I} + O\left(\frac{r^2 \log b}{b}\right)\mathbf{C} = 1 + o(1)$$

which is very good for the practical values involved, though not for very large r.

We compare now the complexity against other parallel algorithms. In [10] a mergesort-based parallel algorithm is proposed, which is $O(b \log n)\mathbf{I} + O(n)\mathbf{C}$ in the average and WAT case. The WAT case is similar to ours, but our average case is much better. In [18], a recursive quicksort-based parallel algorithm is presented, which is $O(b \log n)\mathbf{I} + O(b \log r)\mathbf{C}$ on average and $O(b \log n)\mathbf{I} + O(b \log^2 r)\mathbf{C}$ in the WAT case. Although our average case is better, their WAT case is better than ours. This is because they use a process of pivoting and partitioning by half which allows bad partitions (i.e., one taking more processors than the other). The partition continues until each processor contains a slice. Our present algorithm can be seen as a version of the above procedure in which partitions are built in just one step, losing however the flexibility to handle bad partitions efficiently.

5 A Simpler Algorithm

We show experimentally in Section 6.1 that it is not necessary in practice to compute the exact percentiles. A quick approximation works equally well. This allows to devise a simpler algorithm with the same average case, although the WAT case is worse.

This algorithm replaces Step 2 of the previous one. Instead of engaging in a process of computing the global percentiles, each processor broadcasts its local r uniform percentiles (this costs $O(r^2)\mathbf{C}$). Every processor receives all the percentiles and *estimates* the global percentiles by taking the median of the samples. The rest proceeds in the same way.

We prove in Appendix B that the deviation from the actual values is extremely small on average (i.e., $O(1/\sqrt{n})$), and our experiments in Section 6.1 confirm these assertions. Therefore, the average case cost of this algorithm is

$$O(b \log n)\mathbf{I} + O(r^2 + b)\mathbf{C} \; = \; O(b \log n)\mathbf{I} + O(b)\mathbf{C}$$

To analyze the WAT case, we find the maximum size of an approximated slice. Suppose that we compute an α-percentile. Since every processor broadcasts a value which is larger than αb local items, the median of the values is guaranteed to be larger than $\alpha/2\,n$ elements. With the same argument, it is guaranteed to be smaller than $(1-\alpha)/2\,n$ values. If we take the smallest possible value on an estimated percentile and the largest value in the next percentile, the slice in the middle can be up to $n/2 + b = O(n)$. These $O(n)$ pointers are to be sent to a single processor, which will need $O(n)\mathbf{C}$ time to receive the elements and $O(n \log n)\mathbf{I}$ time to sort them. This is the WAT case of this algorithm: worse than sequential sorting.

6   Simulation Results

The implementation of the proposed algorithm is not concluded yet. However, we performed experiments to validate the most contrived assumptions used in our work.

The first experiment shows that, for a typical text file[6], the distribution of words inside each processor approximately follows that of the whole text, and therefore Steps 2-3 will work well on average text, as well as the simpler algorithm.

The second experiment is related to Step 4. The goal is to find a suitable pruned suffix size ℓ, so that the processors are able to sort locally without normally asking more characters of remote suffixes.

---

[6] In our experiments, the 262,755,189 bytes Wall Street Journal file from TIPSTER/TREC collection [6].

6.1   Word Distribution

In this simulation, the Wall Street Journal (WSJ) file is broken into $r = 16$ blocks of (almost) identical sizes b. For each block, we computed $r - 1$ local percentiles. Next, these percentiles are made available to every simulated processor which computes $r - 1$ medians, each one corresponding to a percentile. Table 1 presents the average and standard deviation of the slice sizes exchanged between any pair of processors. Suffixes were pruned at 48 characters (recall that a large ℓ can be used for Step 2).

p	sent	received	p	sent	received
	$\beta \pm$ stdev	$\beta \pm$ stdev		$\beta \pm$ stdev	$\beta \pm$ stdev
1	$0.94 \pm 2.32\%$	$1.01 \pm 7.68\%$	9	$0.98 \pm 2.40\%$	$1.01 \pm 3.25\%$
2	$0.96 \pm 2.39\%$	$1.00 \pm 2.78\%$	10	$1.01 \pm 1.95\%$	$1.00 \pm 3.01\%$
3	$0.97 \pm 1.46\%$	$1.00 \pm 3.55\%$	11	$1.00 \pm 1.95\%$	$0.99 \pm 2.99\%$
4	$0.99 \pm 1.12\%$	$1.00 \pm 2.89\%$	12	$1.03 \pm 1.73\%$	$1.01 \pm 3.03\%$
5	$0.97 \pm 2.01\%$	$1.01 \pm 3.04\%$	13	$1.00 \pm 1.17\%$	$1.00 \pm 2.98\%$
6	$1.00 \pm 1.02\%$	$0.99 \pm 3.24\%$	14	$1.03 \pm 2.14\%$	$1.00 \pm 2.98\%$
7	$1.00 \pm 1.64\%$	$1.01 \pm 3.21\%$	15	$1.03 \pm 1.67\%$	$1.00 \pm 3.46\%$
8	$1.03 \pm 1.27\%$	$0.99 \pm 3.10\%$	16	$1.07 \pm 1.34\%$	$1.00 \pm 3.83\%$

Table 1. Amount of exchanged data in bytes for $r = 16$ (β is the ratio between the average and the expected b/r). Standard deviation is presented as a percentage of the average.

We remark that the amount of messages sent and received for each pair is approximately the same in most cases. This shows that the partition in words is quite even among processors, and that each processor ends up with a slice of size almost b to perform Step 4.

With regard to the number of bytes transferred during redistribution we observe that the variation among the slices sent by a given processor is rather low ($< 2.5\%$). On the other hand, the variation of the number of bytes received by a given processor from each other processor is higher ($< 8\%$: the higher variation is 7.68% followed by a 3.83%). The largest slice transferred in the whole process is 11% over the expected b/r. This shows that the redistribution of slices is $O(b)\mathbf{C}$ in practice, even with pruned suffixes (of length 48 in this case).

Finally, since we are using estimated percentiles, this shows that the simpler algorithm of Section 5 performs well on natural language texts.

6.2 Suffixes Comparison

In this experiment, we generated sequentially the suffix array for a 100 megabytes subset of the WSJ. We computed for each suffix the number L of identical characters when compared with the previous suffix (given by the sorted suffix array). For example:

```
suffix x   : "A document is a piece of paper..."
suffix x+1: "A document preparation system..."   L=11
suffix x+2: "A dollar in my pocket..."            L=4
```

The purpose of this experiment is to find an ℓ which will work reasonably well even in the final moments of the sorting process, when the algorithm compares suffixes that are almost neighbors in the final suffix array. If each word in the text is considered an index point, we find that the average L is 15.04 with a standard deviation of 10.19. If we consider instead that suffixes do not start with *stop words* (e.g., "a", "the", etc), the average L is 15.45 with a standard deviation of 10.62. The distribution of L is given in the Figure 3.

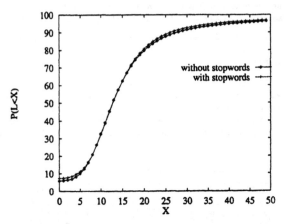

Fig. 3. Distribution of L (probability of a given $L < X$) for 100 megabytes of the WSJ file.

For both cases (suffixes starting and not starting with stop words), the distributions are similar. With $\ell = 30$, 90% of the comparisons are locally solved (i.e., the pruned suffixes differ). In our algorithm, the graph presented in Figure 3 can be considered an upper bound, as explained. Since $L \approx 15$ on average, by using $\ell = 30$ we can save 50% of communication and storage costs by compressing the slices to redistribute (cf. Section 3.4) [7]. For larger texts, the value of ℓ will grow if the same probability of remote access is to be maintained. However, the growing rate is known to be very low (i.e., $O(\log n)$, the average height of a leaf in the suffix trie [19]). The average L will grow at a similar rate, what allows to keep the same compression ratio.

7 Conclusions and Future Work

We have discussed a distributed algorithm for the generation of suffix arrays for large texts. The algorithm is executed on a parallel computer composed of processors connected through a high band-width network. The aggregate memory of the various processors is used as a giant cache for disks.

---

[7] Preliminary more realistic simulations of the sorting process show that this upper bound is pessimistic. With $\ell = 20$ we have 90% of the comparisons decided locally (and therefore compression is 75% effective), and for $\ell = 30$ the probability of a successful local comparison is 95%.

In such aggregate memory, remote accesses are as time consuming as sequential accesses to a local disk. The algorithm quickly splits the problem in one independent subproblem per processor and the rest proceeds locally. The improvement in performance comes from parallelism and from the fact that remote memory can be accessed randomly at uniform cost.

We analyzed the average and worst (on average text) case complexity of our algorithm considering a text of size n and the presence of r processors storing b index points each. Such analysis points out many important advantages over previous work. First, our proposed algorithm has average running time complexity $O(b \log n)$ for computation and $O(b)$ for communication on average, which has optimal speedup over sequential algorithms. Second, it is faster than the previous parallel algorithms that solve this problem. Third, it scales up much nicer than other previous algorithms (e.g., one based on mergesort).

We are currently working on the implementation of the above parallel algorithms. The mergesort implementation is concluded [11] and we compared its performance with that of a local implementation of the sequential algorithm. Besides such implementation efforts, we are investigating the application of our ideas to the generation of the more popular inverted lists [17].

Acknowledgments

We thank the anonymous referees for their useful comments to improve this work.

References

1. T. Anderson, D. Culler, and D. Patterson. A case for NOW (Network of Workstations). *IEEE Micro*, 15(1):54–64, February 1995.
2. A. Apostolico, C. Iliopoulos, G. Landau, B. Schieber, and U. Vishkin. Parallel construction of a suffix tree with applications. *Algorithmica*, 3:347–365, 1988.
3. E. Barbosa and N. Ziviani. From partial to full inverted lists for text searching. In R. Baeza-Yates and U. Manber, editors, *Proc. of the Second South American Workshop on String Processing (WSP'95)*, pages 1–10, April 1995.
4. G. Gonnet. *PAT 3.1: An Efficient Text Searching System - User's Manual*. Centre of the New Oxford English Dictionary, University of Waterloo, Canada, 1987.
5. G. H. Gonnet, R. A. Baeza-Yates, and T. Snider. New indices for text: PAT trees and PAT arrays. In *Information Retrieval - Data Structures & Algorithms*, pages 66–82. Prentice-Hall, 1992.
6. D. Harman. Overview of the third text retrieval conference. In *Proceedings of the Third Text Retrieval Conference - TREC-3*, Gaithersburg, Maryland, 1995. National Institute of Standards and Technology. NIST Special Publication 500-225.
7. J. Jájá. *An Introduction to Parallel Algorithms*. Addison-Wesley, 1992.
8. J. Jájá, K. W. Ryu, and U. Vishkin. Sorting strings and constructing digital search trees in parallel. *Theoretical Computer Science*, 154(2):225–245, 1996.
9. J. Karkkainen. Suffix cactus: A cross between suffix tree and suffix array. In *Proc. CPM'95*, pages 191–204. Springer-Verlag, 1995. LNCS 937.
10. J. P. Kitajima, B. Ribeiro, and N. Ziviani. Network and memory analysis in distributed parallel generation of PAT arrays. In *Fourteenth Brazilian Symposium on Computer Architecture*, pages 192–202, Recife, August 1996.
11. J.P. Kitajima, M.D. Resende, B. Ribeiro, and N. Ziviani. Distributed parallel generation of indices for very large text databases. Technical Report 008/97, Universidade Federal de Minas Gerais - Departamento de Ciência da Computação, Belo Horizonte, Brazil, April 1997. ftp://ftp.dcc.ufmg.br/pub/research/-nivio/papers/.
12. Donald E. Knuth. *The Art of Computer Programming: Sorting and Searching*. Addison Wesley, 1973.
13. U. Manber and G. Myers. Suffix arrays: A new method for on-line string searches. *SIAM Journal on Computing*, 22, 1993.
14. D.R. Morrison. PATRICIA - Practical Algorithm To Retrieve Information Coded In Alphanumeric. *JACM*, 15(4):514–534, October 1968.
15. M. J. Quinn. *Parallel Computing: Theory and Practice*. McGraw-Hill, second edition, 1994.
16. B. Ribeiro, J. P. Kitajima, and N. Ziviani. Distributed parallel generation of PAT arrays. Technical Report 019/96, Universidade Federal de Minas Gerais - Departamento de Ciência da Computação, Belo Horizonte, Brazil, June 1996. ftp://ftp.dcc.ufmg.br/pub/research/nivio/papers/.

17. B. Ribeiro, J.P. Kitajima, G. Navarro, and N. Ziviani. Parallel generation of inverted lists on a network of workstations. Technical Report 009/97, Universidade Federal de Minas Gerais - Departamento de Ciência da Computação, Belo Horizonte, Brazil, April 1997. ftp://ftp.dcc.ufmg.br/pub/research/-nivio/papers/.
18. B. Ribeiro, G. Navarro, J. P. Kitajima, and N. Ziviani. Recursive parallel generation of suffix arrays. Technical Report 010/97, Universidade Federal de Minas Gerais - Departamento de Ciência da Computação, Belo Horizonte, Brazil, April 1997. ftp://ftp.dcc.ufmg.br/pub/research/nivio/papers/.
19. W. Szpankowski. Probabilistic analysis of generalized suffix trees. In Proc. CPM'92, pages 1–14. Springer-Verlag, April 1992. LNCS 644.
20. E. Ukkonen. Constructing suffix trees on-line in linear time. Algorithmica, 14(3):249–260, Sep 1995.

Appendix A: Analysis of Pairwise Exchange

We show that the maximum amount of data exchanged by a pair of processors in the Step 3 of our algorithm is $O(b/r)$ on average.

Since the global index is divided so that an equivalent slice is assigned to each processor, we have r equal-sized slices in the suffix array. The part of the local suffix array to transfer to each processor can be taken as a random sampling over the whole set of suffixes. Therefore, the number of elements of the local suffix array of processor i corresponding to the slice of processor j has a Binomial distribution with parameters $B(b, 1/r)$, since it comes from randomly taking b elements of the global suffix array and observing how many of them correspond to processor j (which occurs with probability $1/r$).

The amount of pointers exchanged between $\lfloor r/2 \rfloor$ pairs of processors can be seen as r independent random variables with the same Binomial distribution (since the pairs exchange data in both ways). The maximum amount of data exchanged in a stage corresponds therefore to the maximum of r independent random variables with distribution $B(b, 1/r)$. Let $X_1, ..., X_r$ be those random variables.

We first show that, for $j > b/r$, $P(X_i \geq j) = O(P(X_i = j))$, i.e. the first term of the summation of probabilities dominates the rest once we passed the mean of the distribution. If we call $p_j = P(X_i = j)$, we have

$$P(X_i \geq j) \; = \; \sum_{k=j}^{b} p_k \; = \; \sum_{k=j}^{b} \binom{b}{k} \frac{(1-1/r)^{b-k}}{r^k}$$

and we observe that

$$\frac{p_{k+1}}{p_k} \; = \; \frac{b-k}{(k+1)r(1-1/r)} \; \leq \; \frac{b}{(k+1)r} \; \leq \; C \; < \; 1$$

where the inequalities come from the fact that $k \geq b/r$. $C = b/(b+r)$ is a new constant introduced to indicate that there is a fixed upper bound for all p_{k+1}/p_k which is independent of k and smaller than 1. Therefore, the terms of the summation decrease at least by a multiplicative constant, what makes their sum a constant proportion of the first summand, i.e. Dp_j, where the constant D is bounded above by $D = 1/(1-C)$.

We now consider the probability of $Y = \max(X_1, ..., X_r) \geq k$. This is equivalent to some X_i being $\geq k$. Bounding again, we have

$$P(Y \geq k) \; \leq \; P(X_1 \geq k) + ... + P(X_r \geq k) \; = \; rP(X \geq k) \; \leq \; Drp_k$$

We find out now how must k be in order for the above probability to be $\leq 1/r$ (we use that result later). That is

$$Drp_k \; = \; Dr \binom{b}{k} \frac{1}{r^k} \; \leq \; 1/r$$

where we pessimistically discarded the factor $(1-1/r)^{b-k}$. Taking logarithms we have

$$b \log b - k \log k - (b-k) \log(b-k) + O(\log b) \; \leq \; k \log r$$

(an $O(\log r)$ error term is discarded assuming $r < b$).

We substitute now $k = \alpha b/r$, for constant α, in the above equation and simplify, to obtain

$$-\frac{\alpha}{r}\log\alpha - \log(1 - \alpha/r) + \frac{\alpha}{r}\log(1 - \alpha/r) + O\left(\frac{\log b}{b}\right) \leq 0$$

which by expanding logarithms yields

$$\log\alpha + \frac{\alpha}{r} \geq 1 + O\left(\frac{\log b}{b} + \frac{1}{r^2}\right)$$

which is clearly achieved by some constant α.

Therefore, we have proved that for $k = O(b/r)$, the probability of the maximum Y among all the random variables X_i being $\geq k$ is $\leq 1/r$. We use it to bound the mean of Y, which is the value we are seeking for:

$$E(Y) = \sum_{j=0}^{b} jP(Y = j) \leq k + (b - k)P(Y > k) \leq k + b/r = O(b/r)$$

what completes the proof.

Appendix B: Average Median by Sampling b Out of n

We prove that by sampling b elements out of n we arrive at the correct median with a relative error of $O(n^{-1/2})$, provided $b > \sqrt{n}$. Since the median is the highest variance percentile, the proof is automatically valid for any percentile. This is stronger than the result we need, since we show that even the median estimation at a single processor is good enough, and therefore doing the same at r processors and combining the results (as done in the paper) is better.

For simplicity, we assume that $b = 2m + 1$. The probability $s(j)$ of our estimated median being the position j in the sorted array is that of, in our sampling, selecting m elements in the range $[1..j - 1]$, m elements in the range $[j + 1..n]$, and of course selecting j. This is

$$s(j) = \frac{\binom{j-1}{m}\binom{n-j}{m}}{\binom{n}{2m+1}}$$

We are interested in the expected proportional size of the larger partition. This is

$$P = \frac{1}{n}\left(\sum_{j=1}^{n/2}(n - j + 1)s(j) + \sum_{j=n/2+1}^{n} js(j)\right) = 2\sum_{j=n/2+1}^{n}(j/n)s(j)$$

We call $t(j) = js(j)$ and $f(x) = t(nx)$ the continuous version of $t(j)$ over the interval $[1/2 .. 1]$. Hence

$$P = \frac{2}{n}\sum_{j=n/2+1}^{n} t(j) = \frac{2}{n}\sum_{j=n/2+1}^{n} f(j/n) \leq 2\int_{1/2}^{1-m/n} f(x)dx$$

(since $t(j)$ is descending for large n). Since $f(x) = t(nx)$, it follows that

$$\frac{f(x + 1/n)}{f(x)} = \frac{t(nx + 1)}{t(nx)} = v(nx)$$

where we have just defined

$$v(j) = \frac{t(j+1)}{t(j)} = \frac{(j+1)(n - m - j)}{(j - m)(n - j)}$$

Taking logarithms and multiplying by n, we have

$$\frac{\ln f(x + 1/n) - \ln f(x)}{1/n} = n \ln v(nx)$$

This last equation defines $(\ln f)'$, hence

$$f(x) = K \; e^{n \int_{1/2}^{x} \ln v(ny) dy}$$

the constant K coming from the integration. We obtain it observing that $f(1/2) = K = t(n/2)$, from where

$$f(x) = \sqrt{\frac{m}{\pi}} \; e^{n \int_{1/2}^{x} \ln v(ny) dy} (1 + O(m/n) + O(1/m))$$

We now solve the integral of $\ln v(ny)$. We have

$$n \int_{1/2}^{x} \ln v(ny) dy = \int_{n/2}^{nx} \ln v(z) dz$$

$$\leq m(2 \ln 2 + \ln x + \ln(1 - x)) + \ln 2 + \ln x + O(1/n)$$

We then rewrite the equation for $f(x)$ as follows

$$f(x) = \sqrt{\frac{m}{\pi}} \; 2^{2m+1} x^{m+1} (1 - x)^m (1 + O(m/n) + O(1/m))$$

and return to our wanted result on P

$$P \leq 2 \int_{1/2}^{1-m/n} f(x) dx \leq \frac{4^{m+1} \sqrt{m}}{\sqrt{\pi}} \int_{1/2}^{1} x^{m+1} (1 - x)^m dx \; (1 + O(m/n) + O(1/m))$$

This last integral is not trivial. We solve it by induction. Let

$$h(d) = \int_{1/2}^{1} x^{m+1+d} (1 - x)^{m-d} dx$$

then

$$h(m) = \frac{1 - \frac{1}{4^{m+1}}}{2m + 2}$$

and our desired result is $h(0)$. Using $fg = \int f'g + \int fg'$, we have

$$h(d) = \int_{1/2}^{1} x^{m+1+d} (1 - x)^{m-d} dx = \frac{1}{(m - d + 1)4^{m+1}} + \frac{m + d + 1}{m - d + 1} \int_{1/2}^{1} x^{m+d} (1 - x)^{m-d+1} dx$$

where the last integral is $h(d - 1)$, hence the recurrence. By using $g(i) = h(m - i)$ we have the more conventional one

$$g(0) = \frac{1 - \frac{1}{4^{m+1}}}{2m + 2} \quad , \quad g(i+1) = \frac{(i+1)g(i) - \frac{1}{4^{m+1}}}{2m - i + 1}$$

which yields

$$g(m) = \frac{1}{8} \frac{\sqrt{\pi/m}}{4^m} (1 + O(1/m))$$

and we have the final result

$$P \leq \frac{1}{2} (1 + O(m/n) + O(1/m)) = \frac{1}{2} (1 + O(b/n) + O(1/b))$$

what shows that the estimated median is very close to the real one for moderately large b.

A question that naturally arises is why the result seems to be worse as b grows (i.e., the $O(b/n)$ error term). This is because we used upper bounds in some parts, hiding factors depending on m that made the error smaller. Since it is clear that, as b grows, the estimation gets better, and that we can assume $b > r$ (i.e., $b > \sqrt{n}$), we have an estimation error independent of b

$$P \leq \frac{1}{2} \left(1 + O\left(1/\sqrt{n}\right)\right)$$

This proves that the largest piece of the partition is $O(1/\sqrt{n})$ in excess over the average, for the median and for every percentile, even if only one processor samples the data.

Direct Construction of
Compact Directed Acyclic Word Graphs

Maxime CROCHEMORE and Renaud VÉRIN

Institut Gaspard Monge
Université de Marne-La-Vallée,
2, rue de la Butte Verte, F-93160 Noisy-Le-Grand.
http://www-igm.univ-mlv.fr

Abstract. The Directed Acyclic Word Graph (DAWG) is an efficient data structure to treat and analyze repetitions in a text, especially in DNA genomic sequences. Here, we consider the Compact Directed Acyclic Word Graph of a word. We give the first direct algorithm to construct it. It runs in time linear in the length of the string on a fixed alphabet. Our implementation requires half the memory space used by DAWGs.

Keywords: pattern matching algorithm, suffix automaton, DAWG, Compact DAWG, suffix tree, index on text.

1 Introduction

In the classical string-matching problem for a word w and a text T, we want to know if w occurs in T, *i.e.*, if w is a factor of T. In many applications, the same text is queried several times. So, efficient solutions are based on data structures built on the text that serve as an index to look for any word w in T. The typical running of various implementations of the search is $\mathcal{O}(|w|)$ (on a fixed alphabet). Among the implementations, the *suffix tree* ([13]) is the most popular. Its size and construction time are linear in the length of the text. It has been studied and used extensively. Apostolico [2] lists over 40 references on it, and Manber and Myers [12] mention several others. Many variants have been developed, like *suffix arrays* [12], *PESTry* [11], *suffix cactus* [10], or *suffix binary search trees* [9]. Besides, the suffix trie, the non-compact version of the suffix tree, has been refined to the *suffix automaton* (*Directed Acyclic Word Graph, DAWG*). This automaton is a good alternative to represent the whole set of factors of a text. It is the minimal automaton accepting this set. It has been fully exposed by Blumer [3] and Crochemore [7]. As for the suffix tree, its construction and size is linear in the length of the text.

In the genome research field, DNA sequences can be viewed as words over the alphabet $\{a, c, g, t\}$. They become subjects for linguistic and statistic analysis. For this purpose, suffix automata are useful data structures. Indeed, the structure is fast to compute and easy to use.

Meanwhile, the length of sequences in databases grows rapidly and the bottleneck to using the above data structures is their size. Keeping the index in main

117

memory is more and more difficult for large sequences. So, having a structure using as little space as possible is appreciable for its construction as well as for its utilization. Compression methods are of no use to reduce the memory space of such indexes because they eliminate the direct access to substrings. On the contrary, the *Compact Directed Acyclic Word Graph* (CDAWG) keeps the direct access while requiring less memory space. The structure has been introduced by Blumer *et al.* [4, 5]). The automaton is based on the concatenation of factors issued from a same context. This concatenation induces the deletion of all states of outdegree one and of their corresponding transitions, excepting terminal states. This saves 50% of memory space. At the same time, the reduction of the number of states (2/3 less) and transitions (about half less) makes the applications run faster. Both time and space are saved.

In this paper, we give an algorithm to build compact DAWGs. This direct construction avoids constructing the DAWG first, which makes it suitable for the actual DNA sequences (more than 1.5 million nucleotides for some of them). The compact DAWG allows to apply standard treatment on sequences twice as long in reasonable time (a few minutes).

In Section 2 we recall the basic notions on DAWGs. Section 3 introduces the compact DAWG, also called compact suffix automaton, with the bounds on its size. We show in Section 4 how to build the CDAWG from the DAWG in time linear in the size of this latter structure. The direct construction algorithm for the CDAWG is given in Section 5. A conclusion follows.

2 Definitions

Let Σ be a nonempty alphabet and Σ^* the set of words over Σ, with ε as the empty word. If w is a word in Σ^*, $|w|$ denotes its length, w_i its ith letter, and $w_{i..j}$ its factor (subword) $w_i w_{i+1} \ldots w_j$. If $w = xyz$ with $x, y, z \in \Sigma^*$, then x, y, and z denote some factors or subwords of w, x is a prefix of w, and z is a suffix of w. $S(x)$ denotes the set of all suffixes of x and $F(x)$ the set of its factors.

For an automaton, the tuple (p, a, q) denotes a transition of label a starting at p and ending at q. A roman letter is used for mono-letter transitions, a greek letter for multi-letter transitions. Moreover, $(p, \alpha]$ denotes a transition from p for which α is a prefix of its label.

Here, we recall the definition of the DAWG, and a theorem about its implementation and its size proved in [3] and [7].

Definition 1. The **Suffix Automaton** of a word x, denoted $DAWG(x)$, is the minimal deterministic automaton (not necessarily complete) that accepts $S(x)$, the (finite) set of suffixes of x.

For example, Figure 1 shows the DAWG of the word **gtagtaaac**. States which are double circled are terminal states.

Theorem 2. *The size of the DAWG of a word x is $\mathcal{O}(|x|)$ and the automaton can be computed in time $\mathcal{O}(|x|)$. The maximum number of states of the automaton is $2|x| - 1$, and the maximum number of edges is $3|x| - 4$.*

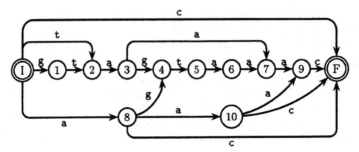

Fig. 1. $DAWG(\texttt{gtagtaaac})$

Recall that the right context of a factor u of x is $u^{-1}S(x)$. The syntactic congruence, denoted by $\equiv_{S(x)}$, associated with $S(x)$ is defined, for $x, u, v \in \Sigma^*$, by:

$$u \equiv_{S(x)} v \iff u^{-1}S(x) = v^{-1}S(x).$$

We call *classes of factors* the congruence classes of the relation $\equiv_{S(x)}$. The longest word of a class of factors is called the *representative* of the class. States of $DAWG(x)$ are exactly the classes of the relation $\equiv_{S(x)}$. Since this automaton is not required to be complete, the class of words not occurring in x, corresponding to the empty right context, is not a state of $DAWG(x)$.

Moreover, we induce a selection among the congruence classes that we call *strict classes of factors* of $\equiv_{S(x)}$ and that are defined as follows:

Definition 3. Let u be a word of C, a class of factors of $\equiv_{S(x)}$. If at least two letters a and b of Σ exist such that ua and ub are factors of x, then we say that C is a **strict class of factors** of $\equiv_{S(x)}$.

We also introduce the function $endpos_x : F(x) \rightarrow \mathbb{N}$, defined, for every word u, by:

$$endpos_x(u) = \min\{|w| \mid w \text{ prefix of } x \text{ and } u \text{ suffix of } w\}$$

and the function $length_x$ defined on states of $DAWG(x)$ by:

$$length_x(p) = |u|, \text{ with } u \text{ representative of } p.$$

The word u also corresponds to the concatenated labels of transitions of the longest path from the initial state to p in $DAWG(x)$. The transitions that belong to the spanning tree of longest paths from the initial state are called *solid transitions*. Equivalently, for each transition (p, a, q) we have the property:

$$(p, a, q) \text{ is solid } \iff length_x(q) = length_x(p) + 1.$$

The function $length_x$ works as well for multi-letter transitions, just replacing 1 in the above equivalence by the length of the label of the transition. This extends the notion of solid transitions to multi-letter transitions:

$$(p, \alpha, q) \text{ is solid } \iff length_x(q) = length_x(p) + |\alpha|.$$

In addititon, we define the *suffix link* for a state of $DAWG(x)$ by:

Definition 4. Let p be a state of $DAWG(x)$, different from the initial state, and let u a word of the equivalence class p. The **suffix link** of p, denoted by $s_x(p)$, is the state q which representative v is the longest suffix z of u such that $u \not\equiv_{S(x)} z$.

Note that, consequently to this definition, we have $length_x(q) < length_x(p)$. Then, by iteration, suffix links induce *suffix paths* in $DAWG(x)$, which is an important notion used by the construction algorithm. Indeed, as a consequence of the above inequality, the sequence $(p, s_x(p), s_x^2(p), ...)$ is finite and ends at the initial state of $DAWG(x)$. This sequence is called the *suffix path of p*.

3 Compact Directed Acyclic Word Graphs

3.1 Definition

The compression of DAWGs is based on the deletion of some states and their corresponding transitions. This is possible using multi-letter transitions and the selection of strict classes of factors defined in the previous section (Definition 3). Thus, we define the Compact DAWG as follows.

Definition 5. The **Compact Directed Acyclic Word Graph** of a word x, denoted by $CDAWG(x)$, is the compaction of $DAWG(x)$ obtained by keeping only states that are either terminal states or strict classes of factors according to $\equiv_{S(x)}$, and by labeling transitions accordingly.

Consequently to Definition 3, the strict classes of factors correspond to the states that have an outdegree greater than one. So, we can delete every state having outdegree one exactly, except terminal states. Note that initial and final states are terminal states too, so they are not deleted.

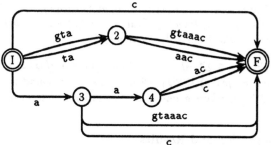

Fig. 2. $CDAWG$(gtagtaaac)

The construction of the DAWG of a word including some repetitions shows that many states have outdegree one only. For example, in Figure 1, the DAWG of the word **gtagtaaac** has 12 states, 7 of which have outdegree one; it has 18 transitions. Figure 2 displays the result after the deletion of these states, using multi-letter transitions. The resulting automaton has only 5 states and 11 edges.

According to experiments to construct DAWGs of biological DNA sequences, considering them as words over the alphabet $\Sigma = \{a, c, g, t\}$, we got that more than 60% of states have an outdegree one. So, the deletion of these states is worth, it provides an important saving. The average analysis of the number of states and edges is done in [5] in a Bernouilly model of probability.

When a state p is deleted, the deletion of outgoing edges is realized by adding the label of the outgoing edge of the deleted state to the labels of its incoming edges. For example, let r, p and q be states linked by transitions (r, b, p) and (p, a, q). We replace the edges (r, b, p) and (p, a, q) by the edge (r, ba, q). By recursion, we extend this method to every multi-letter transition (r, α, p).

In the example (Figure 1), one can note that, inside the word **gtagtaaac**, occurrences of **g** are followed by **ta**, and those of **t** and **gt** by **a**. So, **gta** is the representative of state 3 and it is not necessary to create states for **g** and (**gt** or **t**). Then, we directly connect state I to state 3 with edges (I,gta,3) and (I,ta,3). States 1 and 2 are so deleted.

The suffix links defined on states of DAWGs remain valid when we reduce them to CDAWGs because of the next lemma.

Lemma 6. *If p is a state of $CDAWG(x)$, then $s_x(p)$ is a state of $CDAWG(x)$.*

3.2 Size bounds

By Theorem 2 $DAWG(x)$ is linear in $|x|$. As we shall see below (Section 3.3), labels of multi-letter transitions are implemented in constant space. So, the size of $CDAWG(x)$ is also $\mathcal{O}(|x|)$. Meanwhile, as we delete many states and edges, we review the exact bounds on the number of states and edges of $CDAWG(x)$. They are respectively denoted by $States(x)$ and $Edges(x)$.

Corollary 7. *Given $x \in \Sigma^*$, if $|x| = 0$, then $States(x) = 1$; if $|x| = 1$, then $States(x) = 2$; else $|x| \geq 2$, then $2 \leq States(x) \leq |x| + 1$ and the upper bound is reached when x is in the form $a^{|x|}$, where $a \in \Sigma$.*

Corollary 8. *Given $x \in \Sigma^*$, if $|x| = 0$, $Edges(x) = 0$; if $|x| = 1$, $Edges(x) = 1$; else $|x| \geq 2$, then $Edges(x) \leq 2|x| - 2$ and this upper bound is reached when x is in the form $a^{|x|-1}c$, where a and c are two different letters of Σ.*

3.3 Implementation and Results

Transition matrices and adjacency lists are the classical implementations of automata. Their principal difference lies in the implementation of transitions. The first one gives a direct access to transitions, but requires $\mathcal{O}(States(x) \times \text{card}(\Sigma))$. The second one stores only the exact number of transitions in memory, but needs $\mathcal{O}(\log \text{card}(\Sigma))$ time to access them. When the size of the alphabet is big and the transition matrix is sparse, adjacency lists are preferable. Otherwise, like for genomic sequences, transition matrix is a better choice, as shown by the experiments below. So, we only consider here transition matrices to implement CDAWGs.

We now describe the exact implementation of states and edges. We do this on a four-letter alphabet, so characters take 0.25 byte. We use integers encoded with 4 bytes. For each state, to encode the target state of outgoing edges, transitions matrices need a vector of 4 integers. Adjacency lists need, for each edge, 2 integers, one for the target state and another one for the pointer to the next edge.

The basic information required to construct the DAWG is composed of a table to implement the function s_x and one boolean value (0.125 byte) for each edge to know if it is solid or not. For the CDAWG, in order to implement multi-letter transitions, we need one integer for the $endpos_x$ value of each state, and another integer for the label length of each edge. And that is all.

Indeed, we can find the label of a transition by cutting off the length of this transition from the $endpos_x$ value of its ending state. Then, we got the position of the label in the source and its length. Keeping the source in memory is negligible considering the global size of the automaton (0.25 byte by character). This is quite a convenient solution also used for suffix trees. Figure 3 displays how the

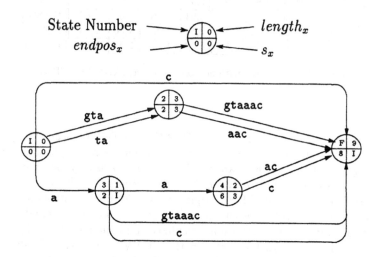

Fig. 3. Data Structure of $CDAWG(\text{gtagtaaac})$

states of $CDAWG(\text{gtagtaaac})$ are implemented.

Then, respectively for transitions matrices and adjacency lists, each state requires 20.5 and 17.13 bytes for the DAWG, and 40.5 and 41.21 bytes for the CDAWG. As a reference, suffix trees, as implemented by McCreight [13], need 28.25 and 20.25 bytes per state. Moreover, for CDAWG and suffix trees the source has to be stored in main memory. Theoretical average numbers of states, calculated by Blumer *et al.* ([5]), are $0.54n$ for CDAWG, $1.62n$ for DAWG, and $1,62n$ for suffix trees, when n is the length of x. This gives respective sizes in bytes per character of the source: 45.68 and 32.70 for suffix trees, 33.26 and 27.80 for DAWGs, and 22.40 and 22.78 for CDAWGs.

Considering the complete data structures required for applications, the function $endpos_x$ has to be added for the DAWG and the suffix tree. In addition, the occurrence number of each factor has to be stored in each state for all the structures. Therefore, the respective sizes in bytes per character of the source become : 58.66 and 45.68 for suffix trees, 46.24 and 40.78 for DAWGs, and 24.26 and 24.72 for CDAWGs.

Source x	$|x|$	Nb states $\frac{}{|x|}$		Nb transitions $\frac{}{|x|}$		Nb transitions $\frac{}{Nb\ states}$		memory gain
		dawg	cdawg	dawg	cdawg	dawg	cdawg	
chro II	807188	1,64	0,54	2,54	1,44	1,55	2,66	50,36%
coli	499951	1,64	0,54	2,54	1,44	1,53	2,66	51,95%
bs 1	183313	1,66	0,50	2,50	1,34	1,50	2,66	54,78%
bs 115	49951	1,64	0,54	2,54	1,44	1,55	2,66	50,16%
random	500000	1,62	0,55	2,54	1,47	1,57	2,68	49,53%
random	100000	1,62	0,55	2,55	1,47	1,57	2,68	49,35%
random	50000	1,62	0,54	2,54	1,46	1,56	2,68	49,68%
random	10000	1,62	0,54	2,54	1,46	1,56	2,68	49,47%
theor. aver. ratios	**1,63**	**0,54**	**2,54**	**1,46**	**1,56**	**2,67**	**50,55%**

Table 1. Statistic table with account between DAWG and CDAWG.

Moreover, Table 1 compares sizes of DAWG and CDAWG meant for applications to DNA sequences. Sizes for random words of different lengths and $|\Sigma| = 4$ are also given. DNA sequences are *Saccharomyces cerevisiae* yeast chromosome II (chro II), a contig of *Escherichia Coli* DNA sequence (coli), and contigs 1 and 115 of *Bacillus Subtilis* DNA sequence (bs). Number of states and edges according to the length of the source and the memory space gain are displayed. Theoretical average ratios are given, calculated from Blumer *et al.* ([5]). First, we observe there are 2/3 less states in the CDAWG, and near of half edges. Second, the memory space saving is about 50%. Third, the number of edges by state is going up to 2.66. With a four-letter alphabet, this is interesting because the transition matrix becomes smaller than adjacency lists. At the same time, we keep a direct access to transitions.

4 Constructing CDAWG from DAWG

The DAWG construction is fully exposed and demonstrated in [3] and [7]. As we show in this section, the CDAWG is easily derived from the DAWG.

Indeed, we just need to apply the definition of the CDAWG recursively. This is computed by the function *Reduction*, given below. Observe that, in this function, $state(p, a]$ denotes the state pointed to by the transition $(p, a]$. The computation is done with a depth-first traversal of the automaton, and runs in time linear in the number of transitions of $DAWG(x)$. Then, by theorem 2, the computation also runs in time linear in the length of the text.

However, this method needs to construct the DAWG first, which spends time and memory space proportional to $DAWG(x)$, though $CDAWG(x)$ is significantly smaller. So, it is better to construct the CDAWG directly.

```
Reduction (state E) returns (ending state, length of redirected edge)
1.  If (E not marked) Then
2.      For all existing edge (E, a] Do
3.          (state(E, a] , |label((E, a])|) ← Reduction(state(E, a]);
4.          mark(E) ← TRUE;
5.      If (E is of outdegree one) Then
6.          Let (E, a] this edge ;
7.          Return (state(E, a] , 1 + |label((E, a])|);
8.      Else
9.          Return (E,1);
```

5 Direct Construction of CDAWG

In this section, we give the direct construction of CDAWGs and show that the running time is linear in the size of the input word x on a fixed alphabet.

5.1 Algorithm

Since the CDAWG of x is a minimization of its suffix tree, it is rather natural to base the direct construction on McCreight's algorithm [13]. Meanwhile, properties of the DAWG construction are also used, especially suffix links (notion that is different from the suffix links of McCreight's algorithm), lengths, and positions, as explained in the previous section.

First, we introduce the notions used by the algorithm, some of them are taken from [13]. The algorithm constructs the CDAWG of the word x of length n, noted $x_{0..n-1}$. The automaton is defined by a set of states and transitions, especially with I and F, the initial and final states. A *partial path* represents a connected sequence of edges between two states of the automaton. A *path* is a partial path that begins at I. The label of a path is the concatenation of the labels of corresponding edges.

The *locus*, or *exact locus*, of a string is the end of the path labeled by the string. The *contracted locus* of a string α is the locus of the longest prefix of α whose locus is defined.

Preliminary Algorithm Basically, the algorithm to build CDAWG inserts the paths corresponding to all the suffixes of x from the longest to the shortest. We define suf_i as the suffix $x_{i..n-1}$ of x. We denote by $\mathcal{A}_i$ the automaton constructed after the insertion of all the suf_j for $0 \leq j \leq i$.

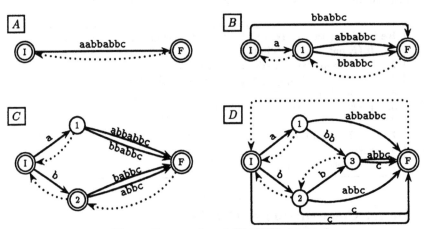

Fig. 4. Construction of $CDAWG(\text{aabbabbc})$

Figure 4 displays four steps of the construction of $CDAWG(\text{aabbabbc})$. In this Figure (and the followings), the dashed edges represent suffix links of states, which are used subsequently. We initialize the automaton $\mathcal{A}_\epsilon$ with states I and F. At step i $(i > 0)$, the algorithm inserts a path corresponding to suf_i in $\mathcal{A}_{i-1}$ and produces $\mathcal{A}_i$. The algorithm satisfies the following invariant properties:

P1: at the beginning of step i, all suffixes suf_j, $0 \leq j < i$, are paths in $\mathcal{A}_{i-1}$.
P2: at the beginning of step i, the states of $\mathcal{A}_{i-1}$ are in one-to-one correspondence with the longest common prefixes of pairs of suffixes longer than suf_j.

We define $head_i$ as the longest prefix of suf_i which is also a prefix of suf_j for some $j < i$. Equivalently, $head_i$ is the longest prefix of suf_i which is also a path of $\mathcal{A}_{i-1}$. We define $tail_i$ as $head_i^{-1} suf_i$. At step i, the preliminary algorithm has to insert $tail_i$ from the locus of $head_i$ in $\mathcal{A}_{i-1}$ (see Figure 5).

To do so, the contracted locus of $head_i$ in $\mathcal{A}_{i-1}$ is found with the help of function *SlowFind* that compares letter-to-letter the right path of $\mathcal{A}_{i-1}$ to suf_i. This is similar to the corresponding McCreight's procedure, except on what is explained below. Then, if necessary, a new state is created to split the last encountered edge, state that is the locus of $head_i$. The automaton B of Figure 4, displays the creation of state 1 during the insertion of suf_1=abbabbc. Note that, if an already existing state matches the strict class of factor of $head_i$, the last

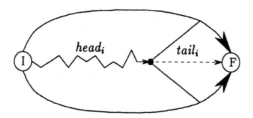

Fig. 5. Scheme of the insertion of a suf_i in $\mathcal{A}_{i-1}$.

encountered edge is split in the same way, but it is redirected to this state. Such an example appears in the same example (case D): the insertion of suf_5=bbc induces the redirection of the edge (2,babbc,F) that becomes (2,b,3). Then, an edge labeled by $tail_i$ is created from the locus of $head_i$ to F. We can write the preliminary algorithm as follows:

```
Preliminary Algorithm
1.   For all suf_i (i ∈[0..n-1]) Do
2.       (q, γ) ← SlowFind(I);
3.       If (γ = ε) Then
4.           insert (q,tail_i,F);
5.       Else
6.           create v locus of head_i splitting (q, γ]
                 and insert (v,tail_i,F);
             or redirect (q, γ] onto v,
                 the last created state;
7.   End For all;
8.   mark terminal states;
```

Note first that *SlowFind* returns the last encountered state. This keeps accessible the transition $(q, \gamma]$ that can be split if this state is not an exact locus.

Second, as in the DAWG construction, if a non-solid edge is encountered during *SlowFind*, its target state has to be duplicated in a clone and the non-solid edge is redirected to this clone. But, if the clone has just been created at the previous step, the edge is redirected to this state. Note that, in the two cases, the redirected transition becomes solid.

Finally, when $tail_i = \varepsilon$ at the end of the construction, terminal states are marked along the suffix path of F.

From the above discussion, a proof of the invariance of properties P1 and P2 can be derived. Thus, at the end of the algorithm all subwords of x and only these words are labels of paths in the automaton (property P1). By property P2, states correspond to strict classes of factors (when the longest common prefix of a pair of suffixes is not equal to any of them) or to terminal states (when the contrary holds). This gives a sketch of the correctness of the algorithm.

The running time of the preliminary algorithm is $\mathcal{O}(|x|^2)$ (with an implementation by transition matrix), like is the sum of lengths of all suffixes of the word x.

Linear Algorithm To get a linear-time algorithm, we use together properties of DAWGs construction and of suffix trees construction. The main feature is the notion of suffix links. They are defined as for DAWGs in Section 2. They are the clue for the linear-running-time of the algorithm.

Three elements have to be pointed out about suffix links in the CDAWG. First, we do not need to initialize suffix links. Indeed, when suf_0 is inserted, x_0 is obviously a new letter, which directly induces $s_x(F)=I$. Note that $s_x(I)$ is never used, and so never defined. Second, traveling along the suffix path of a state p does not necessarily end at state I. Indeed, with multi-letter transitions, if $s_x(p)=I$ we have to treat the suffix $a^{-1}\alpha$ ($a \in \Sigma$) where α is the representative of p. And third, suffix links induce the following invariant property satisfied at step i:

P3: at the beginning of step i, the suffix links are defined for each state of $\mathcal{A}_{i-1}$ according to Definition 4.

The next remark allows redirections without having to search with *SlowFind* for existing states belonging to a same class of factors.

Remark. Let $\alpha\beta$ have locus p and assume that $q = s_x(p)$ is the locus of β. Then, p is the locus of suffixes of $\alpha\beta$ whose lengths are greater than $|\beta|$.

The algorithm has to deal with suffix links each time a state is created. This happens when a state is duplicated, and when a state is created after the execution of *SlowFind*.

In the duplication, suffix links are updated as follows. Let w be the clone of q. In regard to strict classes of factors and Definition 4, the class of w is inserted between the ones of q and $s_x(q)$. So, we update suffix links by setting $s_x(w)=s_x(q)$ and $s_x(q)=w$.

Moreover, the duplication has the same properties as in the DAWG construction. Let (p, γ, q) be the transition redirected during the duplication of q. We can redirect all non-solid edges that end the partial path γ and that start from a state of the suffix path of p. This is done until the first edge that is solid. We are helped in this operation by the function *FastFind*, similar to the one used in McCreight's algorithm [13], that goes through transitions just comparing the first letters of their labels. This function returns the last encountered state and edge. Note that it is not necessary to find each time the partial path γ from a suffix of p, we just need to take the suffix link of the last encountered state and the label of the previous redirected transition.

Let ϑ be the representative of a state of the suffix path of p. Observe that the corresponding redirection is equivalent to insert $suf_{i+|\alpha|-|\vartheta|}$. Indeed, all operations done after this redirection will be the same as for the insertion of suf_i, since they go through the same path.

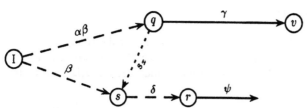

Fig. 6. Scheme of the search using suffix links

After the execution of *SlowFind*, if state v is created, we have to compute its suffix link. Let γ be the label of the transition starting at q and ending at v. To compute the suffix link, the algorithm goes through the path having label γ from the suffix link of q, $s = s_x(q)$. The operation is repeated if necessary. Figure 6 displays a scheme of this search. The thick dashed edges represent paths in the automaton, and the thin dashed edge represents the suffix link of q. This search will allow to insert, as for the duplication, the suffixes suf_j, for $i < j < i+|head_i|$. To travel along the path, we use again the function *FastFind*. Let r and $(r, \psi]$ be the last state and transition encountered by *FastFind*. If r is the exact locus of γ, it is the wanted state, and we set then $s_x(v) = r$. Else, if $(r, \psi]$ is a solid edge, then we have to create a new node w. The edge $(r, \psi]$ is split, it becomes (r, ψ, w), and we insert the transition $(w, tail_i, F)$. Else, $(r, \psi]$ is non-solid. Then, it is split and becomes (r, ψ, v). In the two last cases, since $s_x(v)$ is not found, we run *FastFind* again with $s_x(r)$ and ψ, and this goes on until $s_x(v)$ is eventually found, that is, when $\psi = \varepsilon$.

The discussion shows how suffix links are updated to insure that property P3 is satisfied. The operations do not influence the correctness of the algorithm, sketched in the last section, but yield the following linear-time algorithm. Its time complexity is discussed in the next section.

```
Linear Algorithm
1.   p ← I;   i ← 0;
2.   While not end of x Do
3.       (q, γ) ← SlowFind(p);
4.       If (γ = ε) Then
5.           insert (q, tail_i, F);
6.           s_x(F) ← q;
7.           If (q ≠ I) Then p ← s_x(q) Else p ← I;
8.       Else
9.           create v locus of head_i splitting (q, γ);
10.          insert (v, tail_i, F);
11.          s_x(F) ← v;
12.          find r = s_x(v) with FastFind;
13.          p ← r;
14.      update i;
15.  End While;
16.  mark terminal states;
```

5.2 Complexity

Theorem 9. *The algorithm that builds the CDAWG of a word x of Σ^* can be implemented in time $\mathcal{O}(|x|)$ and in space $\mathcal{O}(|x| \times \text{card}(\Sigma))$ with a transition matrix, or in time $\mathcal{O}(|x| \times \log \text{card}(\Sigma))$ and in space $\mathcal{O}(|x|)$ with adjacency lists.*

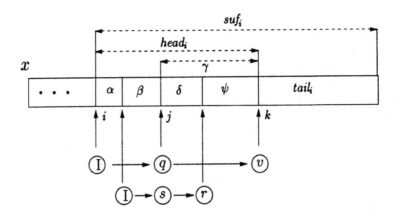

Fig. 7. Positions of labels when suf_i is inserted

Sketch of the proof

It can be proved that each step of the algorithm leads to increase strictly variables j or k in the generic situation displayed in Figure 7. These variables respectively represent the index of the current suffix being inserted, and a pointer on the text. These variables never decrease. Therefore, the total running time of the algorithm is linear in the length of x.

6 Conclusion

We have considered the Compact Direct Acyclic Word Graph, which is an efficient compact data structure to represent all suffixes of a word. There are many data structures representing this set. But, this one allows an interesting space gain compared to the well-known DAWG, which is a reference. Indeed, on the one hand, the upper bounds are of $|x| + 1$ states and $2|x| - 2$ transitions. This saves $|x|$ states and $|x|$ transitions of the DAWG, which leads to faster utilisation. On the other hand, experiments on genomic DNA sequences and random strings display a memory space gain of 50% according to the DAWG. Moreover, when the size of the alphabet is small, transition matrices do not take more space than adjacency lists, keeping direct access to transitions. Thus, we can construct the data structure of twice larger strings, keeping them in main memory, which is actually important to get efficient treatments.

This work shows that the CDAWG can be constructed directly. The algorithm is linear in the length of the text. Of course, it is easier to compute, by reduction, the CDAWG from the DAWG. On the contrary, our algorithm saves time and space simultaneously.

References

1. A. Anderson and S. Nilsson. Efficient implementation of suffix trees. *Software, Practice and Experience*, 25(2):129–141, Feb. 1995.
2. A. Apostolico. The myriad virtues of subword trees. In A. Apostolico & Z. Galil, editor, *Combinatorial Algorithms on Words.*, pages 85–95. Springer-Verlag, 1985.
3. A. Blumer, J. Blumer, D. Haussler, A. Ehrenfeucht, M.T. Chen, and J. Seiferas. The smallest automaton recognizing the subwords of a text. *Theoret. Comput. Sci.*, 40:31–55, 1985.
4. A. Blumer, J. Blumer, D. Haussler, and R. McConnell. Complete inverted files for efficient text retrieval and analysis. *Journal of the Association for Computing Machinery*, 34(3):578–595, July 1987.
5. A. Blumer, D. Haussler, and A. Ehrenfeucht. Average sizes of suffix trees and dawgs. *Discrete Applied Mathematics*, 24:37–45, 1989.
6. B. Clift, D. Haussler, R. McDonnell, T.D. Schneider, and G.D. Stormo. Sequence landscapes. *Nucleic Acids Research*, 4(1):141–158, 1986.
7. M. Crochemore. Transducers and repetitions. *Theor. Comp. Sci.*, 45:63–86, 1986.
8. M. Crochemore and W. Rytter. *Text Algorithms*, chapter 5-6, pages 73–130. Oxford University Press, New York, 1994.
9. R. W. Irving. Suffix binary search trees. *Technical report TR-1995-7, Computing Science Department, University of Glasgow*, April 1995.
10. J. Karkkainen. Suffix cactus : a cross between suffix tree and suffix array. *CPM*, 937:191–204, July 1995.
11. C. Lefevre and J-E. Ikeda. The position end-set tree: A small automaton for word recognition in biological sequences. *CABIOS*, 9(3):343–348, 1993.
12. U. Manber and G. Myers. Suffix arrays: A new method for on-line string searches. *SIAM J. Comput.*, 22(5):935–948, Oct. 1993.
13. E. McCreight. A space-economical suffix tree construction algorithm. *Journal of the ACM*, 23(2):262–272, Apr. 1976.
14. E. Ukkonen. On-line construction of suffix trees. *Algorithmica*, 14:249–260, 1995.

Approximation Algorithms for the
Fixed-Topology Phylogenetic Number Problem

Mary Cryan,[1] Leslie Ann Goldberg,[2] Cynthia A. Phillips,[3]

[1] Department of Computer Science, University of Warwick, Coventry CV4 7AL,
United Kingdom. [†]
[2] Department of Computer Science, University of Warwick, Coventry CV4 7AL,
United Kingdom. [‡]
[3] Sandia National Laboratories, Albuquerque, NM. [§]

Abstract. In the ℓ-phylogeny problem, one wishes to construct an evolutionary tree for a set of species represented by characters, in which each state of each character induces no more than ℓ connected components. We consider the fixed-topology version of this problem for fixed-topologies of arbitrary degree. This version of the problem is known to be $\mathcal{NP}$-complete for $\ell \geq 3$ even for degree-3 trees in which no state labels more than $\ell + 1$ leaves (and therefore there is a trivial $\ell + 1$ phylogeny). We give a 2-approximation algorithm for all $\ell \geq 3$ for arbitrary input topologies and we give an optimal approximation algorithm that constructs a 4-phylogeny when a 3-phylogeny exists. Dynamic programming techniques, which are typically used in fixed-toplogy problems, cannot be applied to ℓ-phylogeny problems. Our 2-approximation algorithm is the first application of linear programming to approximation algorithms for phylogeny problems. We extend our results to a related problem in which characters are polymorphic.

[†] maryc@dcs.warwick.ac.uk. This work was partly supported by ESPRIT LTR Project no. 20244 — ALCOM-IT.

[‡] leslie@dcs.warwick.ac.uk. Part of this work took place during a visit to Sandia National Laboratories which was supported by University of Warwick Research and Teaching Innovations Grant 0951CSA and by the U.S. Department of Energy under contract DE-AC04-94AL85000. Part of this work was supported by ESPRIT LTR Project no. 20244 — ALCOM-IT.

[§] caphill@cs.sandia.gov.caphill@cs.sandia.gov. This work was performed under U.S. Department of Energy contract number DE-AC04-76AL85000.

1   Introduction

The evolutionary biologist collects information on extant species (and fossil evidence) and attempts to infer the evolutionary history of a set of species. Most mathematical models of this process assume divergent evolution, meaning that once two species diverge, they never share genetic material again. Therefore, evolution is modelled as a tree (*phylogeny*), with extant species as leaves and (extant, extinct, or hypothesized) ancestors as internal nodes. Species have been modelled in several ways, depending upon the nature of available information and the mechanism for gathering that information. Based upon these representations, differing measures of evolutionary distance and objective function are used to evaluate the goodness of a proposed evolutionary tree.

In this paper we assume that input data is character-based. Let S be an input set of n species. A *character* c is a function from the species set S to a set R_c of **states**. The set of species in figure 1 have two states. The first character represents skin covering and has three states: h for hair, s for scales, and f for feathers. The second character represents size, where t (tiny) means at most one foot long, m (medium) means one to three feet long and l (large) means greater than three feet long. If we are given a set of characters $c_1, \ldots, c_k$ for S, each species is a vector from $R_{c_1} \times \ldots \times R_{c_k}$ and any such vector can represent a hypothesized ancestor. For example, an anaconda would be represented on these simple characters as (s, l) and a hummingbird would be (f, t). Characters can be used to model biomolecular data, such as a column in a multiple sequence alignment, but in this paper, we think of characters as morphological properties such as coloration or the ability to fly.

Character-based phylogenies are typically evaluated by some *parsimony-like* measure, meaning that the total evolutionary change is somehow minimized. In this paper, we consider the ℓ-*phylogeny* metric introduced in [8]. Given a phylogenetic tree, a character c_i and a state $j \in R_{c_i}$, let ℓ_{ij} be the number of connected components in $c_i^{-1}(j)$ (the subtree induced by the species with state j in character i). A phylogeny is an ℓ-phylogeny if each state of each character induces no more than ℓ connected components. That is, $\max_{c_i, j \in R_{c_i}} \ell_{ij} \leq \ell$. The ℓ-*phylogeny problem* is to determine if an input consisting of a species set S and a set of characters $c_1, \ldots, c_k$ has an ℓ-phylogeny. The *phylogenetic number problem* is to determine the minimum ℓ such that the input has an ℓ-phylogeny.

The classic parsimony problem is to find a tree that minimizes the total number of evolutionary changes: $\sum_{c_i, j \in R_{c_i}} \ell_{ij}$. The compatibility problem is to maximize the number of characters that are *perfect*, meaning that all states of that character induce only one connected component. Thus the compatibility problem is to maximize $|\{c_i : \ell_{ij} = 1 \text{ for all } j \in R_{c_i}\}|$. A 1-phylogeny is called a *perfect phylogeny*. All three problems (ℓ-phylogeny for $\ell \geq 1$, parsimony, compatibility) are $\mathcal{NP}$-complete [1, 8, 4, 6, 13]. Parsimony, ℓ-phylogeny, and compatibility all allow states of a character to evolve multiple times. However, both parsimony and compatibility allow some characters to evolve many times. The ℓ-phylogeny metric requires *balanced* evolution, in that no one character can pay for most of the evolutionary changes. Thus, ℓ-phylogeny is a better measure

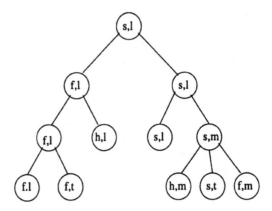

Fig. 1. An example of a 2-phylogeny. States h and f in the first character and state t in the second character are each in two components.

than parsimony or compatibility in biological situations in which all characters are believed to evolve slowly.

In this paper we consider the *fixed-topology* variant of the ℓ-phylogeny problem, where in addition to the species set and characters, we are also given a tree T in which internal nodes are unlabelled, each leaf is labelled with a species $s \in S$ and each species $s \in S$ is the label of exactly one leaf of T. The *fixed-topology ℓ-phylogeny problem* is the problem of determining labels for the internal nodes so that the resulting phylogeny is an ℓ-phylogeny, or determining that such a labelling does not exist. In figure 1, the hypothesized ancestor (s, m) labels one node. This example is a 2-phylogeny.

Fixed-topology algorithms can be used as *filters*. Current phylogeny-producing software can generate thousands of trees which are (approximately) equally good under some metric such as maximum likelihood or parsimony. We can think of these outputs as proposed topologies. One way to differentiate these hypotheses is to see which topologies also have low phylogenetic number. For example, the original trees can be generated by biomolecular sequence data, and they can then be filtered using morphological data with slowly-evolving traits.

It will be convenient to allow a node to remain unlabelled in one or more characters in a fixed topology. In this case, the node disagrees with all of its neighbors on all unlabelled characters. We can easily extend such a labelling to one in which every node is labelled without increasing ℓ_{ij} for any i or j: for any character j, for each connected component of nodes which are not labelled, choose any neighbouring node v which is labelled i_v and label the entire component with the state i_v. This does not introduce any extra component for i_v, nor does it break components of any other state that weren't already broken.

In the fixed-topology setting, optimal trees for the parsimony and compatibility metrics can be found in polynomial time [7]. The fixed-topology ℓ-phylogeny problem can be solved in polynomial time for $\ell \leq 2$, but is $\mathcal{NP}$-complete for $\ell \geq 3$ even for degree-3 trees in which no state labels more than $\ell + 1$ leaves (and therefore there is a trivial $\ell + 1$ phylogeny) [8]. Fitch's algorithm for parsimony uses dynamic programming. Dynamic programming also gives good algorithms in some cases for finding phylogenies when characters are polymorphic [2].

Jiang, Lawler, and Wang [10] consider the fixed-topology tree-alignment problem, where species are represented as biomolecular sequences, the cost of an edge in the tree is the edit distance between the labels at its endpoints, and the goal is to minimize the sum of the costs over all edges. They give a 2-approximation for bounded-degree input topologies and extend this to obtain a polynomial-time approximation scheme (PTAS). In Lemma 3 of [10], they prove that the best lifted tree (in which the label of each internal node is equal to the label of one of its children) is within a factor of 2 of the best tree with arbitrary labels. The proof only uses the triangle inequality (it does not use any other facts about the cost measure). Therefore, the result holds for several other cost measures, including ℓ-phylogeny, parsimony, and the minimum-load cost measure for phylogenies with polymorphic characters which was introduced in [2]. It also holds for the variant of ℓ-phylogeny in which ℓ_i is specified for each character c_i. This variant was introduced in [8]. We refer to it as the *generalized ℓ-phylogeny problem*. In fact, Lemma 3 of [10] holds for the fixed-topology problem with *arbitrary* input topologies, though the authors do not state this fact since they do not use it. Despite the applicability of Lemma 3, the algorithmic method of Jiang, Lawler and Wang does not seem to be useful in developing approximation algorithms for the fixed-topology ℓ-phylogeny problem (or for related problems). Jiang et al. use dynamic programming to find the minimum-cost lifted tree. Dynamic programming is not efficient for the more global metric of ℓ-phylogeny. The dynamic programming proceeds by computing an optimal labelling for a subtree for each possible labelling of the root of the subtree. For metrics where cost is summed over edges (such as parsimony or tree alignment), one only needs to find the lowest-cost labelling for a given root label. For the ℓ-phylogeny problem, the cost of a tree depends upon how many times each state is broken for a given character. One cannot tell *a priori* which state will be the limiting one. Therefore, instead of maintaining a single optimal tree for each root label, we must maintain all trees whose cost (represented as a vector of components for each state) is undominated. This number can be exponential in r, the number of states, even for bounded-degree input trees. This is a common theme in combinatorial optimization: the more global nature of minimax makes it harder to compute than summation objectives, but also more useful.

Gusfield and Wang [14] take the approach of [10] a step further by proving that the best uniform lifted tree (ULT) is within a factor of 2 of the best arbitrarily-labelled tree. In a uniform lifted tree on each level, all internal nodes are labeled by the same child (e.g. all nodes at level one take the label of their leftmost child). This proof also extends to the ℓ-phylogeny metric. If the input tree is a complete binary tree, then there are only n ULTs, and exhaustive search is efficient, giving an algorithm which is faster than ours and has an equivalent performance bound. However, when the input tree isn't complete (even if it is binary), Gusfield and Wang use dynamic programming to find the minimum-cost ULT, and so their method fails when it is applied to the ℓ-phylogeny problem. Wang, Jiang, and Gusfield recently improved the efficiency of their PTAS for tree alignment [15], but still use dynamic programming.

We give a simple 2-approximation for the fixed-topology ℓ-phylogeny problem that works for arbitrary input topologies. It is based on rounding the linear-programming relaxation of an integer programming formulation for the fixed-topology ℓ-phylogeny problem. To our knowledge, this is the first application of linear-programming technology to phylogeny problems.

As we described earlier, ℓ-phylogeny is most appropriate for slowly-evolving characters. It is most restrictive (and hence most different from parsimony) when ℓ is small. Therefore, we look more closely at the first NP-hard case: $\ell = 3$. For this case, we give an algorithm based upon the structure of a 3-phylogeny that will construct a 4-phylogeny if the input instance has a 3-phylogeny.

The remainder of our paper is organized as follows: in section 2, we give the 2-approximation algorithm for the ℓ-phylogeny problem. In section 3 we give the optimal approximation algorithm for inputs with 3-phylogenies. In section 4, we extend the linear-programming-based techniques to finding low-cost labellings for inputs with polymorphic characters.

2 A 2-approximation algorithm for the fixed-topology phylogenetic number problem

The interaction between characters in phylogeny problems affects the choice of the topology, but it does not affect the labelling of the internal nodes once the topology is chosen. Thus, for this problem, we can consider each character separately.

Let $c : S \to \{1, \ldots, r\}$ be a character and let T be a tree with root q and leaves labelled by character states $1, \ldots, r$. For each state i, let T_i be the subtree of T consisting of all the leaves labelled i and the minimum set of edges connecting these leaves. Let $L(T_i)$ be the set of leaves of T_i, and let rt_i, the root of T_i, be the node of T_i closest to the root of T. The *important nodes* of T_i are the leaf nodes and the nodes of degree greater than 2. An *i-path* p of T_i is a sequence of edges of T_i that connects two important nodes of T_i, but does not pass through any other important nodes. The two important nodes are referred to as the *endpoints* of p, and the other nodes along the i-path are said to be *on* p (an i-path need not have any nodes on it). Although the edges of the tree T are undirected, we will sometimes use the notation $(v \to w)$ for an edge or i-path with endpoints v and w, to indicate that v is nearer to the root of T than w (v is the *higher* endpoint and w is the *lower* endpoint); otherwise we will write edges and i-paths as (v, w). If the lower endpoint w is labelled i and the label of the upper endpoint v or some node on the i-path $p = (v \to w)$ is not i, then we say that p *breaks* state i. If an i-path goes through the (degree-2) root of T_i, then *both* endpoints are considered lower endpoints.

Given a tree T with each node labeled from the set $\{1, \ldots, r\}$, we need a way to count the number of components induced by the nodes labeled i. Since the tree is rooted, we can assign each connected component a root, namely the node closest to the root of T. We then count the number of roots for components labelled i. A node is the root of its component if its label differs from that of its parent. The root q, which has no parent, is also the root of its component. Therefore we have the following:

Observation 1 *Let T be a tree with its leaves and internal nodes labelled by elements of $\{1, \ldots, r\}$. For each i, let T_i be defined as above, and let q be the root of tree T. Then the number of connected components induced by the nodes labelled i is $|\{e = (v \to w) : c(v) \neq i, c(w) = i\}| + Y_i$, where $Y_i = 1$ if q is labelled i and 0 otherwise.*

We now define an integer linear program (ILP) which solves the fixed-topology ℓ-phylogeny problem. The linear-programming relaxation of this ILP is the key to our 2-approximation algorithm. The integer linear program $\mathcal{I}$ uses the variables $X_{v,i}$, for each state $i \in \{1, \ldots, r\}$, and each node v in the tree T, the variables $X_{p,i}$ for each state i, and each i-path p of T_i and the variables $cost_{p,v,i}$ for each state i, i-path p in T_i and each lower endpoint v of path p. Recall that each path has one lower endpoint except when there is an i-path through a degree-2 root, in which case both endpoints are lower endpoints. These variables have the following interpretation:

$$X_{v,i} = \begin{cases} 1 \text{ if node } v \text{ is labelled } i \\ 0 \text{ otherwise} \end{cases}$$

$$X_{p,i} = \begin{cases} 1 \text{ if all nodes on } p \text{ are labelled } i \\ 0 \text{ otherwise} \end{cases}$$

$$cost_{p,v,i} = \begin{cases} 1 \text{ if lower endpoint } v \text{ of } p \text{ is the root of a component of state } i \\ 0 \text{ otherwise} \end{cases}$$

ILP $\mathcal{I}$ is defined as follows:

$$\text{minimize } \ell$$

subject to

$$X_{v,i} = 1 \qquad \text{for each leaf } v \in T_i, \, i = 1, \ldots, r \quad (1)$$

$$X_{v,i} = 0 \qquad \text{if } v \notin T_i, \, i = 1, \ldots, r \quad (2)$$

$$\sum_{i=1}^{r} X_{v,i} \leq 1 \qquad \forall v \in T \quad (3)$$

$$X_{p,i} = X_{v,i} \qquad i = 1, \ldots, r, \, \forall p \in T_i, \, \forall v \in p \quad (4)$$

$$X_{p,i} \leq X_{v,i} \qquad i = 1, \ldots, r, \, \forall p \in T_i, \text{ endpoint } v \in p \, (5)$$

$$cost_{p,v,i} \geq X_{v,i} - X_{p,i} \qquad i = 1, \ldots, r, \, \forall p \in T_i \quad (6)$$

$$\sum_{p,v} cost_{p,v,i} + X_{rt_i,i} \leq \quad \ell \qquad i = 1, \ldots, r \qquad (7)$$

$$X_{v,i}, X_{p,i}, cost_{p,v,i} \in \quad \{0, 1\} \qquad (8)$$

Constraint (8) assures that the cost $(cost_{p,v,i})$, i-path $(X_{p,i})$, and vertex $(X_{v,i})$ variables serve as indicator variables in accordance with their interpretation. Constraint (1) labels the leaves in accordance with the input. Constraint (2) prohibits labelling a node v with a state i when v is not in T_i (the number of components labelled i could not possibly be reduced by this labelling). Constraint (3) ensures that each internal node will have no more than one label. Constraints (4) and (5) ensure that for each tree T_i, nodes on paths are taken all-or-none; if any node on an i-path p (including endpoints) is lost to a state i, then it does no good to have any of the other nodes on the path (though it may be beneficial to maintain one or both endpoints). Constraint (6) computes the path costs (counts roots), and constraint (7) ensures that each state has no more than ℓ connected components. This is an implementation of Observation 1. Since there is no i-path in T_i with rt_i as its lower endpoint, we must explicitly check the root of each tree T_i, just as we checked the global root in Observation 1.

Integer program $\mathcal{I}$ solves the fixed-topology ℓ-phylogeny problem. We will now show that the optimal value of ℓ given by $\mathcal{I}$ is a lower bound on the phylogenetic number of tree T with the given leaf labelling.

Proposition 2. *If there exists an ℓ-phylogeny for tree T with a given leaf labelling, then there is a feasible solution for the integer linear program for this value of ℓ.*

Proof. Suppose there exists an ℓ-phylogeny on the tree T with leaves and internal nodes labelled from $\{1, \ldots, r\}$. Consider one particular ℓ-phylogeny, and assume without loss of generality that all node labels are useful for connectivity (i.e. changing the label of node v from i to something else will increase the number of components labelled i). This may require some nodes to be unlabelled. We obtain a feasible solution to $\mathcal{I}$ as follows. Set variable $X_{v,i}$ to 1 if node v is labelled i in this phylogeny and 0 otherwise. Set $X_{p,i}$ to 1 if both endpoints and all internal nodes of i-path p are labelled i and 0 otherwise. Set $cost_{p,v,i} = 1$ if lower endpoint v of p is labelled i and the i-path is not, and set $cost_{p,v,i} = 0$ otherwise. We now show this assignment is a solution to $\mathcal{I}$.

The $X_{v,i}$, $X_{p,i}$, and $cost_{p,v,i}$ variables are binary by construction, thus satisfying Constraint (8). By construction, Constraint (1) will be satisfied by our assignment. Constraint (2) will also be satisfied, because it is never useful to label nodes outside T_i with i, and we have assumed all the labels on nodes are useful for connectivity. Constraint (3) is also satisfied because each node of the phylogeny will be labelled with at most one state. Constraints (4) and (5) are satisfied because the condition that all labelled nodes are necessary for connectivity ensures that a node on an i-path will only be labelled i if all the nodes and endpoints of the i-path are labelled i. Constraint (6) is satisfied by construction.

To show that constraint (7) is satisfied, consider the connected components for i; by our assumption, these all lie in T_i. Let $\gamma = \{e = (v \to w) : c(v) \neq$

i, $c(w) = i$}. By Observation 1 we have $|\gamma| + X_{q,i} \leq \ell$, where q is the root of T. To calculate $(\sum_{p,v} cost_{p,v,i}) + X_{rt_i,i}$, note that $cost_{p,w,i} = 1$ if and only if $X_{w,i} = 1$ for the lower endpoint w and $X_{p,i} = 0$ and otherwise $cost_{p,w,i}$ is 0. By our definitions above, $X_{p,i} = 0$ and $X_{w,i} = 1$ if and only if the edge $(v, w) \in T$ from w's parent (on the i-path p or its upper endpoint) into w has $c(v) \neq i$ and $c(w) = i$. Furthermore, this is the only edge on the i-path with this property (the cost of each other edge is 0) unless path p passes through a degree-2 root and both its endpoints have breaks. In the latter case there is a second endpoint w' such that $cost_{p,w',i} = 1$. Since the i-paths partion T_i, each i-path p and lower endpoint v with $cost_{p,v,i} = 1$ contains one element of γ which is unique to that i-path and lower endpoint. Thus $(\sum_{p,v} cost_{p,v,i}) \leq |\gamma| \leq \ell$. If rt_i is the node q, then $X_{rt_i,i} = X_{q,i}$ and $(\sum_{p,v} cost_{p,v,i}) + X_{rt_i,i} \leq |\gamma| + X_{q,i} \leq \ell$. Otherwise, if rt_i is not the global root q, by our assumption that only useful nodes of T are labelled with i, the ancestor node a_i of rt_i is not labelled i. Then, if $X_{rt_i,i} = 1$ the edge $e = (a_i \rightarrow rt_i)$ contributes 1 to $|\gamma|$, and therefore $(\sum_{p,v} cost_{p,v,i}) + X_{rt_i,i} \leq |\gamma| + X_{q,i} \leq \ell$. Hence constraints (7) are satisfied and we have a solution for the integer program $\mathcal{I}$. □

Integer linear programming in $\mathcal{NP}$-hard in general [3, 9, 11], so we cannot solve it directly in polynomial time. (In fact, doing so would solve the fixed-topology ℓ-phylogeny problem, which we know to be $\mathcal{NP}$-hard for $\ell \geq 3$ from [8].) However, we can solve the linear-programming relaxation $\mathcal{L}$ of $\mathcal{I}$, which consists of all the constraints of $\mathcal{I}$ except that Constraint (8) is replaced by the constraint $0 \leq X_{v,i}, X_{p,i}, cost_{p,v,i} \leq 1$ (8').

Theorem 3. *If there is a solution for the linear program $\mathcal{L}$ for a fixed topology T with leaves labelled with states from $\{1, \ldots, r\}$, then we can assign states to the internal nodes of T such that no state $i \in \{1, \ldots, r\}$ has more than 2ℓ components.*

Proof. The 2ℓ phylogeny for the character $c : S \rightarrow \{1, \ldots, r\}$ on T is constructed by assigning states to the nodes of each tree T_i based on the $X_{v,i}$ values. For each state $i \in \{1, \ldots, r\}$, consider each internal node v of T_i. A node v is labelled i if and only if $X_{v,i} > 1/2$, and there is a path $v, w_1, w_2, \ldots, w_k, v^*$ through tree T_i to a leaf v^* of T_i where $X_{w_j,i} > 1/2$ for all $j = 1, \ldots, k$. If $X_{v,i} > 1/2$, but there is no such path, then node v is *isolated*, and by our procedure remains unlabelled. A node v also remains unlabelled if $X_{v,i} \leq 1/2$ for all states i.

To show that the labelling is a 2ℓ-phylogeny, we show that each component of state i adds at least $1/2$ to the sum $(\sum_{p,v} cost_{p,v,i}) + X_{rt_i,i}$. From Observation 1, the number of connected components for the state i is $|\{e = (v \rightarrow w) : c(v) \neq i, c(w) = i\}| + Y_i$, where Y_i is 1 if q has state i (and therefore $q = rt_i$) and 0 otherwise. Constraints (5) and (4) ensure that if the edge $e = (v \rightarrow w)$ has $c(v) \neq i$ and $c(w) = i$ then either w is the root of T_i, or w must be an endpoint node with $X_{w,i} > 1/2$, and that either $X_{v,i} \leq 1/2$ or v is isolated. However, since w is labelled i, w must not be isolated, and therefore v would not be isolated if $X_{v,i}$ was greater than $1/2$. So $X_{v,i} \leq 1/2$, and $X_{p,i} \leq 1/2$ for the i-path p

with lower endpoint w. Therefore we need only calculate the number of lower endpoints w from i-paths p such that $X_{p,i} \leq 1/2$, $X_{w,i} > 1/2$, and w is not isolated.

Suppose w is a lower endpoint of i-path p. Since w is not isolated and the node above w is not labelled i, there is a sequence $p_1 = (w \to v_1)$, $p_2 = (v_1 \to v_2), \ldots, p_j = (v_{j-1} \to v_j)$ of i-paths of T_i such that $X_{p,i} > 1/2$ for every $p \in \{p_1, \ldots, p_j\}$ and $X_{v,i} > 1/2$ for every $v \in \{v_1, \ldots, v_j\}$, and v_j is a leaf of T_i. Calculating $cost_{p,w,i} + cost_{p_1,v_1,i} + \ldots + cost_{p_j,v_j,i} = (X_{w,i} - X_{p,i}) + (X_{v_1,i} - X_{p_1,i}) + \ldots + (X_{v_j,i} - X_{p_j}) = -X_{p,i} + (X_{w,i} - X_{p_1,i}) + (X_{v_1,i} - X_{p_2,i}) + \ldots + (X_{v_{j-1},i} - X_{p_j,i}) + X_{v_j,i}$, we know by constraints (5) that $X_{w,i} - X_{p_1,i} \geq 0$, $X_{v_1,i} - X_{p_2,i} \geq 0$, $\ldots$, $X_{v_{j-1},i} - X_{p_j,i} \geq 0$. So $cost_{p,w,i} + cost_{p_1,v_1,i} + \ldots + cost_{p_j,vj,i} \geq X_{v_j,i} - X_{p,i} = 1 - X_{p,i} \geq 1/2$.

Note that for any two breaks that appear at lower endpoints w and w' of i-paths p and p' respectively, the i-labelled paths to leaves are disjoint (because they are in separate components of i). Therefore each break of i at a lower endpoint w contributes at least $1/2$ to the sum $(\sum_{p,v} cost_{p,v,i})$. If rt_i is labelled i (and hence the root of a component of i), then $X_{rt_i,i} > 1/2$ (corresponding to an edge $(v \to rt_i)$ in T or to the case $Y_i = 1$). So $2 \times ((\sum_{p,v} cost_{p,v,i}) + X_{rt_i,i}) \geq |\{e = (v \to w) : c(v) \neq i, c(w) = i\}| + Y_i$, and therefore $2\ell \geq |\{e = (v \to w) : c(v) \neq i, c(w) = i\}| + Y_i$.

$\square$

Recall that we have considered each character separately in our 2-approximation algorithm. Thus, our work applies to the generalized ℓ-phylogeny problem (and not just to the ordinary ℓ-phylogeny problem). In particular, we have the following theorem.

Theorem 4. *There is a 2-approximation algorithm for the generalized ℓ-phylogeny problem.*

3 4-phylogeny algorithm

In this section we give an algorithm which takes a fixed-topology phylogeny instance with arbitrary topology and, as long as it has a 3-phylogeny, finds a 4-phylogeny for the instance.

We use the following definitions, in addition to those that we used for the 2-approximation. We will maintain a forest F_i for every state i, which corresponds to the set of nodes that state i is contending for. A **branch point** of F_i is a node in F_i with degree 3. We say that a node $v \in F_i$ is *claimed* by state i if it is not in F_j for any $j \neq i$.

The algorithm generalizes the fixed-topology 2-phylogeny algorithm of [8]. It consists of a *forced phase* and then an *approximation phase*. The forced phase produces a partial labelling (resolution of labels on some subset of the nodes) which can still be extended to a 3-phylogeny; it makes no labelling decisions that are not forced if one is to have a 3-phylogeny. The approximation phase removes all remaining contention for labels, but it can break some states into four pieces. Because finding a fixed-topology 3-phylogeny is $\mathcal{NP}$-complete [8], this is an optimal approximation algorithm for phylogeny instances with 3-phylogenies.

3.1 The Forced Phase of the Algorithm

Initially, for every state i we will have $F_i = T_i$. During the forced phase of the algorithm, nodes will be removed from the forests F_i. The invariant during the forced phase of the algorithm is that there is a 3-phylogeny in which every node v is assigned a state j such that $v \in F_j$. The forced phase applies the following rules in any order until none can be applied. If any forest F_i is broken into more than three components by the application of these rules, then the instance has no 3-phylogeny and the algorithm terminates.

1. For any i-path (v, w), let S be the set containing v and w and the nodes on the i-path. If S contains two or more branch points of F_j (for $i \neq j$) then every node on the i-path is removed from F_i. Note that in the updated copy of F_i (after the rule is applied), v and w will have lower degree than in the original F_i. Furthermore, if v has degree 2 in the updated F_i then the i-path containing it will consist of nodes from two different i-paths in the original F_i. Similarly, i-paths can be merged as a result of the following rules.
2. If F_i has $C(F_i)$ connected components and F_i contains a node v of degree at least $5 - C(F_i)$ then in every forest F_j with $j \neq i$, v and all nodes on j-paths adjacent to v are removed from F_j (i.e., i claims node v).
3. Suppose v is a branch point of F_i but not a branchpoint of F_j, and suppose two i-paths (v, w_1) and (v, w_2) adjacent to v each contain a branch point of F_j. Then in every forest F_k with $k \neq i$, every branch point $w \notin \{v, w_1, w_2\}$ of F_i and every node on every k-path adjacent to w is removed from F_k (i.e., F_i claims all branchpoints except v, w_1, and w_2).

Rule 1 is justified by observing that in any 3-phylogeny, each forest F_i gives up at most two disjoint i-paths, or a single branchpoint with the i-paths adjacent to it. In the setting in which rule 1 is applied, if F_i were to claim the path in question, then F_j would lose two branchpoints and necessarily be in at least four components. Therefore, in any 3-phylogeny for the input, F_i cannot have that i-path. Note that once any node on an i-path is lost to F_i, then F_i has no reason to claim any other nodes on the i-path.

Rule 2 is justified by the following observations. If there is a node of degree at least 4 in tree T_i, then it must be labelled i in any 3-phylogeny (losing it will break state i into at least 4 pieces). Once F_i has been forced to give up an i-path, it cannot give up another branchpoint. Finally, once F_i has been forced into three pieces, then it must claim all remaining nodes in F_i.

Rule 3 is applied when we isolate a region where a break in F_i must occur, but do not yet know exactly where the break will occur. If two paths adjacent to a branchpoint of F_i contain branchpoints of F_j, then by the previous argument for rule 1, F_i cannot keep both of those paths. Therefore, outside of the affected region (those two i-paths), F_i can act as though the forest has been cut into at least two pieces, and can claim all branchpoints.

3.2 The Approximation Phase of the Algorithm

In the following, *releasing* a degree-2 node $v \in F_i$ removes all nodes on its i-path from F_i. Releasing a higher-degree node $v \in F_i$ removes v and all nodes on i-paths adjacent to v from F_i. The approximation phase consists of the following steps.

1. For each connected component C of F_i, if the root of C is unclaimed then F_i releases the root of C. Also, if this root has degree 2, F_i releases any unclaimed branch points at the ends of the i-path through this root.
2. If, after the forced phase, F_i is in a single component with exactly one unclaimed branch point, w, then it releases w.
3. If, after the forced phase, F_i is in a single component with exactly two unclaimed branch points, w_1 and w_2 which are the two endpoints of an i-path, and the path from the root to w_2 passes through w_1, then F_i releases w_2.
4. If, after the forced phase, F_i is in a single component with exactly three unclaimed branch points, w_1, w_2 and w_3 where there is an i-path from w_1 to w_2 and an i-path from w_2 to w_3, then F_i releases w_2.

3.3 The Proof of Correctness

The proof of correctness of the algorithm requires the following observation, and follows from Lemma 6 and Lemma 11.

Observation 5 *If F_i is in one component and it releases two branchpoints w_1 and w_2 which share an i-path, then the resulting forest F_i has at most 4 components.*

Proof. Suppose without loss of generality that branchpoint w_1 is removed first. This leaves F_i in three pieces. Because w_2 shares an i-path with w_1, this operation reduces the degree of w_2 to two, so the two remaining i-paths adjacent to w_2 are merged. Subsequently removing the i-path through w_2 adds only one more component. □

Lemma 6. *At the end of the approximation phase, every forest F_i has at most 4 connected components.*

Proof. The forest F_i can be in at most three components at the end of the forced phase. If F_i is in three components at the end of the forced phase, then, by Rule 2 of the forced phase, every remaining node in F_i is claimed during the forced phase, so nothing is removed from F_i during the approximation phase. If F_i is in two components after the forced phase, then, again by Rule 2, all branch points of F_i are claimed during the forced phase, so no branch points are removed from F_i during the approximation phase. Step 1 of the approximation phase, therefore, will remove at most one path from each component (when the root has degree 2, since degree-3 roots are claimed) and therefore breaks F_i into at most four components. In this case Steps 2–4 of the approximation phase do not apply, and at most one of Steps 2–4 can apply to each of the remaining cases.

If F_i is in one component after the forced phase, and Steps 2–4 do not apply then we have two cases. If the root is degree three, Step 1 results in at most 3 components. If the root is degree two, then F_i could release the two branch-points on either end of this i-path, resulting in at most four components by Observation 5.

Suppose F_i is in one component with exactly one unclaimed branchpoint after the forced phase. If that branchpoint is released by Step 1, then F_i is in at most 3 components after that step (only that branchpoint and its adjacent i-paths are removed from F_i), and Step 2 is redundant. Otherwise, Step 1 only has an effect if the root has degree 2. In this case, Step 1 releases only the i-path through the root, since its endpoints are claimed, resulting in two components, and the subsequent application of Step 2 adds at most two more for a total of four.

Suppose Step 3 can be applied to F_i. If w_1 is not an endpoint of the i-path through the root of F_i (or the root itself), then, as in the previous case, Step 1 results in at most two components. Subsequently removing w_2 by Step 3 results in at most two more components for a total of four. If w_1 is an endpoint of the i-path containing the root of F_i (or the root itself), then both w_1 and w_2 are released (and nothing more). Since they share an i-path, this results in at most four components by Observation 5.

Finally, suppose Step 4 can be applied to F_i. If none of w_1, w_2 or w_3 is the root or is an endpoint of the i-path adjacent to the root, then Step 1 will result in additional components only if the root has degree two. Since both of the endpoints of this i-path are claimed in this case, removing this i-path and w_2 (by Step 4) results in at most four components. Otherwise, the combined application of Steps 1 and 4 requires the release of w_2, and possibly one of w_1 and w_3 as well (but not both, since if w_2 is the root, neither of the other branchpoints will be released). By Observation 5 this will result in at most four components. □

The following lemmas use this fact:

Fact 7 *([8]) The intersection of two subtrees of a tree is connected and contains the root of at least one of the subtrees.*

Lemma 8. *If F_i and F_j are each in two components after the forced phase, then after the approximation phase, there is no node that is in F_i and in F_j.*

Proof. This proof is similar to the correctness proof of the fixed-topology 2-phylogeny algorithm in [8]. Let C_i be a component of F_i after the forced phase, and let C_j be a component of F_j after the forced phase. Since F_i and F_j are both split in two components during the forced phase, all branch points in C_i and C_j are claimed during the forced phase (by Rule 2), and their intersection is a path in the fixed topology (i.e., all nodes are degree 2). Furthermore, the root of C_i or C_j is in the intersection. Therefore, the contention is cleared in Step 1 of the approximation phase of the algorithm. □

Lemma 9. *If F_i is in one component after the forced phase, and F_j is in two components after the forced phase, then there is no node that is in F_i and in F_j after the approximation phase.*

Proof. First note that no branch point of T_j is part of T_i. (Since F_j is in only 2 pieces, it gave up only degree-2 nodes in the forced phase, and subsequently claimed all branch points. None of these are in F_i, since F_i was unbroken in the forced phase). Thus, the intersection of T_i and T_j is a path in T_j. We conclude that the intersection of F_i and F_j is a path in F_j and contains at most one branch point of F_i (otherwise, the path would be removed from F_j during the forced phase by Rule 1). If the intersection contains the root of F_j, then the contention will be removed during Step 1 of the approximation phase. Otherwise, the intersection contains the root of F_i. Thus, the single branch point of F_i that is contained in the intersection of F_i and F_j is either the root of F_i or it is an endpoint of the i-path containing the root of F_i. In either case, the contention will be removed in Step 1 of the approximation phase. □

Lemma 10. *If F_i and F_j are each in one component after the forced phase, then there is no node that is in F_i and in F_j after the approximation phase.*

Proof. We will consider various cases. Case (α, β, γ) will represent the situation in which the intersection of F_i and F_j after the forced phase contains α branch points of F_i and β branch points of F_j, γ of which are shared. Recall that when a branchpoint of F_i is released, all i-paths adjacent to it are released as well.

Case $(0,0,0)$: As in the proof of Lemma 8, the intersection of F_i and F_j is a path containing the root of one of F_i and F_j, so the contention is cleared in Step 1 of the approximation phase.

Case $(1,0,0)$: As in the proof of Lemma 9, the intersection of F_i and F_j is a path of F_j containing one branch point of F_i so the contention is cleared in Step 1 of the approximation phase.

Case $(1,1,\gamma)$: Suppose without loss of generality that the root of F_i is in the intersection of F_i and F_j after the forced phase. Then the branch point of F_i is either the root of F_i or it is the endpoint of an i-path containing the root of F_i. In either case, it is released by F_i during Step 1 of the approximation phase.

Case $(2,0,0)$: This case cannot arise after the forced phase, because it requires two branch points of F_i on a single path of F_j, which is forbidden by Rule 1 of the forced phase.

Case $(2,1,0)$: By Rule 1, after the forced phase, the intersection of F_i and F_j has the branchpoint of F_j (w_2), between the two branchpoints for F_i (w_1 and w_3) as illustrated in Figure 2(a). Let w_4 be the other endpoint of the j-path adjacent to w_2 that contains w_1, and let w_5 be the other endpoint of the j-path adjacent to w_2 that contains w_3. By Rule 3 of the forced phase, all branch points of F_j except w_2, w_4 and w_5 are claimed. If node w_2 is released by F_j during the approximation phase, then the contention is cleared. Otherwise, by Rules 2 and 4 of the approximation phase, exactly one of $\{w_4, w_5\}$ is unclaimed after the forced phase. Suppose without loss of generality this is w_4. Because w_2 is not released by F_j in the approximation phase, the root of F_j is not w_2 or on any j-path adjacent to it. Therefore the root of F_i is in the intersection. If the root of F_i is on the i-path between w_1 and w_3 then by Step 1 of the approximation phase, F_i will release both w_1 and w_3, and the contention will be cleared. If the root of F_i was on the other i-path adjacent to w_1 in the intersection, then w_4 would be the closest to the root of F_j among the three unreleased F_j branchpoints, and F_j would have released w_2 by step 3 of the approximation phase. Finally, if the root is on an i-path adjacent to w_3 (but not w_1), F_i will release w_3 by Step 1 of the approximation phase, clearing the i-path from w_3 to w_1 (not including w_1). By Step 3 of the approximation phase F_j will release w_4, clearing the j-path from w_2 to w_4 (not including w_2). Therefore the contention is removed.

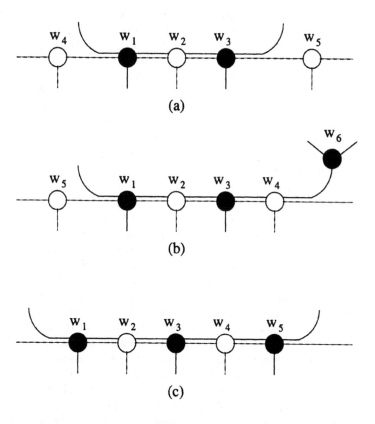

Fig. 2. Cases from Lemma 10. Branchpoints of forest F_i are represented by solid circles, and i-paths are solid lines. Branchpoints of forest F_j are represented as empty circles with j-paths as dashed lines. Dashed and solid together represent shared paths. (a) case $(2,1,0)$, (b) case $(2,2,0)$, and (c) case $(3,2,0)$.

Case $(2, 1, 1)$: This case cannot arise after the forced phase, because it requires two branch points of F_i on a single path (including endpoints) of F_j, which is forbidden by Rule 1 of the forced phase.

Case $(2, 2, 0)$: By Rule 1 of the forced phase, the branch points of F_i and F_j are interleaved and lie on a path, as illustrated in Figure 2(b). Label the four relevant branch points w_1, w_2, w_3, w_4 such that for $k \in \{1, 2, 3\}$, there is a path between w_k and w_{k+1} which does not pass through any $w_\ell \notin \{w_k, w_{k+1}\}$. Without loss of generality, assume that w_1 is a branch point of F_i. Let w_5 be the other endpoint of the j-path adjacent to w_2 that contains w_1, and let w_6 be the other end of the i-path adjacent to w_3 that contains w_4. By repeated applications of Rule 3 of the forced phase, all branch points of F_i and F_j other than w_1–w_6 are claimed.

If F_j does not claim w_5 in the forced phase and F_i does not claim w_6, then by Rule 4 of the approximation phase, w_3 is released by F_i and w_2 is released by F_j, so the contention is cleared.

So, suppose without loss of generality that w_6 is claimed by F_i. Consider the location of the root of F_i relative to w_1 and w_3. If the root of F_i is on the far side of w_6 (down one of the i-paths not adjacent to w_3), or on the (w_3, w_6) i-path between w_4 and w_6, then w_4 is on a j-path adjacent to the root of F_j. Therefore, by Step 1 of the approximation phase, F_j will release w_4, clearing contention up to, but not including w_2, and by Step 3 of the approximation phase, F_i will release w_1, clearing the remaining contention. A similar argument holds when the root is on the far side of w_1. If the root is on the (w_1, w_3) i-path, then by Step 1 of the approximation phase, F_i releases both w_1 and w_3, removing contention. Similarly, if the root is on the (w_2, w_4) j-path, (included in F_i), then F_j releases both w_2 and w_4.

Case $(3, 2, 0)$: By Rule 1 of the forced phase, the branch points of F_i and F_j are interleaved and lie on a path as illustrated in Figure 2(c). Label the five relevant branch points w_1, w_2, w_3, w_4, w_5 such that for $k \in \{1, 2, 3, 4\}$, there is a path between w_k and w_{k+1} which does not pass through any $w_\ell \notin \{w_k, w_{k+1}\}$. By repeated application of Rule 3 of the forced phase, all branch points of F_i and F_j other than w_1–w_5 are claimed. Node w_3 is released by F_i by Rule 4 of the approximation phase. Contention remains at w_1 and w_5. Consider where the global root is with respect to this intersection. If the global root is located down one of the paths adjacent to w_1 (but not adjacent to w_3), then the root of F_j is on an j-path adjacent to w_2. F_j releases w_2 by Step 1 of the approximation phase and w_4 by Step 3 of the approximation phase, removing the remaining contention. A similar argument holds if the root is down a i-path adjacent to w_5 (other than (w_3, w_5)). If the global root is down a j-path adjacent to w_2 (other than (w_2, w_4)), then the root of F_i is on an i-path adjacent to w_1 (or w_1 itself), and therefore F_i will release it by Step 1 of the approximation phase. As before, F_j will release w_4 by Step 3, removing the rest of the contention. A similar argument holds when the global root is down a j-path adjacent to w_4 (other than (w_2, w_4)). If the global root is on j-path (w_2, w_4), or down the i-path adjacent to w_3 which does not intersect this j-path, then by Step 1, F_j releases both w_2 and w_4, removing all contention.

Case $(3,3,0)$: This case cannot arise. By Rule 1 of the forced phase, if it did exist, the branch points of F_i and F_j would be interleaved and lie on a path. But then, by applying Rule 3 of the forced phase, we find that at least one of the relevant branch points of F_i and F_j would have been claimed during the forced phase.

Case $(\alpha > 1, \beta > 1, \gamma > 0)$: This case cannot arise after the forced phase, because it requires two branch points of F_i on a single path of F_j, which is forbidden by Rule 1 of the forced phase.

Case $(3, \beta \leq 1, \gamma)$: This case cannot arise after the forced phase, because it requires two branch points of F_i on a single path of F_j, which is forbidden by Rule 1 of the forced phase. □

Lemma 11. *After the approximation phase, every node is in at most one forest F_i.*

Proof. The lemma follows from Lemmas 8, 9 and 10. □

4 Approximating Polymorphism

A **polymorphic character** (see [12]) allows more than one state per character per species. This type of character has strong application in linguistics [2, 16]. If there are r states, a polymorphic character is a function $c : S \rightarrow (2^{\{1,\ \cdots,\ r\}} - \emptyset)$, where $2^{\{1,\ \cdots,\ r\}}$ denotes the power set (set of all subsets) of $\{1, ..., r\}$. For a given set of species, the *load* is the maximum number of states for any character for any species.

Often the evolution of biological polymorphic characters from parent to child is modelled by mutations, losses and duplications of states between species (see [12]). A **mutation** changes one state into another; a **loss** drops a state from a polymorphic character from parent to child; and a **duplication** replicates a state which subsequently mutates. We associate a cost with each mutation, duplication and loss between a pair of species. In the state-independent model, which we will consider, a loss costs c_l, a mutation costs c_m and a duplication costs c_d, regardless of which states are involved. Following the justification in [2], we insist $c_l \leq c_m \leq c_d$. Let $s_1, s_2 \in S$ and assume s_1 is the parent of s_2. In the state-independent model, we first look at the differences in cardinality of the parent and child sets. If the parent has fewer states, then we pay the appropriate duplication costs to account for the increased size of the child. If the parent is bigger, then we pay the loss cost. Then we match up as many elements as possible, and pay for the remaining changes as mutations. More specifically, as given in [2], we define $X = c(s_1) - c(s_2)$, and $Y = c(s_2) - c(s_1)$. Then the cost for the character c from s_1 to s_2 is $c_m * |X|$ if $|X| = |Y|$, and is $c_l * [|X| - |Y|] + c_m * |Y|$ if $|X| > |Y|$ and is $c_d * [|Y| - |X|] + c_m * |X|$ if $|Y| > |X|$.

As input we are given a fixed-topology T which has a unique species from S associated with each of its leaves, and label the leaf associated with $s \in S$ with the set of states $c(s)$. The **parsimony problem** is the problem of extending the function c to the internal nodes of T so that the sum of the costs over all edges of

T is minimised. In the monomorphic case (one state per character per species), as discussed earlier, this problem can be solved in polynomial time [7], though the problem of finding a mimimum cost labelling is NP-hard if the input does not include a topology [4, 5]. We will consider the **load problem**, introduced in [2]; calculate a labelling of the internal nodes of a fixed topology T with load at most ℓ and cost at most p. This problem was shown to be NP-hard in [2], even when $c_l = 0$ and the topology T is a binary tree. An (α, β)-approximation algorithm for the load problem computes a phylogeny with load at most $\alpha\ell$ and cost at most βc provided there is a load-ℓ cost-c phylogeny. Note that this is a pseudoapproximation algorithm, since the cost of the best $\alpha\ell$-load phylogeny may be significantly lower than the cost of the best ℓ-load phylogeny. In this section of the paper, we consider the load problem when $c_l = 0$ and the topology is arbitrary. We extend the results of section 2 to obtain, for any $\alpha > 1$, an $(\alpha, \frac{\alpha}{\alpha-1})$-approximation algorithm for the problem. (Note that taking $\alpha = 2$ gives a $(2, 2)$-approximation algorithm.)

We first quote the following observation, which was first noted in [2]:

Observation 12 *If $c_l = 0$, then if there is a labelling for the topology T which has load ℓ and cost p, then there is also a labelling for T with load ℓ and cost p such that each internal node contains all the states in the subtree rooted at it or else has load ℓ.*

Therefore to approximate the load we only need to consider the labellings where each internal node contains all the states in the subtree rooted at it or else has load ℓ. We begin by presenting an ILP which provides an exact solution for the load problem. We then use the solution to the linear-programming relaxation of this ILP to compute an $(\alpha, \frac{\alpha}{\alpha-1})$-approximation for the problem. The integer program $\mathcal{P}$ uses the variables $X_{v,i}$, for each node v of the fixed-topology T and each state $i \in \{1, \ldots, r\}$, cost variables $cost_{e,i}$ for each edge $e \in E(T)$ and each state $i \in \{1, \ldots, r\}$ and the total cost variable $cost_e$ for each edge e. These variables have the following interpretation:

$$X_{v,i} = \begin{cases} 1 \text{ if state } i \text{ is in } c(v) \\ 0 \text{ otherwise} \end{cases}$$
$$cost_{e,i} = \begin{cases} 1 \text{ if } i \in c(v) \text{ and } i \notin c(u), \text{ for } e = (u \to v) \\ 0 \text{ otherwise} \end{cases}$$
$$cost_e = \sum_{i=1}^{r} cost_{e,i}$$

The ILP $\mathcal{P}$ is then defined as:

$$\text{minimize } p$$

subject to

$$X_{v,i} = 1 \qquad \text{for each leaf } v \in V(T), \forall i \in c(v) \qquad (9)$$

$$\sum_{i=1}^{r} X_{v,i} \leq \quad \ell \qquad \forall v \in V(T) \tag{10}$$

$$cost_{e,i} \geq \quad 0 \qquad \forall e \in E(T),\ i = 1,\ldots,r \tag{11}$$

$$cost_{e,i} \geq X_{v,i} - X_{u,i} \qquad \forall e = (u \to v) \in E(T),\ i = 1,\ldots,r \tag{12}$$

$$cost_e = \sum_{i=1}^{r} cost_{e,i} \qquad \forall e = (u \to v) \in E(T) \tag{13}$$

$$\sum_{e \in E(T)} cost_e \leq \quad p/c_m \tag{14}$$

$$X_{v,i}, cost_{e,i} \in \quad \{0,1\} \tag{15}$$

The integer program $\mathcal{P}$ solves the load problem. However, we require only that it provides a lower bound on the best cost. We now show that when we solve $\mathcal{P}$ with parameter ℓ, the optimal value of p is a lower bound on the cost of the best load-ℓ solution to the fixed topology problem.

Lemma 13. *Let S be a species set, T be a fixed topology and $c : S \to (2^{\{1,\ldots,r\}} - \emptyset)$ be a polymorphic character on S. If the internal nodes v of T can be labelled with subsets of $\{1,\ldots,r\}$ to create a phylogeny for c with load ℓ and cost p, then there is a feasible solution for the linear program for this value of ℓ and p.*

Proof. Because of Observation 12, we can assume that in the load-ℓ, cost-c phylogeny, for each internal node v in $V(T)$, either $c(v') \subset c(v)$ for every child v' of v, or else $|c(v)| = \ell$. Therefore the cost of this phylogeny is $\sum_{(u \to v) \in E(T)} (c_m * |c(v) - c(u)|) = p$. Assign values to the $X_{v,i}$ variable for each internal node v and to the $cost_{e,i}$ variable for each edge $e = (u \to v)$ as follows:

$$X_{v,i} = \begin{cases} 1 \text{ if } i \in c(v) \\ 0 \text{ otherwise} \end{cases}$$

$$cost_{e,i} = \begin{cases} 1 \text{ if } i \in c(v) - c(u) \\ 0 \text{ otherwise} \end{cases}$$

This assignment satisfies constraints (15), (10), (11) and (12) of $\mathcal{P}$. Constraint (9) is automatically satisfied, and constraint (13) is definitional. Also, $c_m * cost_e = c_m * |c(v) - c(u)|$ for every $e = (u \to v)$ by definition of the $cost_{e,i}$, and therefore $\sum_{e \in E(T)} c_m * cost_e = p$, and constraint (14) is satisfied. □

Once again, since integer linear programming is $\mathcal{NP}$-hard, we solve the linear-programming relaxation $\mathcal{LP}$ of $\mathcal{P}$, which consists of all the constraints of $\mathcal{P}$ except that Constraint 15 is replaced with the constraint $0 \leq X_{v,i}, cost_{e,i} \leq 1$ (15').

Theorem 14. *Suppose there is a solution for the linear program $\mathcal{LP}$. Then we can assign states to the internal nodes of input tree T such that the resulting phylogeny for c has load $\alpha\ell$ and cost no more than $\left(\frac{\alpha}{\alpha-1}\right) p$.*

Proof. We assign states to the internal nodes of the fixed topology from the leaves upwards. For each internal node $v \in V(T) - L(T)$, consider the set $R(v) = \cup_{(v \to v') \in E(T)} c(v')$. If $|R(v)| \leq \alpha\ell$, then define $c(v) = R(v)$. If $|R(v)| > \alpha\ell$ then choose the $\alpha\ell$ states i of $R(v)$ which have the greatest $X_{v,i}$ values. By definition, this assignment of states to the internal nodes of T has load at most $\alpha\ell$.

To show that the cost of this assignment is no more than $\left(\frac{\alpha}{\alpha-1}\right) p$, note that the cost on an edge $e = (u \to v) \in E(T)$ is $c_m * |c(v) - c(u)|$, as $|c(v) - c(u)|$ is the number of mutations on e. Our assignment guarantees that if $|c(u)| < \alpha\ell$ then $c(u) \supset c(v)$, which implies $c_m * |c(v) - c(u)| = 0$, so we need only consider edges whose upper endpoint has full load. Suppose $|c(u)| = \alpha\ell$ and $i \in c(v) - c(u)$. Then, by construction of the phylogeny, there is a downwards path from v to some leaf w which has $i \in c(v')$ at every node along the path, including w. Suppose this path is $e_1 = (v \to v_1)$, $e_2 = (v_1 \to v_2)$, ..., $e_j = (v_{j-1} \to w)$. By the constraints of the linear program, $cost_{e,i} + cost_{e_1,i} + \ldots + cost_{e_j,i} \geq (X_{v,i} - X_{u,i}) + (X_{v_1,i} - X_{v,i}) + \ldots + (X_{w,i} - X_{v_{j-1},i}) = X_{w,i} - X_{u,i}$, and as w is a leaf and $i \in c(w)$, this is $1 - X_{u,i}$. Then, since $i \notin c(u)$, and the $\alpha\ell$ states in $c(u)$ were chosen to have the greatest $X_{u,j}$ values, we know $X_{u,i} < \ell/(\alpha\ell + 1)$. The worst case is achieved when there are $\alpha\ell + 1$ positive $X_{u,j}$ values all equal. They sum to at most ℓ from constraint 10, and therefore the smallest one, which cannot be included in the set, has value at most $\ell/(\alpha\ell+1)$. Therefore $cost_{e,i} + cost_{e_1,i} + \ldots + cost_{e_j,i} > ((\alpha - 1)\ell + 1)/(\alpha\ell + 1)$. Furthermore, the costs $cost_{e,i}$, $cost_{e_1,i}$, $\ldots cost_{e_j,i}$ will not be allocated to any other mutation to i, because any mutation occurring above u will not have an unbroken path in i intersecting with any of the edges $e, e_1, \ldots, e_j$. So every mutation along an edge $e = (u \to v) \in T$ with $|c(u)| = \alpha\ell$ contributes at least $((\alpha - 1)\ell + 1)/(\alpha\ell + 1)$ to the sum $\sum_{e \in E(T)} cost_e$ in our linear program. Hence $p/c_m \geq \sum_{e \in E(T)} cost_e \geq \sum_{e=(u \to v) \in E(T)} |c(v) - c(u)| \left(\frac{(\alpha-1)\ell+1}{\alpha\ell+1}\right)$, so the cost $\sum_{e=(u \to v) \in E(T)} c_m * |c(v) - c(u)| \leq (\alpha/(\alpha - 1)) * p$.

$\square$

References

1. H. Bodlaender, M. Fellows, T. Warnow, "Two Strikes Against Perfect Phylogeny", *procs. of the 19th International Congress on Automata, Languages and Programming (ICALP)*, pp. 273-287, Springer-Verlag Lecture Notes in Computer Science, 1992.
2. M. Bonet, C. Phillips, T.J. Warnow and S. Yooseph, Constructing Evolutionary Trees in the Presence of Polymorphic Characters, *Proceedings of the 28th Annual ACM Symposium on the Theory of Computing* (1996).
3. I. Borosh and L.B. Treybig, Bounds on positive integral solutions of linear Diophantine equations, *Proceedings of the American Mathematical Society*, Vol 55 (1976).
4. W.H.E. Day, Computationally difficult parsimony problems in phylogenetic systematics, *Journal of Theoretical Biology*, Vol 103 (1983).
5. W.H.E. Day, D.S. Johnson and D. Sankoff, The computational complexity of inferring phylogenies by parsimony, *Mathematical biosciences*, Vol 81 (1986).
6. W.H.E. Day and D. Sankoff, "Computational complexity of inferring phylogenies by compatibility", *Systematic Zoology*, 35(2): 224-229, 1986.
7. W. Fitch, Towards defining the course of evolution: minimum change for a specified tree topology, *Systematic Zoology*, Vol 20 (1971).
8. L.A. Goldberg, P.W. Goldberg, C.A. Phillips, E. Sweedyk and T. Warnow, Minimizing phylogenetic number to find good evolutionary trees, *Discrete Applied Mathematics*, to appear.
9. M.R. Garey and D.S. Johnson, *Computers and Intractability: A Guide to the Theory of NP-completeness*, W.H. Freeman and Company (1979).
10. T. Jiang, E.L. Lawler and L. Wang, Aligning Sequences via an Evolutionary Tree: Complexity and Approximation, *Proceedings of the 26th Annual ACM Symposium on the Theory of Computing* (1994).
11. R.M. Karp, Reducibility among combinatorial problems, *Complexity of Computer Computations*, eds. R.E. Miller and J.W. Thatcher, Plenum Press (1972).
12. M. Nei, *Molecular Evolutionary genetics*, Columbia University Press, New York (1987).
13. M.A. Steel, "The complexity of reconstructing trees from qualitative characters and subtrees", *Journal of Classification*, 9 91-116, 1992.
14. L. Wang and D. Gusfield, Improved Approximation Algorithms for Tree Alignment, Proceedings of CPM 1996, 220-233.
15. L. Wang, T. Jiang, and D. Gusfield, "A more efficient approximation scheme for tree alignment", To appear in *Proceedings of the of First Annual International Conference on Computational Molecular Biology*, Jan. 1997.
16. T. Warnow, D. Ringe and A. Taylor, *A character based method for reconstructing evolutionary history for natural languages*, Tech Report, Institute for Research in Cognitive Science, 1995, and *Proceedings 1996 ACM/SIAM Symposium on Discrete Algorithms*.

A New Algorithm for the
Ordered Tree Inclusion Problem*

Thorsten Richter

Department of Computer Science IV, University of Bonn
Roemerstr. 164, 53117 Bonn, Germany
e-mail: richter@cs.uni-bonn.de

Abstract. In the problem of *ordered tree inclusion* two ordered labeled trees P and T are given, and the *pattern tree* P matches the *target tree* T at a node x, if there exists a one-to-one map f from the nodes of P to the nodes of T which preserves the labels, the ancestor relation and the left-to-right ordering of the nodes. In [7] Kilpeläinen and Mannila give an algorithm that solves the problem of ordered tree inclusion in time and space $\Theta(|P| \cdot |T|)$. In this paper we present a new algorithm for the ordered tree inclusion problem with time complexity $O(|\Sigma_P| \cdot |T| + \#matches \cdot \text{DEPTH}(T))$, where Σ_P is the alphabet of the labels of the pattern tree and $\#matches$ is the number of pairs $(v, w) \in P \times T$ with $\text{LABEL}(v) = \text{LABEL}(w)$. The space complexity of our algorithm is $O(|\Sigma_P| \cdot |T| + \#matches)$.

1 Introduction and Motivation

The problem of *ordered tree inclusion* [6] can be seen as an extension of the classic problem of *tree pattern matching* [3], [9], [1]. In the latter problem two ordered labeled trees P and T are given, and the *pattern tree* P matches the *target tree* T at a node x, if there exists a one-to-one map f from the nodes of P to the nodes of T , such that

(1) the root of P maps to x,
(2) $\forall v \in P$: v and $f(v)$ have the same labels,
(3) $\forall v \in P$: if v is not a leaf, then the i-th child of v maps to the i-th child of $f(v)$.

In the problem of ordered tree inclusion the third condition is replaced by

(3') $\forall v_1, v_2 \in P$:
    (a) v_1 is an ancestor of v_2 $\iff$ $f(v_1)$ is an ancestor of $f(v_2)$,
    (b) v_1 is to the left of v_2 $\iff$ $f(v_1)$ is to the left $f(v_2)$.

This is obviously a relaxation. Another formulation of this problem is given by Knuth in [8], Exercise 2.3.2-22.

    One motivation for considering the ordered tree inclusion problem comes

---

* This work was supported by the DFG under grant Bl 320/2-1

from the concept of structured text databases. One can use context-free grammars to describe the structure of natural language sentences in terms of their parse trees. Hence a structured text database can be realized as a collection of parse trees (see [2], [4]). Then one can use tree inclusion as a means of retrieving information from documents stored in such a database [5].

In [7] Kilpeläinen and Mannila give an algorithm that solves the problem of ordered tree inclusion in time and space $\Theta(|P| \cdot |T|)$. In that paper it is also shown that the tree inclusion problem becomes $\mathcal{NP}$-complete when considering unordered trees.

In this paper we present a new algorithm for the ordered tree inclusion problem with time complexity $O(|\Sigma_P| \cdot |T| + \#matches \cdot \text{DEPTH}(T))$, where Σ_P is the alphabet of the labels of the pattern tree and $\#matches$ is the number of pairs $(v, w) \in P \times T$ with $\text{LABEL}(v) = \text{LABEL}(w)$. This complexity beats the complexity of the algorithm of [7] if the number of matches is relatively small, i. e. $\#matches = o(|P| \cdot |T| / \text{DEPTH}(T))$. Furthermore, the time bound of our algorithm is not a tight one, but an upper bound. The space complexity of our algorithm is $O(|\Sigma_P| \cdot |T| + \#matches)$.

The main idea of our algorithm is to construct an inclusion map f by considering and mapping the nodes of P in ascending preorder, either until the pattern tree is completely mapped, or until it arrives at a point where it is impossible to continue with the construction of the inclusion map. In the latter case the algorithm returns to a node of P that has already been mapped, and maps it to another candidate. To avoid duplicate work, it derives as much information as possible from such a "dead end". Then it considers and maps the remaining nodes of P again in ascending preorder. Hence our algorithm uses some kind of *backtracking*.

2 Definitions and Basic Facts

Let $T = (V, E)$ be an ordered labeled tree. Then we use for a node $u \in V$ the following notations. $\text{LABEL}(u)$ is the label of u; we assume that the labels of the pattern and target tree are chosen from a finite alphabet Σ. $T[u]$ denotes the subtree of T with root u and $\text{RIGHTMOST\_LEAF}(u)$ is the rightmost leaf of the subtree $T[u]$. When we use in the following the terms *ancestor* and *successor*, we mean *proper* ancestors and proper successors. Analogously, we mean by *to the left of* and *to the right of* always *properly* left and *properly* right. Now the problem of ordered tree inclusion can be formally defined as follows.

Definition 1. In the *problem of ordered tree inclusion* there are given two ordered labeled trees P and T with $|P| \leq |T|$. P is the *pattern tree* and T is the *target tree*. Sought is a one-to-one map f from the nodes of P to the nodes of T, such that $\forall v, v_1, v_2 \in P$ the following conditions hold

- *label condition:* $\text{LABEL}(v) = \text{LABEL}(f(v))$;
- *ancestor condition:* v_1 is ancestor of $v_2 \iff f(v_1)$ is ancestor of $f(v_2)$;
- *order condition:* v_1 is to the left of $v_2 \iff f(v_1)$ is to the left of $f(v_2)$.

We call such a map f an *inclusion map* from P to T. □

Note that there may be *exponentially* many inclusion maps from a pattern tree to a target tree in the general case. Thus it is not feasible to look for *all* inclusion maps. Hence we look for only *one* inclusion map in the following.

In the remainder of this paper $P = (V_P, E_P)$, $V_P = \{v_1, \ldots, v_n\}$, is the given pattern tree and $T = (V_T, E_T)$, $V_T = \{w_1, \ldots w_m\}$, is the given target tree. $V_P[i]$ denotes the set of the first i nodes of the pattern tree in preorder. A pair $(v, w) \in V_P \times V_T$ with $\text{LABEL}(v) = \text{LABEL}(w)$ is called a *match*; then w is called a *candidate* for v.

To access the candidates efficiently in the algorithm, we use the NEXT **array**: if $\Sigma_P = \{s_1, \ldots, s_t\}$ is the alphabet of the labels of the nodes of the pattern tree, then we denote, for all $1 \le i \le t$ and $1 \le j \le m$, by $\text{NEXT}(s_i, w_j)$ the first node of T with label s_i and a preorder number greater than j, if there is one. If there is none, the value of $\text{NEXT}(s_i, w_j)$ is NIL, i. e. it is undefined.

Now suppose that we have already mapped the first i, $1 \le i < |V_P|$, nodes of the pattern tree to nodes of the target tree. Then we call a map f from $V_P[i]$ to the nodes of T a *partial inclusion map for* $V_P[i]$, if f satisfies for all nodes of $V_P[i]$ the conditions of ordered tree inclusion. If we want to extend this map f to $V_P[i + 1]$, we have to consider only candidates for v_{i+1} which are compatible with this partial inclusion map. We call a candidate w for v_{i+1} *feasible with respect to f*, if the map f' from $V_P[i+1]$ to the nodes of T, defined by $f'(u) = f(u)$, $u \in V_P[i]$, and $f'(v_{i+1}) = w$, is a partial inclusion map for $V_P[i + 1]$. Hence to extend a partial inclusion map f for $V_P[i]$ to $V_P[i + 1]$ it suffices to consider only feasible candidates for v_{i+1}. The next lemma specifies their position in the target tree.

Lemma 2. *Let f be a partial inclusion map for $V_P[i]$. A candidate w for v_{i+1} is feasible with respect to f, if and only if*

(a) w is successor of $f(\text{PARENT}(v_{i+1}))$;
(b) w is to the right of $f(\text{LEFT\_SIBLING}(v_{i+1}))$, if v_{i+1} has a left sibling. □

This lemma implies that a feasible candidate for v_{i+1} must have a greater preorder number than $f(\text{PARENT}(v_{i+1}))$ or than $\text{RIGHTMOST\_LEAF}(f(\text{LEFT\_SIBLING}(v_{i+1})))$, if v_{i+1} has a left sibling. Together with the definition of the NEXT array, we have the following corollary.

Corollary 3. *Let f be a partial inclusion map for $V_P[i]$, and let there be a feasible candidate for v_{i+1}. Then we have that the feasible candidate for v_{i+1} with the smallest preorder number is*

(a) $\text{NEXT}(\text{LABEL}(v_{i+1}), f(\text{PARENT}(v_{i+1})))$, if v_{i+1} has no left sibling;
(b) $\text{NEXT}(\text{LABEL}(v_{i+1}), \text{RIGHTMOST\_LEAF}(f(\text{LEFT\_SIBLING}(v_{i+1}))))$,
*   otherwise.* □

From this result we can derive a criterion for the existence of a feasible candidate.

Corollary 4. *Let f be a partial inclusion map for $V_P[i]$. Then there is no feasible candidate for v_{i+1}, if and only if*

(a) v_{i+1} *has no left sibling and*
$$\text{NEXT}(\text{LABEL}(v_{i+1}), f(\text{PARENT}(v_{i+1})))$$
is not successor of $f(\text{PARENT}(v_{i+1}))$;

(b) v_{i+1} *has a left sibling and*
$$\text{NEXT}(\text{LABEL}(v_{i+1}), \text{RIGHTMOST\_LEAF}(f(\text{LEFT\_SIBLING}(v_{i+1}))))$$
is not successor of $f(\text{PARENT}(v_{i+1}))$. □

Hence to extend a partial inclusion map f for $V_P[i]$ to $V_P[i+1]$ we consider the node
$$w_k =$$
$$\left\{ \begin{array}{l} \text{NEXT}(\text{LABEL}(v_{i+1}), f(\text{PARENT}(v_{i+1}))), \text{ if } v_{i+1} \text{ has no left sibling;} \\ \text{NEXT}(\text{LABEL}(v_{i+1}), \text{RIGHTMOST\_LEAF}(f(\text{LEFT\_SIBLING}(v_{i+1})))), \text{ otherwise.} \end{array} \right\}$$

If w_k is not successor of $f(\text{PARENT}(v_{i+1}))$, we know, by Corollary 4, that there is no feasible candidate for v_{i+1} at all. If w_k is successor of $f(\text{PARENT}(v_{i+1}))$, we can map v_{i+1} to it to get a partial inclusion map for $V_P[i+1]$. By Corollary 3 we know that we have not skipped any eligible candidate for v_{i+1}.

During the following construction of the inclusion map it may happen that there is no feasible candidate for a node v_j, $j > i+1$, if v_{i+1} is mapped to w_k. That is, the partial inclusion map constructed for $V_P[i+1]$ cannot be extended to an inclusion map for the whole pattern tree P: although w_k is a feasible candidate for v_{i+1}, it is not suited for the construction of an inclusion map for P. We call a node w_k *not suitable with respect to* f, if there is a $j > i+1$ such that there is no partial inclusion map f' for $V_P[j]$ with $f'(u) = f(u)$, $u \in V_P[i]$, and $f'(v_{i+1}) = w_k$. If there is no such j, then w_k is called *suitable with respect to* f. If it turns out that the candidate w_k to which v_{i+1} has been mapped is not suitable, this candidate is dismissed and another candidate for v_{i+1} is chosen. The following corollary generalizes the Corollaries 3 and 4 to this situation.

Corollary 5. *Let f be a partial inclusion map for $V_P[i]$, and let u be a node of T which is successor of $f(\text{PARENT}(v_{i+1}))$ and to the right of $f(\text{LEFT\_SIBLING}(v_{i+1}))$, if v_{i+1} has a left sibling. That is, if u has the label $\text{LABEL}(v_{i+1})$, then it is a feasible candidate for v_{i+1}. Let all feasible candidates for v_{i+1} whose preorder numbers are less than or equal to that of u be not suitable, as shown in Figure 1. Then we have*

(a) *if there are suitable candidates for v_{i+1} whose preorder numbers are greater than that of u, then $\text{NEXT}(\text{LABEL}(v_{i+1}), u)$ is the one with the smallest preorder number;*

(b) *if $\text{NEXT}(\text{LABEL}(v_{i+1}), u)$ is not successor of $f(\text{PARENT}(v_{i+1}))$, then there are no suitable candidates for v_{i+1}.* □

Hence, after dismissing the candidate w_k chosen first, we consider the candidate $\widetilde{w_k} = \text{NEXT}(\text{LABEL}(v_{i+1}), u)$ for v_{i+1}. If it is feasible, we map v_{i+1} to it. Otherwise, we know that there is no suitable candidate for v_{i+1}. From the latter we

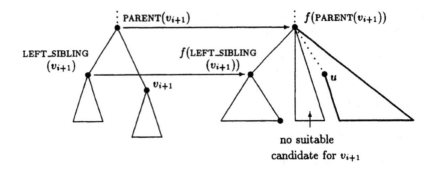

Fig. 1. *The situation of Corollary 5.*

can conclude that the candidate chosen for v_i is not suitable either.

It later turns out that we can conclude from a feasible candidate being not suitable that some following feasible candidates are also not suitable. Consequently, we can skip them. However, for the correctness of our algorithm it is necessary that we skip only feasible candidates which we know for certain are not suitable. If we have no such information, we have to go from one feasible candidate to the next.

Definition 6. Let f a partial inclusion map for $V_P[i]$. We call a candidate w for v_{i+1} the next eligible candidate for v_{i+1}, if

- the preorder number of w is greater than that of $f(\text{PARENT}(v_{i+1}))$ or than that of $\text{RIGHTMOST\_LEAF}(f(\text{LEFT\_SIBLING}(v_{i+1})))$, if v_{i+1} has a left sibling;
- all feasible candidates for v_{i+1} with a smaller preorder number than w are not suitable;
- whether w is suitable or not is still open. □

Note that the next eligible candidate for v_{i+1} is feasible if and only if it is successor of $f(\text{PARENT}(v_{i+1}))$. But if it is not feasible, we know by Corollary 5 that there is no feasible candidate for v_{i+1} at all. Which candidate the next eligible candidate is depends on the "knowledge" of the algorithm, but is unequivocal at any time. Hence we can always choose the next eligible candidate for the node of the pattern tree under consideration.

3 The New Algorithm

The **input** to our new algorithm *OrderedTreeInclusion* for the ordered tree inclusion problem consists of two ordered labeled trees, the pattern tree $P = (V_P, E_P)$ and the target tree $T = (V_T, E_T)$, where $V_P = \{v_1, \ldots, v_n\}$, $V_T = \{w_1, \ldots, w_m\}$, and $n \leq m$. In our description we assume that the subscripts of the nodes correspond to their preorder number. We further assume without loss of generality that the roots of the pattern and target tree have the same label, this is the root of the target tree is a candidate for the root of the pattern tree.

After some preprocessing our algorithm begins to construct an inclusion map f from the pattern tree P to the target tree T iteratively. We start the construction of the inclusion by mapping v_1 to v_2 and choosing the next eligible candidate for v_2 Then we are in the start state of the first iteration of the algorithm.

In the **start state** of an iteration of the algorithm we have already constructed a partial inclusion map for $V_P[next - 1]$. Now we are considering the node v_{next}, for which we have chosen the next eligible candidate w_{next}. We then check whether w_{next} is a feasible candidate for v_{next}. If this is the case, we can map v_{next} to w_{next}: we carry out a **forward step**. If not, then there is no suitable candidate for v_{next}, and hence we have come to a dead end: we carry out a **backward step**.

In a **forward step** we map v_{next} to w_{next}, i. e. we extend the partial inclusion map for $V_P[next - 1]$ to $V_P[next]$. Then we choose the next eligible candidate for v_{next+1}.

In a **backward step** the current candidate w_{next} is not feasible. Hence we know that there is no suitable candidate for v_{next} at all. So we have come to a dead end in trying to map $P[\text{PARENT}(v_{next})]$ to $T[f(\text{PARENT}(v_{next}))]$. Hence we choose another candidate for either the left sibling of v_{next} or its parent.

In either case we proceed with a new iteration of the algorithm.

By mapping v_{next} to w_{next} in a forward step, we begin to construct an inclusion map from $P[v_{next}]$ to $T[w_{next}]$. This construction is called a **phase of the algorithm** for the match (v_{next}, w_{next}). At the end of this phase we have either constructed an inclusion map from $P[v_{next}]$ or $T[w_{next}]$ or ascertained, that there is none. To store the results of the phases, we use an **array STATE**, which is defined for the matches. Its fields can have three values:

$$\text{STATE}(v, w) = \begin{cases} \text{NIL}, & \text{if the match } (v, w) \text{ has not been considered yet;} \\ \text{TRUE}, & \text{if } P[v] \text{ can be completely mapped to } T[w]; \\ \text{FALSE}, & \text{if } P[v] \text{ cannot be completely mapped to } T[w] \end{cases}.$$

Within a phase for the match (v_{next}, w_{next}) we may discover that there is no candidate for a child v_k of v_{next} to the right of a node w_l of T whose choice would let us map $P[v_{next}]$ completely to $T[w_{next}]$. Hence there is no suitable candidate for v_k. This information restricts the mapping range of v_k within this phase. To use this information, we store it in an **array RANGE** that is defined for the nodes of P. First, the entries of this array all have the value NIL, meaning that there is no restriction on the mapping range of the corresponding nodes. If we get within a phase of the algorithm the information that there is no suitable candidate for the node v_k to the right of the node w_l, we set the value of $\text{RANGE}(v_k)$ to w_l. At the end of this phase, we reset all entries of the RANGE array for the children of v_{next} that have been set in this phase.

We take the RANGE array into account by checking in the start state of an iteration described above not ony if the current candidate w_{next} is feasible, but also if it is within the range specified by RANGE(v_{next}).

If the algorithm has been successful in constructing an inclusion map from P to T, i. e. if it has set STATE(v_1, w_1) to TRUE, its **output** is the constructed inclusion map f. Otherwise, i. e. if it has set STATE(v_1, w_1) to FALSE, it returns a message that it is not possible to map the pattern tree completely to the target tree.

3.1 Preprocessing

In the preprocessing part of the algorithm we compute some information on the pattern and target tree that is used in the main part of the algorithm. In our description we assume that for every node u of P and T, respectively, the values of PARENT(u), LEFT_SIBLING(u), RIGHT_SIBLING(u) and RIGHTMOST_CHILD(u) are already known. If the trees are given by adjacency lists, for example, these values can obviously be computed in linear time.

We start preprocessing by traversing both trees in preorder, storing the preorder numbers of the nodes of the pattern and target tree in the arrays PRE$_P$ and PRE$_T$, respectively. The target tree is also traversed in postorder in order to store in the array POST$_T$ the postorder numbers of its nodes. The preorder and postorder numbers can be used to determine the relative position of two nodes in a tree.

For each node w of the target tree T we compute the value of RIGHTMOST_LEAF(w) and the values of NEXT(s, w) for all $s \in \Sigma_P$. This can be done in time $O(|\Sigma_P| \cdot |T|)$ by traversing T in descending preorder. We omit the details here.

To implement the STATE array efficiently, we proceed as follows. First we traverse the pattern tree P in preorder, where we store for every node v with label s its preoder number in a list OCCURENCES$_P(s)$. Then we traverse the target tree T in preorder to build analogous lists OCCURENCES$_T$. Here we maintain for every label $s \in \Sigma_P$ a counter OCCURENCE_NUMBER(s), which is initially zero for every $s \in \Sigma_P$. If we visit a node w with label s during the traversal of T, we increase OCCURENCE_NUMBER(s) by one, store its new value in LABEL_NUMBER(w), and add it to the list OCCURENCES$_T(s)$.

After the traversal of T we can define for every node $v \in P$ an array STATE(v)() with exactly OCCURENCE_NUMBER(LABEL(v)) entries, which are initialized with the value NIL. During the main part of the algorithm we can access the STATE value of a match (v, w) via STATE(v, LABEL_NUMBER(w)).

Since a traversal of a tree can be done in linear time, the initialization of the STATE array takes the time $O(\#matches)$.

Altogether, the preprocessing part of the algorithm has a time complexity of $O(|\Sigma_P| \cdot |T| + \#matches)$.

3.2 Main Part of the Algorithm

Since we have assumed that the root of the target tree is a candidate for the root of the pattern tree, i. e. $\text{LABEL}(v_1) = \text{LABEL}(w_1)$, we can start the construction of the inclusion map f by mapping v_1 to w_1. Next we consider the node v_2 of P. The next eligible candidate for v_2 is the node with the smallest preorder number that has the same label as v_2, but is not the root of T. This node is obviously successor of $f(\text{PARENT}(v_2)) = f(v_1)$. Hence we can set

$$v_{next} := v_2$$
$$w_{next} := \text{NEXT}(\text{LABEL}(v_2), w_1)$$

and proceed with the first iteration of the algorithm.

In the **start state** of an iteration of the algorithm we have already constructed a partial inclusion map for $V_P[next - 1]$, and are now considering the node v_{next}, for which we have chosen the next eligible candidate w_{next}. Then we check whether w_{next} is successor of $f(\text{PARENT}(v_{next}))$, and whether it is within the range specified by $\text{RANGE}(v_{next})$. If it is not successor of $f(\text{PARENT}(v_{next}))$, it is not feasible. Hence by Corollaries 4 and 5, respectively we have that there is no feasible candidate at all. If w_{next} is out of the range specified by $\text{RANGE}(v_{next})$, it is not suitable. In both cases we proceed with a backward step; otherwise we proceed with a forward step.

The **forward step.** When we carry out a forward step, w_{next} is successor of $f(\text{PARENT}(v_{next}))$, hence a feasible candidate for v_{next}, and it is within the range specified by $\text{RANGE}(v_{next})$. Now we have to distinguish the following cases depending on the value of $\text{STATE}(v_{next}, w_{next})$.

(1) If $\text{STATE}(v_{next}, w_{next}) = \text{NIL}$, we have not tried to map $P[v_{next}]$ to $T[w_{next}]$ yet. Consequently we map v_{next} to w_{next} now:

$$f(v_{next}) := w_{next}.$$

In choosing the next eligible candidate for v_{next+1}, we have to distinguish the following cases.

(1.a) If v_{next} *has a child*, the next node v_{next+1} is the leftmost child of v_{next}. Then the next eligible candidate for v_{next+1} must only have a greater preorder number than $f(v_{next})$ (cf. Corollary 3.(a)). Hence, we set

$$w_{next} := \text{NEXT}(\text{LABEL}(v_{next+1}), f(v_{next})),$$
$$v_{next} := v_{next+1}$$

in this case.

(1.b) If v_{next} *has no child, but a right sibling*, then v_{next} is a leaf. Hence we can set

$$\text{STATE}(v_{next}, w_{next}) := \text{TRUE}.$$

The next node v_{next+1} is the right sibling of v_{next}, and the next eligible candidate for v_{next+1} has to be to the right of $f(v_{next})$ (cf. Corollary 3.(b)). Consequently, we set

$$w_{next} := \text{NEXT}\big(\text{LABEL}(v_{next+1}), \text{RIGHTMOST\_LEAF}(f(v_{next}))\big),$$

$$v_{next} := v_{next+1}$$

in this case.

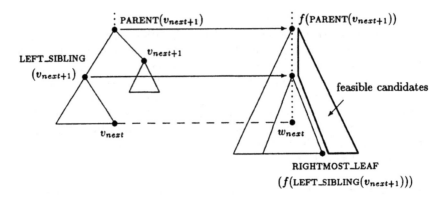

Fig. 2. *Forward step, Case (1.c), v_{next} is a rightmost leaf.*

(1.c) If v_{next} *is a rightmost leaf*, we have mapped a subtree of P completely to a subtree of T, as shown in Figure 2. To store this, we go up the path from v_{next} to its first ancestor with a right sibling. Thereby we set for every node v on this path

$$\text{STATE}(v, f(v)) := \text{TRUE},$$

and, for each of its children u whose the RANGE-value has been set, we reset the value of $\text{RANGE}(u)$ to NIL.

If v_{next} has no ancestor with a right sibling, v_{next} is the rightmost leaf of the whole pattern tree. This means that we have mapped the pattern tree completely to the target tree. Hence we can finish the algorithm.

If v_{next} is not the rightmost leaf of the pattern tree, the next node v_{next+1} is the right sibling of the first ancestor of v_{next} with a right sibling, and the next eligible candidate for v_{next} has to be to the right of $\text{RIGHTMOST\_LEAF}(f(\text{LEFT\_SIBLING}(v_{next+1})))$ (cf. Corollary 3.(b)). Consequently, we set

$$w_{next} := \text{NEXT}\big(\text{LABEL}(v_{next+1}), \text{RIGHTMOST\_LEAF}(f(\text{LEFT\_SIBLING}(v_{next+1})))\big),$$

$$v_{next} := v_{next+1}$$

in this case.

(2) If $\text{STATE}(v_{next}, w_{next}) = \text{FALSE}$, we already know that $P[v_{next}]$ cannot be completely mapped to $T[w_{next}]$. Hence w_{next} is not a suitable candidate for v_{next}, and we do not map v_{next} to w_{next}, but immediately choose a new candidate for v_{next}. Since successors of w_{next} with label $\text{LABEL}(v_{next})$ are also not suitable candidates for v_{next}, the next eligible candidate for v_{next} has to be to the right of w_{next}. Hence we set

$$w_{next} := \text{NEXT}\big(\text{LABEL}(v_{next}), \text{RIGHTMOST\_LEAF}(w_{next})\big)$$

in this case (cf. Corollary 5.(a)).

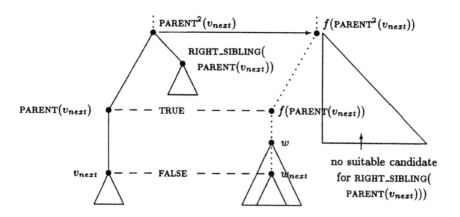

Fig. 3. *A match* (v_{next}, w_{next}) *with* $\text{STATE}(v_{next}, w_{next}) = \text{FALSE}$ *is considered again.*

Remark. Note that this case can actually occur. In the situation shown in Figure 3, $\text{STATE}(v_{next}, w_{next})$ has been set to FALSE. Nevertheless, it has been possible to map $P[\text{PARENT}(v_{next})]$ completely to $T[f(\text{PARENT}(v_{next}))]$ afterwards with another mapping of v_{next}. Hence $\text{STATE}(\text{PARENT}(v_{next}), f(\text{PARENT}(v_{next})))$ has been set to TRUE. But now it is impossible to map $\text{RIGHT\_SIBLING}(\text{PARENT}(v_{next}))$. Hence the algorithm dismisses the match $(\text{PARENT}(v_{next}), f(\text{PARENT}(v_{next})))$ and seeks for another candidate w for $\text{PARENT}(v_{next})$ among the successors of $f(\text{PARENT}(v_{next}))$ (cf. Case (1) of the backward step). If it finds one, and if this is also ancestor of w_{next}, it may happen that the algorithm considers the match (v_{next}, w_{next}) again. □

(3) If STATE(v_{next}, w_{next}) = TRUE, we have already mapped $P[v_{next}]$ completely to $T[w_{next}]$. Nevertheless, we again map v_{next} to w_{next}:

$$f(v_{next}) := w_{next}.$$

But we do not have to map the entire subtree $T[v_{next}]$ again; instead we can implicitly make use of the inclusion map from $P[v_{next}]$ to $T[w_{next}]$ constructed before. Consequently the constructed inclusion map f from P to T may be incomplete. Hence we have to reconstruct the missing parts after (successfully) finishing the algorithm. Nevertheless, we can immediately go to the next node of P that is not successor of v_{next}. Thereby we have to distinguish two cases.

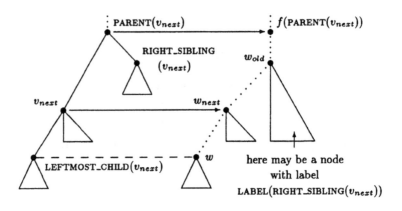

Fig. 4. *The inclusion map of a subtree can be taken over completely.*

(3.a) In the case when v_{next} *has a right sibling*, we consider this right sibling next. The next eligible candidate for it must be to the right of w_{next} (cf. Corollary 3.(a)). Hence we set

$$w_{next} := \text{NEXT}\big(\text{LABEL}(\text{RIGHT\_SIBLING}(v_{next})), \text{RIGHTMOST\_LEAF}(w_{next})\big),$$
$$v_{next} := \text{RIGHT\_SIBLING}(v_{next})$$

in this case.

Remark. Note that this situation can actually occur. In Figure 4 v_{next} had been mapped first to w_{old}, but with this choice it was not possible to map RIGHT_SIBLING(v_{next}). Subsequently another feasible candidate w_{next} for v_{next} was found in a backward step, so that v_{next} was mapped to it in the following forward step. But the subtree $T[w]$, on which the subtree

$P[\text{LEFTMOST\_CHILD}(v_{next})]$ has been mapped previously, is not only sub-tree of $T[w_{old}]$, but also subtree of $T[w_{next}]$. Hence the inclusion map from $P[\text{LEFTMOST\_CHILD}(v_{next})]$ to $T[w]$ can be taken over completely. However, note that there may be new possibilities to map $\text{RIGHT\_SIBLING}(v_{next})$ through the change from w_{old} to w_{next}. $\qquad\square$

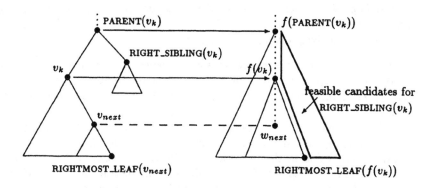

Fig. 5. *Forward step, Case (3.b), v_{next} has no right sibling.*

(3.b) In the case when v_{next} *has no right sibling*, we have mapped a subtree of P completely to a subtree of T, as shown in Figure 5. In this case, the values of STATE of the nodes on the path from v_{next} to v_k, its first ancestor with a right sibling, have already been correctly set. Hence we consider the right sibling of v_k (note that this is the node which succeeds $\text{RIGHTMOST\_LEAF}(v_{next})$ in preorder) next. The next eligible candidate for it has to be to the right of $f(v_k)$ (cf. Corollary 3.(b)). Hence, we set

$$v_{next} := v_{\text{PRE}(\text{RIGHTMOST\_LEAF}(v_{next}))+1},$$
$$w_{next} := \text{NEXT}(\text{LABEL}(v_{next}), \text{RIGHTMOST\_LEAF}(f(\text{LEFT\_SIBLING}(v_{next}))))$$

in this case.

In every case of the forward step, we go with the new match (v_{next}, w_{next}) to a new iteration of the algorithm. Note that the candidate w_{next} chosen in the forward step is always the next eligible candidate for v_{next}.

The backward step. When we carry out a backward step, w_{next} is either not successor of $f(\text{PARENT}(v_{next}))$ or it is out of the range specified by $\text{RANGE}(v_{next})$. In the first case w_{next} is not feasible and in the second case it is not suitable. In either case we know that there is no suitable candidate for v_{next} , so that

$f(\text{PARENT}(v_{next}))$ or $f(\text{LEFT\_SIBLING}(v_{next}))$, if v_{next} has a left sibling, are not suitable either. Depending on whether v_{next} has a left sibling, we have to distinguish two cases in the backward step.

(1) If v_{next} *has a left sibling*, there is no candidate for v_{next} to the right of $f(\text{LEFT\_SIBLING}(v_{next}))$ whose choice enables us to map $P[\text{PARENT}(v_{next})]$ completely to $T[f(\text{PARENT}(v_{next}))]$.

Lemma 7. *Let w_{next} be no suitable candidate for v_{next} and let v_{next} have a left sibling. Then there is no suitable candidate for v_{next} to the right of $f(\text{LEFT\_SIBLING}(v_{next}))$.*
PROOF. We inductively distinguish two cases.

- If we have entered the backward step, because w_{next} is not successor of $f(\text{PARENT}(v_{next}))$, then there is no suitable candidate for v_{next} to the right of $f(\text{LEFT\_SIBLING}(v_{next}))$, because w_{next} is the next eligible candidate for v_{next}.
- If we have entered the backward step, because w_{next} is to the right of $\text{RANGE}(v_{next})$, then there is no suitable candidate for v_{next} to the right of $\text{RANGE}(v_{next})$ due to the induction hypothesis. As w_{next} is the next eligible candidate for v_{next}, there is no suitable candidate for v_{next}, neither between $f(\text{LEFT\_SIBLING}(v_{next}))$ and $\text{RANGE}(v_{next})$ nor among the successors of $\text{RANGE}(v_{next})$. □

From this lemma we can conclude that $P[\text{PARENT}(v_{next})]$ cannot be completely mapped to $T[f(\text{PARENT}(v_{next}))]$, if $\text{LEFT\_SIBLING}(v_{next})$ is mapped to $f(\text{LEFT\_SIBLING}(v_{next}))$ or to a node that is to the right of $f(\text{LEFT\_SIBLING}(v_{next}))$. Hence we have restricted the mapping range of v_{next} and $\text{LEFT\_SIBLING}(v_{next})$. For both nodes there is no suitable candidate to the right of $f(\text{LEFT\_SIBLING}(v_{next}))$. Consequently we set

$$\text{RANGE}(v_{next}) := f(\text{LEFT\_SIBLING}(v_{next}))$$

and

$$\text{RANGE}(\text{LEFT\_SIBLING}(v_{next})) := f(\text{LEFT\_SIBLING}(v_{next})).$$

Note that the value of $\text{RANGE}(\text{LEFT\_SIBLING}(v_{next}))$ does not exclude that $\text{LEFT\_SIBLING}(v_{next})$ is mapped to $f(\text{LEFT\_SIBLING}(v_{next}))$ again. But we can prove that the algorithm does not do this again. We omit this technical detail here.

However, to check if $P[\text{PARENT}(v_{next})]$ can be completely mapped to $T[f(\text{PARENT}(v_{next}))]$, i.e. if $f(\text{PARENT}(v_{next}))$ is a suitable candidate for $\text{PARENT}(v_{next})$, we must still consider other candidates for the left sibling of v_{next}. Thereby only those candidates are in question which are successors of its current image $f(\text{LEFT\_SIBLING}(v_{next}))$. Candidates which are to the left of this current image or ancestors of it are either not feasible or not suitable (note that we have always chosen the next eligible candidate). Furthermore, candidates

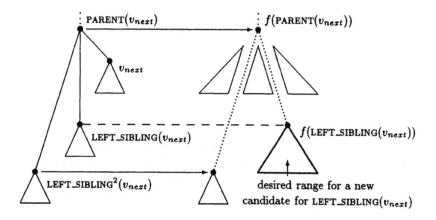

Fig. 6. *The desired range for* LEFT_SIBLING(v_{next}).

which are to the right of this image are either not feasible or they are excluded by the corollary above (see Figure 6). Hence we set

$$v_{next} := \text{LEFT\_SIBLING}(v_{next})$$
$$w_{next} := \text{NEXT}(\text{LABEL}(v_{next}), f(v_{next}))$$

in this case (cf. Corollary 5.(a)).

(2) If v_{next} *has no left sibling*, there is no suitable candidate for v_{next} in the whole subtree $T[f(\text{PARENT}(v_{next}))]$. Hence $P[\text{PARENT}(v_{next})]$ cannot be completely mapped to $T[f(\text{PARENT}(v_{next}))]$.

Lemma 8. *Let* w_{next} *be no suitable candidate for* v_{next}, *and let* v_{next} *have no left sibling. Then* $P[\text{PARENT}(v_{next})]$ *cannot be completely mapped to* $T[f(\text{PARENT}(v_{next}))]$.
PROOF. The proof uses an argument that is very similar to that of Lemma 7 and is hence omited here. □

Therefore we set

$$\text{STATE}(\text{PARENT}(v_{next}), f(\text{PARENT}(v_{next}))) := \text{FALSE}.$$

If $\text{PARENT}(v_{next})$ is the root of P, we have found that the pattern tree P cannot be completely mapped to the target tree T. Hence we can finish the algorithm. Otherwise we reset the RANGE-values of the children of $\text{PARENT}(v_{next})$ which have been set. Then we go to the next eligible candidate for $\text{PARENT}(v_{next})$. Since $f(\text{PARENT}(v_{next}))$ and successors of it with label $\text{LABEL}(\text{PARENT}(v_{next}))$

are not suitable candidates for PARENT(v_{next}), the next eligible candidate has to be to the right of $f(\text{PARENT}(v_{next}))$. Hence, we set

$$v_{next} := \text{PARENT}(v_{next}),$$
$$w_{next} := \text{NEXT}(\text{LABEL}(v_{next}), \text{RIGHTMOST\_LEAF}(f(v_{next})))$$

in this case.

When we leave the backward step, we have dismissed a mapping of a node of the pattern tree to a node of the target tree, and chosen a new candidate w_{next} for this node of P – either for PARENT(v_{next}) or for LEFT_SIBLING(v_{next}). Note further that the candidate w_{next} chosen in the backward step is always the next eligible candidate for v_{next}.

4 Analysis of the Algorithm

In this section we first prove the **correctness** of our algorithm.

Theorem 9. *The algorithm* OrderedTreeInclusion *is correct.*
PROOF. A candidate w_j chosen for a node v_i within the algorithm is always the next eligible one. Furthermore, v_i is only mapped to w_j, if w_j is feasible. Hence any (partial) inclusion map constructed by the algorithm satisfies the condiditons of the ordered tree inclusion. Hence the algorithm has constructed an inclusion map from the pattern tree P to the target tree T that satisfies the conditions of ordered tree inclusion, if it sets the value of STATE(v_1, w_1) to TRUE.

A value of STATE(v, w) can be set to FALSE only in Case (2) of the backward step. Hence it follows from Lemma 8, that there exists no inclusion map from $P[v]$ to $T[w]$ in this case. Hence there is no inclusion map from the pattern tree P to the target tree T that satisfies the conditions of ordered tree inclusion, if the algorithm set the value of STATE(v_1, w_1) to FALSE. □

Now we analyze the **time complexity** of our algorithm OrderedTreeInclusion. This is composed of the time complexity of preprocessing and the time spent in the forward and backward steps of the main part.

A forward step can be carried out in constant time, with the exception of *going up the path* in Case (1.c) of the forward step. During *going up the path* the value of STATE(v, w) for a match (v, w) is set from NIL to TRUE in every step. Since the value of STATE(v, w) is not changed afterwards, the total time spent in all forward steps for *going up the path* is at most $O(\#\text{matches})$. Furthermore the value of STATE(v, w) for a match (v, w) is also not changed after it has been set from NIL to FALSE.

Next we prove an upper bound for the number of forward steps carried out per match.

Lemma 10. *A match* (v_k, w_l) *is considered in at most*

$$|\{w \mid w \text{ is on the path from } w_l \text{ to } \text{ROOT}(T) \land \text{LABEL}(w) = \text{LABEL}(\text{PARENT}(v_k))\}|$$

forward steps.

PROOF. The key argument of the proof is that the match (v_k, w_l) can only be considered, if PARENT(v_k) has been mapped to an ancestor of w_l, i. e. a node that is on the path from w_l to ROOT(T) and that has the label LABEL(PARENT(v_k)). By a careful analysis we can show that the use of the RANGE array guarantees that the match (v_k, w_l) is considered only once per such a mapping of PARENT(v_k) in a forward step. We omit the details here. □

$|\{w \mid w \text{ is on the path from } w_l \text{ to ROOT}(T) \land \text{LABEL}(w) = \text{LABEL}(\text{PARENT}(v_k))\}|$ is bounded by DEPTH(T). Note that this bound is generally not very tight. It follows that at most $O(\#\text{matches} \cdot \text{DEPTH}(T))$ forward steps are carried out. in the algorithm. The number of forward steps is also an upper bound for the total number of single mappings of nodes of the pattern tree to nodes of the target tree. It follows that it is also an upper bound for the number of backward steps, since in every backward step one mapping of a node of the pattern tree to a node of the target tree is dismissed. Furthermore, a backward step can be carried out in constant time.

Since the **space complexity** of the algorithm is dominated by the STATE and the NEXT arrays, we have the following results for the complexity of the algorithm.

Theorem 11. *The algorithm* OrderedTreeInclusion *has a running time of*

$$O(|\Sigma_P| \cdot |T| + \#matches \cdot \text{DEPTH}(T)),$$

where Σ_P *is the alphabet of the labels of the pattern tree, and a space complexity of*

$$O(|\Sigma_P| \cdot |T| + \#matches).$$

□

5 Conclusion and Further Work

We have presented a new algorithm for the ordered tree inclusion problem that is better than the previous ones if the number of matches is relatively small. Next we would like to eliminate the factor DEPTH(T) in the time complexity of our algorithm. Furthermore, we would like to implement our algorithm to see how competetive it is in practice. In this implementation some heuristics could be applied; for example, we can map a subtree of the pattern tree to a subtree of the target tree, if the latter is at least as high and as large as the former.

Next we would like to apply the techniques we have used for the ordered tree inclusion to other tree inclusion problems, for example the ordered path and the ordered region inclusion problems (see [6]). We conjecture that we can solve this problem within the same time and space bounds by modifying our algorithm for the ordered tree inclusion problem appropriately.

Finally we would like to attack the largest common substructure problem in order to beat the complexity resulting from the application of the algorithm of [10] for the tree editing problem to this problem.

Acknowledgment. I would like to thank Prof. Dr. N. Blum for helpful discussions on this work.

References

1. M. Dubiner, Z. Galil and E. Magen, *Faster Tree Pattern Matching*, Proc. 31st FOCS (1990), pp. 145 - 150.
2. G. H. Gonnet and F. Wm. Tompa, *Mind your Grammar - a New Approach to Text Databases*, Proc. of the Conf. on Very Large Databases 1987 (VLDB'87), pp. 339 - 346.
3. C. M. Hoffman and M. J. O'Donnell, *Pattern Matching in Trees*, JACM **29** (1982), pp. 68 - 95.
4. P. Kilpeläinen, G. Linden, H. Mannila and E. Nikunen, *A Structured Document Database System*, in R. Furuta (ed.), *EP'90 - Proc. of the Int. Conf. on Electronic Publishing, Document Manipulation & Typography*, The Cambridge Series on Electronic Publishing, Cambridge University Press, 1990.
5. P. Kilpeläinen and H. Mannila, *Retrieval from Hierarchical Texts by Partial Patterns*, in R. Korfhage, E. Rasmussen and P. Willet (eds.), SIGIR '93 - Proc. of the 16th Ann. Int. ACM SIGIR Conf. on Research and Development in Informational Retrieval 1993, pp. 214 - 222.
6. P. Kilpeläinen and H. Mannila, *Query Primitives for Tree-Structured Data*, Proc. 5th CPM (1994), pp. 213 - 225.
7. P. Kilpeläinen and H. Mannila, *Ordered and Unordered Tree Inclusion*, SIAM J. Comput. **24** (1995), pp. 340 - 356.
8. D. E. Knuth, *The Art of Computer Programming*, Vol. 1, Addison-Wesley, Reading, MA, 1969, p. 347.
9. S. R. Kosaraju, *Efficient Tree Pattern Matching*, Proc. 30th FOCS (1989), pp. 178 - 183.
10. K. Zhang and D. Shasha, *Simple Fast Algorithms for the Editing Distance between Trees and Related Problems*, SIAM J. Comput. **18** (1989), pp. 1245 - 1262.

On Incremental Computation of Transitive Closure and Greedy Alignment

Saïd Abdeddaïm

BGBP - UMR CNRS 5558
Université Claude Bernard (LYON 1)
43, Bd. du 11 Novembre 1918
69622 Villeurbanne Cedex - FRANCE
E-mail: Said.Abdeddaim@biomserv.univ-lyon1.fr

Abstract. Several algorithms based on heuristics have been proposed for the multiple alignment of sequences. The most efficient in time computation are often *greedy* algorithms. At each step a greedy alignment algorithm must know if two characters are *alignable* or not, regarding to the characters definitely aligned before. We show that this problem is reducible to find paths in a directed graph. We give an *incremental* algorithm that maintains the transitive closure of a graph for which we know a *spanning* set of k disjoined paths. Our algorithm maintains the transitive closure of a graph of n vertices and m edges (in the final state) in $O(k^2m + n\min\{m, n\})$ time and $O(kn)$ space. We show that this algorithm can be used by any greedy alignment algorithm to know in constant time if two characters are alignable or not, by maintaining the transitive closure of an *alignment* graph in $O(k^2n + n^2)$ time and $O(kn)$ space, for k sequences whose total length is n. As an example of application we have implemented TwoAlign a efficient multiple alignment program based on greedy computation of pairwise local alignments.

1 Introduction

The computation of a minimal cost alignment of an indefinited number of sequences is known as an NP-hard problem for the costs usually used [1, 2, 3]. Solutions thus necessarily rely on the choice of heuristics. Several approaches have been proposed, the most efficient in time computation are often *greedy* algorithms. In a greedy alignment algorithm the process of alignment is irreversible. When two characters are aligned ("put together" in the same column of the final alignment) this decision cannot be reconsidered, that explain the efficiency of such algorithms. The most known greedy alignment approaches are *progressive* alignment algorithms [4, 5, 6, 7, 8, 9]. These algorithms are based on iterative selection of pairwise alignments. Other approaches are based on greedy computation of *blocks* [10, 11].

At each step a greedy alignment algorithm must know if two characters x and y are *alignable* or not, regarding to the set of pairs of characters which are already aligned. In a progressive alignment algorithm if x or y is a character of a sequence that is not yet aligned, then necessarily x and y are alignable.

However the problem of deciding if x and y are alignable or not is not such simple in the general case. We show that this problem is reducible to find paths in a directed graph, in which the vertices correspond to the characters of the sequences, and such that the vertices corresponding to each sequence form a path in this graph. When a new pair of aligned characters is added, edges are added in this directed graph. To maintain the *alignability* relation between characters, when new characters are aligned, a greedy algorithm must then maintains the *transitive closure* of the directed graph.

An *incremental* algorithm that maintains the transitive closure of a graph must be able to answer at each step and in constant time, if there is a path or not between two vertices, when additions of edges are allowed. The efficiency of an incremental algorithm is generally measured by the global number of operations required for maintaining the solution in a graph G that has n vertices and m edges (in the final state). Ibaraki and Katoh [12] have described an algorithm that maintains the transitive closure of a graph in $O(nm^*)$ time and $O(n^2)$ space, where m^* is the number of edges in the transitive closure of G. This result was improved by Italiano [13] and by La Poutre and van Leeuwen [14], their algorithms maintain the transitive closure of a graph in $O(nm)$ time and $O(n^2)$ space.

We give an algorithm that maintains the transitive closure of a graph G for which we *know* a *spanning* set of k disjoined paths (each vertex is in one and only one path). Our algorithm maintains the transitive closure of a graph in $O(k^2 m + n \min\{m, n\})$ time and $O(kn)$ space. So this algorithm use less space than the incremental algorithms that do not use the knowledge of the k paths, and is faster when the number of paths is lower than the average length of these paths $(k < n/k)$.

We show that this algorithm can be used by any greedy alignment algorithm

1. to know in constant time if two characters are alignable or not,
2. and to compute the set C of characters of a sequence which are alignable with a character c of an other sequence in $O(|C|)$ time.

This is done by maintaining the transitive closure of an *alignment* graph in $O(k^2 n + n^2)$ time and $O(kn)$ space, for k sequences whose total length is n. The saving of memory space ($O(kn)$ rather than $O(n^2)$) is of crucial importance in the case of multiple alignment. If, for example, 100 sequences of average length 2000 are aligned, the space required in our case is 20 millions of memory words rather than 40 billions.

As an example of application of this approach we have implemented TwoAlign a multiple alignment program, based on greedy computation of pairwise local alignments. We show on real data the efficiency (in computation time and alignments quality) of TwoAlign comparing to CLUSTAL W[15] a well know tool based on a progressive alignment approach.

2 Incremental computation of transitive closure of directed graphs

Let $G = (V, \to)$ be a *directed graph*, where V is a set of *vertices* and $\to$ a binary relation in V ($\to \subset V \times V$). A pair (x, y) such that $x \to y$ is called *edge*. In the following a *graph* is always a directed graph.

A *path* P in G is an ordered set $(x_1, x_2, \ldots, x_p)$ of vertices such that $x_i \to x_{i+1}$ for $i = 1, 2, \ldots, p-1$. We tell that the path P *joins* x_1 with x_p and *contains* the vertices x_i ($i = 1, 2, \ldots, p$). We also tell that the vertex x_i ($i = 1, 2, \ldots, p$) is in the path P at the *position* i.

The *transitive closure* of G is the graph $G^* = (V, \overset{*}{\to})$, where $\overset{*}{\to}$ is the relation that contain all pairs joined by a path in G; $x \overset{*}{\to} y$ if there is a path between x and y. Notice that $\overset{*}{\to}$ is reflexive. If $x \overset{*}{\to} y$ then x is called *predecessor* of y, and y called *successor* of x.

For a set E, $|E|$ is the number of elements in E. In the following n is the number of vertices of G ($n = |V|$), m its actual number of edges ($m = |\to|$) and m^* the number of edges in G^* ($m^* = |\overset{*}{\to}|$). We denote by m_0 the number of edges *initially* in G, before the process of additions of edges in the graph.

We are interested by the case where a set $\mathcal{P} = \{P_1, P_2, \ldots, P_k\}$ of k paths is known in the initial graph, such that each vertex of G is in one and only one of these paths P_i. Such a set of paths is called a *spanning set of disjoined paths* (SSDP) in G (see fig. 1).

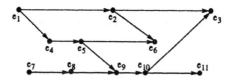

Fig. 1. $\{(e_1, e_2, e_3), (e_4, e_5, e_6), (e_7, e_8, e_9, e_{10}, e_{11})\}$ is a SSDP of the graph

Let $m_{\mathcal{P}}$ be the number of edges traversed by a path of $\mathcal{P}$, and let m_0 be the number of edges in the initial state of G ($m_{\mathcal{P}} \leq m_0 \leq m$).

Theorem 1. *For a spanning set of k disjoined paths, the transitive closure of a graph G can be maintained, when edges are added, in $O(k^2(m - m_0) + n \min\{n, (m - m_{\mathcal{P}})\})$ time and $O(kn)$ space.*

This can be done by an incremental algorithm we give in the following.

2.1 Compact representation of the transitive closure when a spanning set of disjoined paths is known

For a vertex x, let $P(x)$ be the vector where each entry $P(x)[i]$ ($i = 1, \ldots, k$) is the number of vertices of the path P_i which are predecessors of x. In other words, $P(x)[i]$ is the maximal position of a vertice of P_i that is predecessor of x if there is one, otherwise it is equal to 0. $P(x)$ is called *predecessor frontier* of the vertex x. For a vertex x in the path P_i we denotes by $pos(x)$ the position of the vertex x in P_i.

Observation 1 *For x in P_i and y in P_j, $x \overset{*}{\to} y$ iff $pos(x) \leq P(y)[i]$*

This observation shows that it is possible to represent the transitive closure of a graph G in $O(kn)$ space if a spanning set of k disjoined paths is known. A similar observation was first proposed for particular graphs to represent the *causality* relation in *distributed computations* [16, 17].

Symmetrically to the predecessor frontier we define the *successor frontier* of a vertex x, as a vector $S(x)$ of length k in which every position $S(x)[i]$ is equal to $|P_i| - \mathrm{Suc}_i(x) + 1$, where $\mathrm{Suc}_i(x)$ is the number of vertices in P_i which are successors of x. So $S(x)[i]$ is the minimal position of a vertice of P_i that is successor of x if there is one, otherwise it is equal to the length of P_i plus one. The figure 2 shows the predecessor frontier and the successor frontier of the vertex e_5.

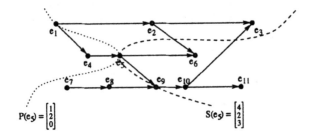

Fig. 2.

The next observation is equivalent to the observation 1 for the successor frontiers.

Observation 2 *For x in P_i and y in P_j, $x \overset{*}{\to} y$ iff $pos(y) \geq S(x)[j]$*

2.2 Algorithm

The addition of a new edge modifies the transitive closure and therefore the frontiers of at least some of the vertices. Rather than computing all frontiers after each insertion of an edge, we give an incremental algorithm that computes only the positions of the frontiers that are changed by the addition of the new edge.

We suppose that the transitive closure of the initial graph is known and known the frontiers of each vertex initially. The purpose is to compute the frontiers of the vertices of G after the addition of an edge (x, y), using the frontiers before this addition.

Let $P^{(1)}(u)$ (resp. $S^{(1)}(u)$) be the predecessor (resp. successor) frontier of a vertex $u \in V$, before the addition of $x \to y$, and $P^{(2)}(u)$ (resp. $S^{(2)}(u)$) be the predecessor (resp. successor) frontier after the addition of $x \to y$. The following observation shows the relation between the frontiers of each u before and after the addition of (x, y) (see an illustration fig. 3).

Observation 3 *For u in P_i and for each path P_j :*

1. $P^{(2)}(u)[j] = \begin{cases} \max\{P^{(1)}(u)[j], P^{(1)}(x)[j]\} & \text{if } u \text{ was successor of } y \\ P^{(1)}(u)[j] & \text{otherwise.} \end{cases}$

2. $S^{(2)}(u)[j] = \begin{cases} \min\{S^{(1)}(u)[j], S^{(1)}(y)[j]\} & \text{if } u \text{ was predecessor of } x \\ S^{(1)}(u)[j] & \text{otherwise.} \end{cases}$

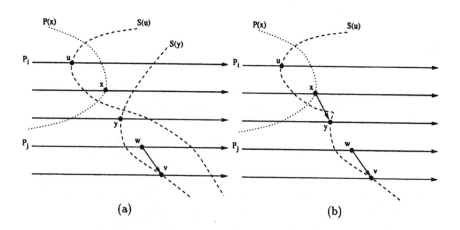

Fig. 3. (a) the successor frontiers of u and y *before* the addition of (x, y), and (b) the successor frontier of u *after* the addition of (x, y)

The procedure EdgeAddition (showed in figure 4), we have designed, enables to compute the new frontiers after the addition of an edge (x, y).

If there is already a path between x and y the frontiers are not modified by the addition of (x, y). Otherwise, the procedure EdgeAddition considers each path P_i and each path P_j and for each u in P_i that is predecessor of x (the position of u in P_i is lower then or equal to $P(x)[i]$) the value of $S(u)[j]$ is updated when it is greater than $S(y)[j]$ (observation 3). This can be done following the decreasing positions of u because $S(u)[j]$ cannot increase when the position of u

EdgeAddition(x,y)

1. If $x \not\xrightarrow{A} y$ then
2.   For each path P_i do
3.   For each path P_j do
4.   For each u in P_i from the position $P(x)[i]$ in the decreasing positions and While $S(u)[j] > S(y)[j]$ do
5. $S(u)[j] \longleftarrow S(y)[j]$
6.   For each path P_i do
7.   For each path P_j do
8.   For each u of P_i from the position $S(y)[i]$ in the increasing positions and While $P(u)[j] < P(x)[j]$ do
9. $P(u)[j] \longleftarrow P(x)[j]$

Fig. 4. Algorithm for the maintenance of transitive closure when an edge is added

decreases, so it cannot be greater than $S(y)[j]$ after it had been equal or lower. The predecessor frontiers are computed in the same way.

2.3 Complexity analysis

Each time a new edge (x, y) is added the procedure considers at most all pairs of paths (P_i, P_j), it means that the steps 4 and 8 requires at least $O(k^2(m - m_0))$ time in the wost case.

We must also add the time computation due to the modifications of the frontiers for all vertices $u \in V$, which is the number of times the steps 5 and 9 are done.

Let us consider the modifications of the successor frontiers. As the paths of $\mathcal{P}$ are disjoined, a vertex $v \in V$ can modify an entry of the successor frontier of u at most once. This can happen when v is successor of y but not successor of u before the addition of an edge (x, y) such that u is a predecessor of x (see fig. 3). It means that for each u the successor frontier of u cannot be modified more than n times. It follows that the step 5 cannot be done more than n^2 times.

Let (w, v) be an edge of G that is not traversed by a path of $\mathcal{P}$. In the procedure EdgeAddition an edge (w, v) can modify an entry of the successor frontier of u after the addition of (x, y) only if w is successor of y but not successor of u before the addition of (x, y) (see fig. 3). After that (w, v) has modified the successor frontier of u, it cannot modify it again. So for each u the successor frontier of u cannot be modified more than $m - m_\mathcal{P}$ times. That means that the step 5 cannot be executed more than $n(m - m_\mathcal{P})$ times.

It follows that the entries of the successor frontiers cannot be modified more than $n \min\{n, (m - m_\mathcal{P})\}$ times. For similar reasons this is also the case for the predecessor frontiers. The theorem 1 follows.

3 Application for greedy alignment of sequences

Let A be an alphabet and $\Delta \notin A$ called *null letter*. A word in A^* is called a *sequence*. A word in $(A \cup \Delta)^*$ is called a *padded* sequence. A *character* of a (padded) sequence S is a position in S together with the letter at that position. In a padded sequence S, the occurrence of a factor of S is called a *segment*. A *gap* is a maximal segment of null letters.

Let $\mathcal{S} = \{S_1, S_2, \ldots, S_k\}$ be a set of sequences. A *pseudo-alignment* of the sequences $S_1, S_2, \ldots, S_k$ is an ordered set of padded sequences $(S'_1, S'_2, \ldots, S'_k)$ of equal length such that removing the null letters from S'_i gives the sequence S_i. The pth *column* of an ordered pseudo-alignment α is the ordered set of the pth character from each padded sequence of α. An *alignment* α is a pseudo-alignment all of whose columns contain a non-null letter (see an example figure 5).

$$S_1 \quad \text{--TCGTC-TGCACGC--GCTCTGCGAT}$$
$$S_2 \quad \text{AGTCGTC-TGCACG-G-GATCTGCGA-}$$
$$S_3 \quad \text{AATAGTCATGGACGCGTGCTCT---A-}$$
$$S_4 \quad \text{-ATAGTCATGGACGCGTGCGC---GAT}$$

Fig. 5. Alignment of four sequences. Δ is denoted by the symbol –

We consider the graph $G = (V, \rightarrow)$ *associated* to $\mathcal{S}$ such that: i) each vertex x in V represents a character $\phi(x)$ in a sequence of $\mathcal{S}$, where ϕ is a mapping, ii) $x \rightarrow y$ if the character $\phi(x)$ precedes immediately the character $\phi(y)$ in a sequence of $\mathcal{S}$.

The graph G contains k paths, such that each path P_i corresponds to a sequence S_i $(i = 1, 2, \ldots, k)$. The greedy alignment of the sequences of $\mathcal{S}$ is a process where pairs of characters are aligned definitely. Each time two characters corresponding to two vertices x and y are aligned, two edges (x, y) and (y, x) are added in the graph G. The graph G is called an *alignment graph* [18] of the sequences of $\mathcal{S}$.

We denote by $x \overset{*}{=} y$ the fact that x and y are joined by a path in both directions $(x \overset{*}{\rightarrow} y$ and $y \overset{*}{\rightarrow} x)$. In this case we tell that x *coincides* with y or that x and y coincide.

Definition 2. Let G be a alignment graph for a set S of sequences, and let $\mathcal{P}$ be the SSDP of G corresponding to $\mathcal{S}$. $(G, \mathcal{P})$ is *regular* if there is an alignment $\mathcal{A}$ of the sequences of $\mathcal{S}$ such that two vertices which coincide in G correspond to characters that are in the same column of $\mathcal{A}$.

A maximal set of vertices which coincide defines what we call an *anchor*. The figure 6 shows an example of four sequences associated to twelve anchors. The frontiers of the vertex corresponding to the character in position 10 of the sequence S_3 are represented.

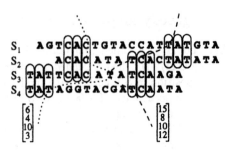

Fig. 6.

Proposition 3. *Let G be an alignment graph and P it associated SSDP. (G, P) is regular if and only if each anchor has at most one vertex for each path of P.*

Proof (if) For (G, P) regular there is, by definition, an alignment A such that each anchor of G is contained in a column of A, as a column cannot not contain two distinct characters of a same sequence of S, then an anchor cannot not contain two distinct vertices of a same path of P.

(only if) Suppose that each anchor of an alignment graph G does not contain two distinct vertices of a same path of P. We consider the partial order relation $\prec$ between the anchors of G, such that $\alpha_1 \prec \alpha_2$ if and only if there is a path between two vertices u and v that are respectively in the anchors α_1 and α_2. As all the anchors of G cannot contain two distinct vertices of the same path, and as they are partially ordered, we can always built an alignment such as each column contains the characters corresponding to one and only one anchor, and where the columns order is consistent with the partial order $\prec$ (topological order). It follows that (G, P) is regular. $\qquad\qquad\square$

From now, we consider that (G, P) is regular. When the sequences of S are greedily aligned, two characters a and b, corresponding to two vertices x and y, can be aligned if and only if (G, P) remains regular when (x, y) and (y, x) are added. If so, a and b are said *alignable*. The next proposition gives a necessary and sufficient condition that enables to know if two characters are alignable or not.

Proposition 4. *For (G, P) regular, two characters a and b, corresponding to two vertices x and y, are alignable if and only if one of these two conditions is true:*

(1) *there is no path between x and y ($x \not\rightarrow y$ and $y \not\rightarrow x$),*

(2) *x and y are joined in the two directions ($x \overset{*}{\rightleftharpoons} y$).*

Proof For (G, P) regular, let G' be the graph G after the addition of (x, y) and (y, x). We must show that (G', P) is regular if and only if (1) or (2).

(if) Suppose that x and y are joined only in one direction in G, for example that $x \overset{*}{\to} y$ and $y \overset{*}{\not\to} x$. As G is a alignment graph, there is necessarily an edge (u, v) in the path from x to y that is traversed by a path P_i of $\mathcal{P}$. In the resulting alignment graph G' there is also a path between v and u (as $v \overset{*}{\to} y$, $y \to x$ and $x \overset{*}{\to} u$). So u coincides with v, then $(G', \mathcal{P})$ is not regular (proposition 3).

(only if) Suppose that $x \overset{*}{=} y$ in G, it means that the anchors of G' are the same as those of G so $(G', \mathcal{P})$ is regular (proposition3). Suppose now that there is no path between x and y in G, let α_1 and α_2 be the two anchors of G that contains respectively x and y. Let $\alpha = \alpha_1 \cup \alpha_2$ be the new anchor of G', here again there is no pair (u, v) of distinct vertices of α that are in the same path in G. Indeed, if $u \in \alpha_1$ and $v \in \alpha_2$ and u and v are in a same path in G, then x is joined with y in only one direction, which is inconsistent with the hypothesis. The other anchors of G' (other than α) are anchors of G, so $(G', \mathcal{P})$ is regular. $\square$

As it is possible to know in constant time if two vertices are joined by a path or not (observations 1 or 2), we can then (using the proposition 4) verify in constant time if two characters are alignable or not.

Furthermore, it is possible to compute the set C of characters of a sequence which are alignable with a character c of an other sequence in $O(|C|)$ time using the frontiers. Indeed:

Observation 4 *Let c be a character of a sequence S_i, and x be the vertex corresponding to c ($c = \phi(x)$). The characters of the sequence S_j which are alignable with the character c have their positions comprised between $P(x)[j]$ and $S(x)[j]$ in the sequence S_j. Furthermore, the characters of the sequence S_j of position comprised between $P(x)[j] + 1$ and $S(x)[j] - 1$ are necessarily alignable with c.*

The next proposition gives the complexity of the maintenance of the transitive closure for a alignment graph G when $(G, \mathcal{P})$ is regular.

Proposition 5. *For $(G, \mathcal{P})$ regular the transitive closure of G can be maintained in $O(k^2 n + n^2)$ time and $O(kn)$ space.*

Proof As $(G, \mathcal{P})$ is regular, each anchor has at most one vertex for each path of $\mathcal{P}$ (proposition 3). So each anchor correspond to at most $2k(k-1)$ edges, but only $2(k-1)$ of them verify the condition of the step 1 when they are added in the procedure EdgeAddition and give rise to new paths. For the $2(k-1)^2$ other edges the condition of the step 1 in the procedure EdgeAddition is not verified. It follows that the number of edges that verify the condition of the step 1 when they are added is upper-bounded by $2n$ ($\geq \frac{2(k-1)}{k} \times n$) when $k > 1$. For these edges the time complexity due to EdgeAddition is then upper-bounded by $O(k^2 n + n^2)$ (theorem 1). The number of remaining edges is upper-bounded by $2n(k-1)$ so they cost at worst $O(kn)$ time. $\square$

4 TwoAlign

We propose **TwoAlign** a new program that compute multiple alignment of biological sequences. The main idea in **TwoAlign** consists in building a multiple alignment by clustering *pairwise local alignments*.

4.1 Definitions and algorithm

A *pairwise local alignment* is a alignment of two segments (occurrences of factors) taken in two sequences S_i and S_j. A pairwise local alignment λ is *inscribed* in an alignment α of the sequences of S, if the columns of β are subsets of columns of α. The pairwise local alignments of a set Λ are *compatible* if all the alignments of Λ can be inscribed in a same alignment. For set Λ of compatible pairwise local alignments, two characters are *aligned* if they are necessarily contained in a column of an alignment where the alignment of Λ are inscribed. A pairwise local alignment λ is *independent* with a set of pairwise local alignments Λ if λ contains no column with a pair of aligned characters.

Given a score matrix between pairs of letters taken in A and a gap score, the *score* of a pairwise (local) alignment is the sum of the scores of the pairs of non null letters of its columns plus the sum of the gap scores.

TwoAlign is greedy and compute at each step a maximal score pairwise local alignment (MSPLA) among all pairs of sequences. This new local alignment must be compatible and independent from those already computed. The process is repeated until the maximal score of a pairwise local alignment (that verify the two conditions) is lower than a fixed score that is considered not significant. The second condition (independent local alignments) avoid to take always the same local alignment at each step.

The global multiple alignment is "finished" between the anchors (corresponding to the pairwise local alignments computed) using a dynamic programming algorithm [11]. This algorithm is based on the assumption that the SP gap cost [19] is minimal between two successive anchors.

4.2 Implementation

In the implementation of **TwoAlign**, each time a new local alignment λ is selected, the edges corresponding to the aligned characters of each column of λ are added in a alignment graph G.

Observation 5 *Let Λ be a set of compatible pairwise local alignments, and let G be the alignment graph corresponding to Λ:*

- *A pairwise local alignment λ is compatible with Λ if and only if the characters of each column of λ are alignable regarding to G.*
- *A pairwise local alignment λ is compatible and independent from Λ if and only if the characters of each column of λ correspond to vertices that are not joined by a path in G.*

So to compute MSPLA of two sequences S_i and S_j that is compatible with and independent from the computed local alignments, TwoAlign considers only the pairs of characters which correspond to vertices that are alignable. This is done easily by an adaptation of the algorithm of Smith and Waterman [20] algorithm. This adaptation is implemented efficiently as we can access easily to the characters of S_j that are alignable with each character c of S_i (see observation 4). TwoAlign also avoid the pairs of characters (c, c') such that the vertices corresponding to c and c' are joined by a path (can be tested in constant time) to consider only independent local alignments.

At the first step, to compute the MSPLA among all pairs of sequences, all the MSPLA of all pairs of sequences S_i and S_j $(1 \le i < j \le k)$ are computed, but when a MSPLA λ is chosen it is not necessary (in order to choose the next MSPLA) to compute again all MSPLA of all pairs of sequences S_i and S_j. To be efficient, we compute the MSPLA of a pair (S_i, S_j) if and only if the addition of λ in the previous step has changed the alignability relationship between two characters a in S_i and b in S_j. This happen when a and b are not aligned before the computation of λ and are aligned after it addition. This condition is equivalent as to tell that the entry $S(a)[j]$ of successor frontier of a is changed when λ is added. So when a new MSPLA λ is added we can easily know for which pairs (S_i, S_j) TwoAlign has to compute the MSPLA.

4.3 Results

We did a comparison of TwoAlign with Clustal W[15] a well know tool based on a progressive alignment approach. For this comparison we used 552 sets of protein sequences taken in the PRINTS database [21]. On this data TwoAlign runs 2.7 times faster than Clustal W (13283" for TwoAlign and 35296" for CLUSTAL W on a DEC Alpha 3000 station).

We used the same score matrix (BLOSUM 30) for the two programs TwoAlign and Clustal W. We also used in TwoAlign the same gap penalties that are used in Clustal W at its first step, namely the pairwise alignment. At this step of Clustal W the gap score gs(l) of a gap of length l is computed as follow:

$$gs(l) = -(g + gh \times l)$$

where g and gh depends on the sequence type (protein or dna), on the length of the two sequences, and on the score matrix used. For two amino-acid sequences of length l_1 and l_2 that are aligned using BLOSUM 30:

$$g = 2 \times AVS \times (go + \min(l_1, l_2))$$
$$gh = 100 \times ge$$

where AVS depends on the BLOSUM 30 matrix, and where go and ge are parameters given by the program user. Notice that in the second step of Clustal W (progressive alignment) the treatment of the gap penalties is more complicated [15]. So any comparison between TwoAlign and Clustal W has a significance limited by the fact that the two programs use the parameters differently.

The quality of the alignments computed by the two programs TwoAlign and Clustal W is measured by there SP cost [19]. The SP cost is a sum of pairwise alignment costs. The pairwise alignment cost parameters we used to estimate the quality are equivalent to the alignment score parameters given in input to the programs [22].

The figure 7 shows the distribution of the 552 sets of sequences according to the improvement of the SP cost that TwoAlign obtain comparing to CLUSTAL W. We used the default parameters used by CLUSTAL W for amino-acid sequences, which are go = 10 and g = 0.1. We observe that for 71% of the 552 sets the results are equivalent (between -2% and 2% of improvement). CLUSTAL W computes better alignment for 3% of the sets, and TwoAlign do better for 26% (more than 5% better for 8% of the sequences). TwoAlign gives particularly better quality when the lengths of the sequences are very different, for example for 5 *glutelin* sequences of lengths varying between 58 and 223, TwoAlign gives an alignment that is 8.4 % better than the alignment given by Clustal W.

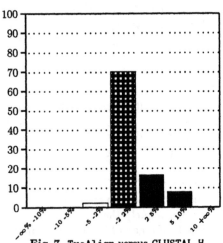

Fig. 7. TwoAlign versus CLUSTAL W

References

1. M. R. Garey and D. S. Johnson. *Computers and intractability; a guide to the theory of NP-completeness.* Freeman, 1979.
2. L. Wang and T. Jiang. On the complexity of multiple sequence alignment. *J. Comput. Biol.*, 1:337–348, 1994.
3. T. Jiang, E. L. Lawler, and L. Wang. Aligning sequences via an evolutionary tree: complexity and approximation. In *Proc. 26-th Annual ACM Symp. Theory of Comput.*, pages 760–769, 1994.

4. P. Hogeweg and B. Hesper. The alignment of sets of sequences and the construction of phyletic trees: an integrated method. *J. Mol. Evol.*, 20:175–186, 1984.

5. D-F. Feng and R. F. Doolittle. Progressive sequence alignment as a prerequisite to correct phylogenetic trees. *J. Mol. Evol.*, 25:351–360, 1987.

6. W. R. Taylor. Protein structure prediction. In M. J. Bishop and C. J. Rawlings, editors, *Nucleic Acid and Protein Sequence Analysis, a Practical Approach.*, pages 285–323. IRL Press, 1987.

7. F. Corpet. Multiple sequence alignment with hierarchial clustering. *Nucleic Acids Research*, 16(22):10881–10890, 1988.

8. D.G. Higgins and P.M. Sharp. Fast and sensitive multiple sequence alignments on a microcomputer. *CABIOS*, 5:151–153, 1989.

9. O. Gotoh. Further improvement in methods of group-to-group sequence alignment with generalized profile operations. *CABIOS*, 10(4):379–387, 1994.

10. A. M. Landraud, J. F. Avril, and P. Chrétienne. An algorithm for finding a common structure shared by a family of strings. *IEEE Transactions on Pattern Analysis and Machine Intelligence*, 11:890–895, 1989.

11. Saïd Abdeddaïm. Fast and sound two-step algorithms for multiple alignment of nucleic sequences. In *Proceedings of the IEEE International Joint Symposia on Intelligence and Systems*, pages 4–11, 1996.

12. T. Ibaraki and N. Katoh. On-line computation of transitive closure for graphs. *Inform. Proc. Lett.*, 16:95–97, 1983.

13. G. F. Italiano. Amortized efficiency of a path retrieval data structure. *Theor. Comput. Sci.*, 48:273–281, 1986.

14. J. A. La Poutré and J. van Leeuwen. Maintenance of transitive closure and transitive reduction of graphs. In *Proc. Workshop on Graph-Theoretic Concepts in Computer Science*, pages 106–120. Lecture Notes in Computer Science 314, Springer-Verlag, 1988.

15. J. D. Thompson, D. G. Higgins, and T. J. Gibson. CLUSTAL W: improving the sensitivity of progressive multiple sequence alignment through sequence weighting, position specific gap penalties and weight matrix choice. *Nucleic Acids Research*, 22:4673–4680, 1994.

16. F. Mattern. Virtual time and global states of distributed systems. In *Proc. Workshop on Parallel and Distributed Algorithms*, pages 215–226, 1989.

17. C. J. Fidge. Timestamps in message-passing systems that preserve the partial ordering. In *11-th Australian Computer Science Conference*, pages 55–66, 1988.

18. J. Kececioglu. The maximum weight trace problem in multiple sequence alignment. In *4-th Annual Symp. Combinatorial Pattern Matching*, volume 684 of *LNCS*, pages 106–119. 1993.

19. S. F. Altschul. Gap costs for multiple sequence alignment. *J. Theor. Biol.*, 138:297–309, 1989.

20. T. F. Smith and M. S. Waterman. Identification of common molecular subsequences. *J. Mol. Biol.*, 147:195–197, 1981.

21. T. K. Attwood, M. E. Beck, A. J. Bleasby, and D. J. Parry-Smith. PRINTS - a database of protein motif fingerprints. *Nucleic Acids Research*, 22:3590–3596, 1994.

22. M. S. Waterman. *Mathematical Methods for DNA Sequences*. C.R.C. Press, 1989.

Aligning Coding DNA in the Presence of Frame-Shift Errors

Lars Arvestad *

Department of Numerical Analysis and Computing Science
Royal Institute of Technology
S-100 44 Stockholm, Sweden.

Abstract. The problem of aligning two DNA sequences with respect to the fact that they are coding for proteins is discussed. Criteria for a good alignment of coding DNA, together with an algorithm that satisfies them, are presented. The algorithm is robust against frame-shifts and forgiving towards silent substitutions. The important choice of objective function is examined and several variants are proposed.

1 Introduction

In this paper we discuss how to align two DNA sequences that come from a coding region, i.e. the DNA is translated to an amino acid sequence, which is something we should take note of when aligning the sequences.

The traditional pairwise sequence alignment algorithm, as found by Sellers in 1974 [13] and independently by others (see Waterman [15] and Sankoff and Kruskal [12] for a thorough treatment), aims at minimizing the amount of *change* (substitutions/replacements, insertions and deletions) between two biological sequences. However, change in DNA does not always have an obvious interpretation.

If the sequences are DNA coding for proteins, we do not necessarily want to count silent substitutions that are often numerous in the third position of a codon. Also, some amino acids are coded by codons that differ in each position. Matching two such codons looks bad on the DNA level, but should not result in a poor score since the proteins may be identical when the DNA is translated to amino acids.

Another common problem [9] with implementations of the traditional aligning algorithm, when applied to DNA, is that stretches of insertions/deletions, gaps, are not constrained to biologically reasonable lengths. Since gaps of a length that is not 0 modulo 3 change the reading-frame, they are very unlikely to occur. If the algorithms do not take this into account, unsatisfactory alignments are computed.

Frame-shift errors further complicate our task at hand. They typically occur from bad gel-readings (compressions) or other sequencing problems. Thus,

---

* *E-mail:* `arve@nada.kth.se`.

aligning newly sequenced DNA often involves investigating the source of error, correcting, and computing a new alignment. It would be desirable to do this automatically. Not only off-the-gel sequences contain frame-shift errors; it has been observed [8] that many sequences found in databases (e.g. EMBL) are faulty. An algorithm that is robust against these problems is therefore useful to many researchers.

Frame-shift errors also invalidate a natural line of attack, namely to translate the DNA to amino acids for each combination of reading frames. Because the translation will soon be obscured by a frame-shift, such an algorithm is very sensitive.

A related problem, the question of how to align a coding DNA sequence with a protein, has been discussed in a range of papers [4, 8, 11, 14] and mentioned applications are database searching as well as error-checking sequences. The problem we focus on in this paper has been addressed by Hein in [5], where an $O(N^4)$ algorithm is given. In this paper, we present a quadratic time algorithm for aligning protein-coding DNA in the presence of frame-shift errors.

Section 2 introduces notation and requirements for a good objective function. The new algorithm is explained in section 3 and variation on the objective function is found in section 4. To conclude, experimental results and a discussion are given in sections 5 and 6.

2 A new scoring scheme

In this section we address the question of how to choose a better scoring scheme for coding DNA sequences. Our intention is to introduce language needed for the new approach and give an abstract scoring scheme in order to present the new algorithm. Especially, notice that we do not yet make adjustments for frame-shifts, but leave that for section 3.

2.1 Definitions and notation

Definition 1. A *frame-shift* is an insertion or deletion of length one or two.

Definition 2. A *gap* is an insertion or deletion whose length is a multiple of three.

If a gap has a length that is not 0 modulo 3, then we regard it as a combination of a frame-shift and a gap. The intuition is that a frame-shift corresponds to something less likely such as a sequencing error or a rare evolutionary event, while a gap typically corresponds to lost or inserted amino acids.

The terms "cost" and "score" are used intermixed in this paper. We want to minimize the cost but maximize the score. Higher scores are assumed to be used for preferred matches.

Denote the cost of inserting a frame-shift γ. In this paper we only consider affine gap costs where opening a gap is associated with a cost α and every triplet of three indels has an additional cost β. So for a gap of length l the cost is $\alpha + l\beta$. We can now make a new definition, slightly different from current practice, of an alignment.

Definition 3. An alignment of two sequences a and b is a pair (a', b') where a' and b' are derived by inserting frame-shifts and gaps in a and b.

Let $\mathcal{A}_{a,b}$ be the set of alignments of sequences a and b. (This notation is also abused to allow a and b to be sets of sequences.) The quality of an alignment is captured in the score of the alignment, thus we need to produce a reasonable scoring function. Let $\mathbf{N} = \{A, C, G, T\}$, the set of nucleotides and $\mathbf{N}^+ = \mathbf{N} \cup \{-\}$, the set of nucleotides together with the indel symbol. Similarly, let $\mathbf{A}$ be the set of amino acids and let $\mathbf{A}^+ = \mathbf{A} \cup \{-\}$ be the same set with indels included. The set of r-tuples over a set X is written X^r.

Definition 4. A function $S_N : (\mathbf{N}^+ \times \mathbf{N}^+)^* \mapsto \mathcal{R}$ is called a *nucleotide scoring function* and $S_A : (\mathbf{A}^+ \times \mathbf{A}^+)^* \mapsto \mathcal{R}$ is an *amino acid scoring function*.

We write the translation of a DNA sequence x, whose length is a multiple of three, to amino acids as $aa(x)$. If x contains frame-shifts or ambiguity symbols, the result may be a set of translations. Writing $aa(a, b)$ is short for $(aa(a), aa(b))$.

2.2 Requirements for a good alignment

Since we want to align coding DNA, we argue that a nucleotide scoring function should have the property that for an alignment (a', b') of a and b,

$$S_N(a', b') = \max_{(\tilde{a}, \tilde{b}) \in \mathcal{A}_{a,b}} (S_N(\tilde{a}, \tilde{b})) \iff S_A(aa(a', b')) = \max_{(\tilde{a}, \tilde{b}) \in \mathcal{A}_{aa(a,b)}} (S_A(\tilde{a}, \tilde{b})). \quad (1)$$

That is, the optimal nucleotide sequence alignment is also, when translated to amino acids, the optimal amino acid sequence alignment, and *vice versa*.

We want the scoring function to work with codons, so let

$$G = \{(x, y) : x, y \in \mathbf{N}^3 \cup \{---\}\}$$

be the set of codon matchings and introduce the following to make S_N easier to define in terms of codons.

Definition 5. A function $s : G \mapsto \mathcal{R}$ is called a *codon scoring function*.

In words, s takes two triplets and assigns the matching of them a score. For equation (1) to be fulfilled, we choose s such that $s(x_1, x_2) = s_A(aa(x_1, x_2))$, where s_A is an amino acid scoring scheme, e.g. from PAM or BLOSSUM matrices, $s(---, x) = s(x, ---) = \beta$. Matchings of type gap-gap, $(---,---)$, are not of interest to us and hence they have an associated score of $-\infty$.

We can now define an appropriate nucleotide scoring function.

Definition 6. Let x be the number of gaps. The score of an alignment (a', b') is then

$$S_N(a', b') = x\alpha + \sum_{i=0}^{|a'|/3} s(a'_{3i}a'_{3i+1}a'_{3i+2}, b'_{3i}b'_{3i+1}b'_{3i+2}). \qquad (2)$$

This function is easy to optimize, and the only things that differ from traditional nucleotide alignment are that we are inserting indel triplets (instead of single indels) and assigning scores based on triplets.

3 Aligning in the presence of frame-shifts

The above scoring function only works when we have correctly sequenced sequences and when we disregard evolutionary relations that have come from accidental frame-shifts during evolution. For maximal flexibility, we want to be able to insert and delete frame-shifts. Inserting frame-shifts is simply a matter of inserting indels and deleting frame-shifts is to be interpreted as ignoring nucleotides.

Therefore, we change the definition of an alignment slightly.

Definition 7. An alignment of two sequences a and b is a pair (a', b') where a' and b' are derived by inserting and removing frame-shifts and adding gaps in a and b.

Observe that removing frame-shifts does not necessarily mean that nucleotides are removed from the sequences or even from the presentation of the alignment, but simply that they are ignored in the objective function.

Adapting to this definition, define G' as

$$G' = \left\{ (x, y) : x, y \in \mathbf{N}^{+^3} \right\}.$$

In this set of matchings, we have all possible ways of constructing codons from full nucleotide triplets as well as from codon fragments and frame-shift symbols. Also included are all matchings against gap symbols.

It is noteworthy that contrary to common practice, columns in our alignments may actually contain only indels. The approach taken here is that the sequences are tried to be reconstructed in parts where the reading-frame is confused. This is more natural when frame-shifts are thought to be sequence errors, than when they are evolutionary events. However, since we may look at alignments as tools for reconstructing sequence ancestors, frame-shifts have the advantage that we are able to guess the dropped nucleotide(s).

There is now an immediate extension of the nucleotide scoring function.

Definition 8. Let x and z be the number of gaps and frame-shifts, respectively. The score of an alignment (a', b') is then

$$S_N(a', b') = x\alpha + z\gamma + \sum_{i=0}^{|a'|/3} s(a'_{3i}a'_{3i+1}a'_{3i+2}, b'_{3i}b'_{3i+1}b'_{3i+2}).$$

In the former, how to choose s was quite immediate, but it is less clear how to do that now. We defer that discussion to section 4.

3.1 A new requirement

Requirement (1) is hard to relate to the current version of the problem since we had not defined a scoring function that could incorporate frame-shifts when it was formulated. The requirement can be restated to include frame-shifts in a slightly more complex way.

$$S_N(a', b') = \max_{(\tilde{a}, \tilde{b}) \in \mathcal{A}_{a,b}} (S_N(\tilde{a}, \tilde{b})) \Longleftrightarrow$$

$$\Longleftrightarrow S_A(aa(a', b')) = \max_{\substack{z \in \mathcal{N} \\ (\tilde{a}, \tilde{b}) \in \mathcal{A}_{aa(a,b,z)}}} (S_A(\tilde{a}, \tilde{b}) + z\gamma). \quad (3)$$

We have added translations from the DNA pair to amino acid sequences which may include frame-shifts; The set $aa(a, b, z)$ is the set of translations of the DNA sequences a and b using z frame-shifts. The effect is that we recognize, both in the DNA alignment and in the DNA to amino acid translation, that our sequences might have frame-shifts.

3.2 The new algorithm

We can now present the new algorithm and we state it in a theorem.

Theorem 9. *We can find the optimal nucleotide alignment with frame-shifts in time $O(|a||b|)$.*

Proof. The result follows by applying generalized substitutions (see Sankoff-Kruskal [12]) and Gotoh's technique for linear gap functions [3]. Let $\mathbf{C} = \mathbf{N}^{1,2,3,4,5}$ be the set of codon fragments and let $M = \{(x, y) : x, y \in \mathbf{C}\}$ be the set of generalized substitutions. That is, M could be derived from G' by removing any gap-matchings (on the form $(---, x)$ for any x) and for each element in G' removing the indel symbols and create copies with one and two extra frame-shifting nucleotides inserted. Let $d_{a,b}$ denote the score of the optimal alignment for the sequences a and b ending with a (reconstructed) codon-matching. Let $d^-_{a,b}$ and $d^+_{a,b}$ denote the optimal score of the alignments on a and b that ends with a gap in a' and b' respectively.

To score substitutions from M, we define $\tilde{s}$ from s as

$$\tilde{s}(x, y) = \max_{\substack{x' \in I(x) \\ y' \in I(y)}} s(x', y') \quad (4)$$

where I is defined as

$$I(a) = \{-a, -a-, a-\} \qquad I(ab) = \{-ab, a-b, ab-\}$$
$$I(abc) = \{abc\} \qquad I(abcd) = \{abc, abd, acd, bcd\} \quad (5)$$
$$I(abcde) = \{abc, abd, acd, bcd, abe, ace, bce, ade, bde, cde\}$$

are the sets of three-letter strings, codons, created from inserting indels into, or removing nucleotides from, the arguments. We are assuming that s is defined on codons containing frame-shifts (with the associated cost γ included), s on a codon-gap matching scores β (unless the codon contains frame-shifts), and that $s(---,---) = -\infty$.

Consider now the following recursion with $\circ$ as the concatenation operator.

$$d_{\epsilon,\epsilon} = d^-_{\epsilon,\epsilon} = d^+_{\epsilon,\epsilon} = 0 \tag{6a}$$

$$d_{a,b} = \max_{\substack{\hat{a}\circ x = a \\ \hat{b}\circ y = b \\ (x,y)\in M}} \begin{cases} d^-_{\hat{a},\hat{b}} + \tilde{s}(x,y) \\ d_{\hat{a},\hat{b}} + \tilde{s}(x,y) \\ d^+_{\hat{a},\hat{b}} + \tilde{s}(x,y) \end{cases} \tag{6b}$$

$$d^-_{a,b} = \max_{\substack{\hat{b}\circ y = b \\ y\in C}} \begin{cases} d^+_{a,\hat{b}} + \alpha + \tilde{s}(---,y) \\ d_{a,\hat{b}} + \alpha + \tilde{s}(---,y) \\ d^-_{a,\hat{b}} + \tilde{s}(---,y) \end{cases} \tag{6c}$$

$$d^+_{a,b} = \max_{\substack{\hat{a}\circ x = a \\ x\in C}} \begin{cases} d^+_{\hat{a},b} + \tilde{s}(x,---) \\ d_{\hat{a},b} + \alpha + \tilde{s}(x,---) \\ d^-_{\hat{a},b} + \alpha + \tilde{s}(x,---) \end{cases} \tag{6d}$$

The recursion can be solved in the usual way with dynamic programming and therefore uses a matrix of size $\Theta(|a||b|)$ which is completed in the same time complexity.

For the number of comparisons, we see that the number of previous cells (a cell being a variable within a matrix element) for each element in the dynamic programming matrix for the above recursion is $3 \times 5 \times 5 = 75$ from (6b) plus $3 \times 5 = 15$ each from (6b) and (6d), giving a total of 105 precursors and thus 104 comparisons. However, we can make a significant speed-up by also introducing $\hat{d}_{a,b} = \max\{d_{a,b}, d^-_{a,b}, d^+_{a,b}\}$, since in equation (6b), this comparison will be made (implicitly) once for each reference to an element (a, b). If this comparison is pre-computed, only 25 instead of 75 cells are considered in (6b), giving 24 comparisons. Also note that a similar comparison is done five times for each (a, b) in equations (6b) and (6d). If we pre-compute $\hat{d}^-_{a,b} = \max\{d_{a,b} + \alpha, d^-_{a,b}, d^+_{a,b} + \alpha\}$ and $\hat{d}^+_{a,b} = \max\{d_{a,b}+\alpha, d^-_{a,b}+\alpha, d^+_{a,b}\}$, only four comparisons in each of the two cases are needed. If we also note that $\hat{d}_{a,b}$, $\hat{d}^-_{a,b}$, and $\hat{d}^+_{a,b}$ can be computed with five comparisons, the number of comparisons have been reduced to $5+24+4+4 = 37$.

4 On choosing a codon scoring function

A first approach to choosing s is to find the best "interpretation" of codon fragments. This is achieved by first mapping a codon fragment x to possible amino acids,

$$C(x) = \{x' : \exists y \in I(x), x'_i = y_i \text{ if } y_i \neq - \text{ and } x'_i \in N \text{ otherwise.}\} \tag{7}$$

and then choosing

$$s(x, y) = \max_{\substack{x' \in C(x) \\ y' \in C(y)}} \left(s_A(aa(x', y')) + |3 - |x||\gamma + |3 - |y||\gamma \right). \tag{8}$$

In practice, it is probably desirable to extend C to work with ambiguity codes; The natural extension is to map a codon (fragment) to all possible interpretations of the ambiguity and choose the most favorable.

There are adjustments we could make to improve this scoring scheme and we now discuss a few suggestions.

4.1 Silent substitutions

If several alignments are possible that give the same score if they are translated to amino acid sequences, we still want to minimize the amount of nucleotide substitutions. That is, between two matches of codons equivalent with respect to amino acids, we choose the one with less substitutions. It therefore seems reasonable to add the cost for nucleotide mismatches in codons coding for the same amino acid. Let δ_s be the cost for nucleotide transitions and δ_v the cost for nucleotide transversions. (More elaborate scoring schemes could be considered to account for the different nucleotide substitution rates.) We adjust s to

$$s'(x, y) = \max_{\substack{x' \in C(x) \\ y' \in C(y)}} \left(s_A(aa(x', y')) + |3 - |x||\gamma + |3 - |y||\gamma + n_s\delta_s + n_v\delta_v \right) \tag{9}$$

for n_s and n_v being the number of silent transitions and transversions, respectively. An unfortunate effect with this is that we now have given up the nucleotide scoring function requirement from equation (3). However, in practice there is little reason to expect a big difference in the end-result.

If the requirement is imperative and special scoring of silent mutations is only to be used to distinguish equivalent (under requirement (3)) alignments, the following method is suggested:

1. Compute all alignments with the optimal score. This is easily done by adjusting the current algorithm to use a technique like e.g. what Chao [1] propose to compute all alignments within ϵ of the optimal value (in this application $\epsilon = 0$). The set of alignments is stored in a graph such that any path in the graph is an alignment.
2. In this graph, use dynamic programming to find the optimal alignment using an adjusted scoring function that penalize silent mutations as in equation (9) above.

Only alignments optimal under requirement (3) are computed in step 1, and the best alignment with respect to silent mutations is computed in step 2, and hence the method computes what we wanted.

4.2   Context-dependent frame-shift costs

One of the sources for frame-shifts are sequencing errors and it is well known [9] that such errors typically come from compression effects in the sequencing gel; for a longer stretch of the same nucleotide it is difficult to correctly determine how many bases we have. Thus, the likelihood of a frame-shift is different in different positions in a sequence. The propensity for a frame-shift to occur in nature might also vary over the sequences, which would be nice to account for if it could be quantified. To that purpose we add *context-dependent frame-shift costs*.

Let $\gamma(i,j,l)$ be the cost of putting a frame-shift of length l (possibly zero) between positions j and $j+1$ of sequence i and let $l_{i,j} \in \{0,1,2\}$ give the length of a frame-shift after position j of sequence i. We may now write a new version of S_N.

$$S_N(a',b') = x\alpha + \sum_{j=0}^{|a|}\gamma(a,j,l_{a,j}) + \sum_{j=0}^{|b|}\gamma(b,j,l_{b,j})$$
$$+ \sum_{i=0}^{|a'|/3} s(a'_{3i}a'_{3i+1}a'_{3i+2}, b'_{3i}b'_{3i+1}b'_{3i+2}) \quad (10)$$

Letting the cost of a frame-shift vary across the sequence has the effect that the choice of codon-fragment interpretation, as given in equations (8) and (9), is no longer valid. This is rectified by pre-computing the codon scoring function for each possible pair of contexts, either by looking at each pair of positions in the two sequences or by simply enumerating possible pair of contexts from knowing which contexts can occur (the latter method is then sequence independent).

4.3   Out-of-frame gaps

So far we have assumed that gaps occur in the reading-frame. If they were not in frame, frame-shifts would have to be inserted in both ends of the gap to compensate and this is probably quite costsome since frame-shifts should be expensive in order to produce reasonable alignments.

Out-of-frame gaps has been observed in HIV [9] and they should not be unexpected since they do not have a much larger impact on the interpretation than in-frame gaps. In particular, an out-of-frame insertion basically exchange one codon c for an inserted string of codons with the end-codons biased by the remaining, split, fragments of c. An out-of-frame deletion may be seen as a deletion followed by an insertion of one codon, because the two affected codons at the ends of the deletion join.

There are two out-of-frame gaps to be distinguished and we call them ± 1 and ± 2, with the sign telling us in which sequence they occured. If a gap is -1, then the gap is in the first sequence delayed with one nucleotide with respect to the reading frame. A type $+2$ gap occurs in the second sequence and is delayed by two nucleotides, see also figure 1. Note that we can consider in-frame gaps to be of type ± 0.

```
...xxx|x--|---|-xx... ...xxx|xxx|xxx|xxx...
...xxx|xxx|xxx|xxx... ...xxx|xx-|---|--x...
```

Fig. 1. *The two types of out-of-frame gaps. To the left, a type −1 gap, while on the right a type +2 gap.*

As with the in-frame gaps, this is solved by the technique Gotoh suggested [3]. For each gap we introduced a variable, d^{-1}, d^{-2}, d^{+1} and d^{+2}, that contains the cost of ending the alignment with a gap of the respective type. We omit the details since they are more lengthy than informative.

4.4 Starting and ending with frame-shifts

It is often desirable to let alignments start and end with gaps without any cost, a so called end-free alignment. This is convenient, for instance, if the sequences have been unevenly sequenced. For the same reason, it is interesting to allow end-free frame-shifts. If we for some reason don't have the sequences starting in the correct reading frame, we don't want any strange behavior with insertion of frame-shifts to try to compensate. Instead, any frame-shifts in the beginning or end of an alignment should be for free.

This is easily achieved by using context-dependent frame-shift costs and choosing γ appropriately for the starting and ending positions.

5 Experimental results

We have implemented a simple version of the described algorithm, only considering lost nucleotides. s is chosen in the most direct way, based on a user-chosen amino acid scoring-scheme (e.g. PAM- or BLOSSUM-matrices [2, 6])). No concern about the improvements to the scoring scheme mentioned in section 4 is given, but ambiguity codes are honored.

HIV data from the ENV-V3 region, 13 sequences kindly supplied by Thomas Leitner [9], was used to test the robustness of the algorithm. Aligning any two of these sequences yields an alignment that is about 261 bases long. About 72 % of the columns are mismatches and about 9 columns contains indels.

Four tests was set up by randomly introducing errors in pairs of sequences and the algorithms ability to find the position of the removed nucleotide(s) was then tested. The codon scoring function s was based on PAM-100 [2], gap-cost and gap extension-cost was set to 20, and single-indel frame-shifts cost 40 while double-indel frame-shifts cost 50. The results are shown in table 5. As seen, the guessed position of the frame-shift was usually only a few bases from the correct position.

Table 1. The average displacement, i.e. the number average of bases away from the known position of the removed nucleotide(s) that the algorithm put a frameshift. The four tests was set up by removing one nucleotide or two adjacent nucleotides from one or both sequences. The positions of the removed nucleotides was chosen at random for each pair sequences. In one case, having deleted a pair of nucleotides in both sequences, the algorithm chose to insert two single indel frame-shifts.

Displacement	Removed nucleotides	
	1	2
In one sequence	3.2	5.8
In both sequences	2.4	5.7

6 Discussion

As has been shown, our algorithm is very robust. However, the price paid is in computation time, with 37 comparisons for each element in in the dynamic programming matrix (with none of the improvements from section 4).

As presented, the algorithm has a quadratic space need. A linear space approach, similar to Hirschberg's technique [7] (popularized by [10]), is straightforward to use also on this algorithm.

In the above, we have used amino acid scoring schemes to base our nucleotide scoring function on for the simple reason that they are well understood and well investigated. But instead of relying on statistics on amino acid substitution probabilities, we could make the same statistics on codons. Basically, this would imply using a 64 by 64 scoring matrix instead of a 23 by 23 matrix. It is noteworthy that codons have different frequencies, which further justifies this approach.

7 Acknowledgments

Thanks to John Kececioglu for helping me understand the problem. Also thanks to Bill Bruno, Aaron Halpern, Johan Håstad and David Torney for discussions and suggestions.

References

1. K.-M. Chao. Computing all suboptimal alignments in linear space. In *5th Symposium on Combinatorial Pattern Matching*, pages 31–42. Springer-Verlag LNCS 807, 1994.
2. M. O. Dayhoff, R. M. Schwartz, and B. C. Orcott. A model of evolutionary change in proteins. *Atlas of Protein Sequence and Structure*, 5:345–352, 1978. National Biomedical Research Foundation, Silver Spring, Maryland, USA.

3. O. Gotoh. An improved algorithm for matching biological sequences. *Journal of Molecular Biology*, 162:705–708, 1982.

4. X. Guan and E. C. Uberbacher. Alignments of DNA and protein sequences containing frameshift errors. *Comp. Appl. Bio. Sci.*, 12(1):31–40, 1996.

5. J. Hein. An algorithm combining DNA and protein alignment. *Journal of Theoretical Biology*, 167:169–174, 1994.

6. S. Henikoff and J. G. Henikoff. Amino acid substitution matrices from protein blocks. *Proc. Natl. Acad.Sci.*, 89:10915–10919, 1992.

7. D. S. Hirschberg. A linear space algorithm for computing longest common subsequences. *Communications of the ACM*, 18:341–343, 1975.

8. L. J. Knecht. *Alignment and Analysis of Genes Coding for Proteins*. PhD thesis, Swiss Federal Institute of Technology, 1996.

9. T. Leitner. Personal communication. Until recently at the Department of Biochemistry, Royal Institute of Technology, Stockholm, now at Los Alamos National Laboratory, USA, Theoretical Biology and Biophysics Group.

10. E. W. Myers and W. Miller. Optimal alignments in linear space. *Comp. Appl. Bio. Sci.*, 4(1):11–17, 1988.

11. H. Peltola, H. Söderlund, and E. Ukkonen. Algorithms for the search of amino acid patterns in nucleic acid sequences. *Nuclear Acids Research*, 14(1):99–107, 1986.

12. D. Sankoff and J. Kruskal. *Time warps, string edits, and macromolecules: The theory and practice of sequence comparison*. Addison-Wesley, 1983.

13. P. H. Sellers. On the theory and computation of evolutionary distances. *SIAM Journal on Applied Mathematics*, 26:787, 1974.

14. D. J. States and D. Botstein. Molecular sequence accuracy and the analysis of protein coding regions. *Proc. Natl. Acad.Sci.*, 88:5518–5522, July 1991.

15. M. S. Waterman. *Introduction to computational biology: Maps, sequences and genomes*. Chapman & Hall, 1995.

A Filter Method for the Weighted Local Similarity Search Problem

Enno Ohlebusch

University of Bielefeld, Technische Fakultät
P.O. Box 100131, 33501 Bielefeld, Germany
enno@TechFak.Uni-Bielefeld.DE

Abstract. In contrast to the extensively studied k-differences problem, in the weighted local similarity search problem one searches for approximate matches of *subwords* of a pattern and subwords of a text whose lengths exceed a certain threshold. Moreover, arbitrary gap and substitution weights are allowed. In this paper, two new prefilter algorithms for the weighted local similarity search problem are presented. These overcome the disadvantages of a similar filter algorithm devised by Myers.

1 Introduction

Given a relatively short pattern (query) $P = P[1 \ldots m]$ of length m, a long text (database) $T = T[1 \ldots n]$ of length n, threshold $k \in \mathbb{R}$, and cost (or weight) function δ, the *weighted approximate string search problem* is the problem of finding all subwords t of the text for which the edit distance of P and t w.r.t. the cost function δ is below threshold k (or $edist_\delta(P, t) \leq k$ for short). Sellers [Sel80] provided the classical dynamic programming solution to the problem; its time complexity is $O(mn)$.

The extensively studied *k-differences problem* is a special case of the weighted approximate string search problem, where the cost function corresponds to the Levenshtein distance (that is, every insertion, deletion, and substitution is assigned the cost 1). In this case, Sellers' [Sel80] dynamic programming algorithm computes an $m + 1$ by $n + 1$ matrix whose entry $D(i, j)$ is the minimum number of edit operations necessary to transform the length i prefix of the pattern into some text fragment ending at the j-th character. Taking advantage of the properties of this matrix, $O(kn)$ worst case dynamic programming algorithms have been devised; see e.g. [LV88, GG88, GP90]. Moreover, Ukkonen's [Ukk85] cut-off variation of Sellers' algorithm has an $O(kn)$ average case time complexity; see [CL92]. It is unknown whether this algorithm has the same complexity for arbitrary cost functions. The above algorithms, however, are impractical for large-scale applications. Recent filter techniques (e.g. [Ukk92, TU93, CL94, CM94, Mye94a, Tak94, ST95, ST96]) reduce the average case complexity to at least linear time, provided k is small enough. The basic idea of a (pre-)filtering algorithm is to scan the database and quickly eliminate regions that can't possibly match via some easily computed criterion. For any region not eliminated, a more expensive dynamic programming computation is applied to determine if there is a match.

For some applications, it is necessary to deal with the more general *weighted local similarity search problem*: Given P, T, k, δ as above and minimum subword length l, find all quadruples (b', b, e', e) such that $edist_\delta(P[b' \ldots e'], T[b \ldots e]) \leq k$ and $e - b + 1 \geq l$. In other words, one has to find every local similarity between P and T whose length is at least l. As usual, it is sufficient to find the end-points e' and e of a local similarity. We quote Myers [Mye94b], page 123: "..., the essence of the generalization required in going from the approximate keyword search problem[1] to the biologically relevant search problem are (1) to allow arbitrary gap and substitution weights, and (2) to consider substrings of the query." One can solve this problem by applying Sellers' algorithm with cost function $\bar{\delta}$, defined by $\bar{\delta}(a \to b) = \delta(b \to a)$, to P (as text) and every subword t of T with $|t| = l$ (as pattern). Since T has $n - l + 1$ subwords of length l, this yields an $O(lmn)$ solution to the problem. Consequently, for the Levenshtein distance this problem can be solved in $O(kmn)$ time in the worst case. This brute force approach seems to be the best dynamic programming solution known so far.

To the best of our knowledge, Myers [Mye94b] filter algorithm is the only known algorithm (aside from classical dynamic programming) for the weighted local similarity search problem. For every q-gram[2] x it precomputes the value $Best(P, x) = \min\{edist_\delta(y, x) \mid y$ is a subword of $P\}$; in [CM94], this value is called $asm(x, P)$. The precomputation needs $O(|\mathcal{A}|^q m)$ time and $O(|\mathcal{A}|^q)$ space, where $|\mathcal{A}|$ is the size of the underlying alphabet $\mathcal{A}$. Then, for every position i in T, the value $Bound(i) = \sum_{j=1}^{l/q} Best(P, T[i - jq + 1 \ldots i - (j-1)q])$ is computed (for simplicity, assume $l \bmod q = 0$). If $Bound(i) \leq k$, then Sellers' algorithm is applied to P and $T[i - l + 1 \ldots i]$. Myers recognized two disadvantages of the algorithm: (i) the measure $Best(P, x)$ is based on the whole pattern P, (ii) the ordering requirement of sequence alignments is ignored. So, the longer P gets, the worse the algorithm will perform (because the more chance there is that $Best(P, x)$ becomes very small). Note moreover that (iii) for any region not eliminated, a dynamic programming computation is applied to this region and the *whole* pattern P. In this paper, an algorithm is given which overcomes all three disadvantages of Myers' algorithm by dividing the pattern into blocks (this was first suggested in [ST95] for the k-differences problem). In addition, a variation of our new filter method is given which needs much less time and space in the preprocessing phase. It will rigorously be proven that, for a certain range of parameters, the algorithms run in linear expected time. This is an enormous improvement over the $O(lmn)$ time requirement of the brute force dynamic programming solution. Moreover, the filtration efficiency of Myers' algorithm has been experimentally compared with the new algorithms. Our experiments confirm that the effectiveness of his algorithm crucially depends on m, whereas this is not true for ours. For instance, for the Levenshtein distance, $|\mathcal{A}| = 20$, $|T| = 1\,000\,000$, and $l = 40$ our best algorithm is effective (filtration efficiency $\geq 95\%$) for the mismatch ratio $k/l = 42.5\%$. If $m = 100$, then Myers' algorithm is effective for a ratio of 35%. If $m = 4000$, however, then $k/l = 10\%$.

---

[1] A synonym for the k-differences problem.
[2] A q-gram is just a word of length q.

2 A Filter Technique

We assume that the reader is familiar with the standard terminology on strings, as used e.g. in [CL94]. A *cost function* δ assigns to every edit operation $a \to b$ a positive real-valued cost $\delta(a \to b)$. The cost of $a \to a$ is 0. If $\delta(a \to b) = 1$ for all edit operations, then δ is the *unit cost function* . The cost function δ is extended to alignments A by: $\delta(A) = \sum_{a \to b \in A} \delta(a \to b)$. The *edit distance* of two strings u and v is $edist_\delta(u, v) = \min\{\delta(A) \mid A$ is an alignment of u and $v\}$. An alignment A of u and v is *optimal* if $\delta(A) = edist_\delta(u, v)$. If δ is the unit cost function, then $edist_\delta(u, v)$ is the *Levenshtein distance* between u and v.

In order to obtain a filter algorithm which can deal with arbitrary cost functions, we need the following simple observations. Let c_{sub} (c_{del}, c_{ins}, respectively) denote the minimum cost of all possible substitutions (deletions, insertions, respectively) and define $c_{eop} = \min\{c_{sub}, c_{del}, c_{ins}\}$. If p is a subword of P, t is a subword of T, and A is an alignment of p and t such that $\delta(A) \leq k$, then A contains at most $max_{sub} = \lfloor \frac{k}{c_{sub}} \rfloor$ substitutions, $max_{del} = \lfloor \frac{k}{c_{del}} \rfloor$ deletions and $max_{ins} = \lfloor \frac{k}{c_{ins}} \rfloor$ insertions, respectively. Furthermore, A contains at most $max_{eop} = \lfloor \frac{k}{c_{eop}} \rfloor$ edit operations. For the Levenshtein distance, we have $max_{sub} = max_{del} = max_{ins} = max_{eop} = k$. The next simple facts will be used frequently.

Lemma 1. *(i) Let $A = (a_1 \to b_1, \ldots, a_c \to b_c)$ be an alignment of two strings p and t, and let A_{del} and A_{ins} be the number of deletions and insertions, respectively, in A. Then $c = |p| + A_{ins} = |t| + A_{del}$ holds.*
(ii) If $edist_\delta(p, t) \leq k$, then $|t| - max_{ins} \leq |p| \leq |t| + max_{del}$.

As in Takaoka's method [Tak94], we take samples from T.

Definition 2. For all $j \in \{1, 2, \ldots, n'\}$, where $n' = \lfloor \frac{n-q}{h} \rfloor + 1$, let

$$Tsam(j) = T[(j-1)h + 1 \ldots (j-1)h + q].$$

The $Tsam(j)$ are called *q-samples* or simply *samples*. So every h-th q-gram is a q-sample in T (see Figure 1).

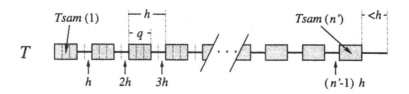

Fig. 1. Locations of the q-samples in T

The following lemma was proven in [ST95] for the case $|t| = m - k$.

Lemma 3. *Let t be a subword of T. Let $q, r \in \mathbb{N}_+$. If the sampling step $h \in \mathbb{N}$ satisfies $h \geq q$ (so that q-samples do not overlap) and*

$$h \leq \frac{|t| - q + 1}{r},$$

then at least r contiguous q-samples occur in t.

Proof. Let us consider the situation where the leftmost q-sample of t starts as right as possible: in this case the first $q - 1$ characters of t belong to a q-sample that started directly left to t. Thus t contains at least one q-sample if $|t| \geq q - 1 + h$. In general, t contains at least r samples if $|t| \geq q - 1 + rh$. This is equivalent to $h \leq \frac{|t| - q + 1}{r}$.

In Sutinen and Tarhio's algorithms LEQ [ST95] and LAQ [ST96] (filter algorithms for the k-differences problem), the location of q-samples is taken into account by dividing the pattern into $k + s$ overlapping blocks of length $h + k + q - 1$. In the sequel we are going to make use of their idea. Note that we have to take care of (i) arbitrary cost functions and (ii) the fact that we are considering subwords of the pattern instead of the pattern itself. The extension to (i) is fairly simple: blocks of length $h + max_{indel} + q - 1$ have to be used, where $max_{indel} = \max\{max_{ins}, max_{del}\}$. A natural extension to (ii) would be to divide the pattern from left-to-right into overlapping blocks such that the end-positions of the blocks do not exceed m. However, this does not work. Instead, the pattern must be "completely covered with blocks". Formally, for all $j \in \mathbb{N}$

1. $P[jh + 1 \ldots max_{indel} + q - 1 + (j + 1)h]$
2. $P[1 \ldots max_{indel} + q - 1 - jh]$

is a block if it contains at least one q-gram of P. Let

$$i_{max} = \lfloor \tfrac{m-q}{h} \rfloor \quad \text{and} \quad i_{min} = \lfloor \tfrac{max_{indel} - 1}{h} \rfloor.$$

(1) is satisfied for all $0 \leq j \leq i_{max}$, so there are $i_{max} + 1$ blocks of the first kind in P. Analogously, (2) is satisfied for all $0 \leq j \leq i_{min}$, so there are $i_{min} + 1$ blocks of the second kind in P.

B_1

P abcdefghijklmnopqrstuvwxyz

B_{11}

Fig. 2. Blocks in P: $m = 26$, $q = 2$, $h = 3$, $k = 4$, δ = Levenshtein distance

Definition 4. Pattern P is divided into blocks (where $1 \leq i \leq i_{min} + i_{max} + 2$)

$$B_i = P[(i - i_{min} - 2)h + 1 \ldots (i - i_{min} - 1)h + max_{indel} + q - 1].$$

Moreover, define $Q_i = \{u \mid u \text{ is a subword of } B_i, |u| = q\}$.

We observe that (i) $q \leq |B_i| \leq h + max_{indel} + q - 1$, (ii) two blocks B_i and B_{i+1} overlap by at most $max_{indel} + q - 1$ positions, and (iii) the first block of size $h + max_{indel} + q - 1$ is $B_{i_{min}+2} = P[1 \ldots h + max_{indel} + q - 1]$.

The next theorem forms the basis of our algorithms.

Theorem 5. *Suppose $h \leq \lfloor \frac{l-q+1}{r} \rfloor$ satisfies $h \geq q$, where $q, r \in \mathbb{N}_+$. Let $p = P[b' \ldots e']$ and $t = T[b \ldots e]$ be given with $edist_\delta(p, t) \leq k$ and $|t| \geq l$. Moreover, let $Tsam(j+1), Tsam(j+2), \ldots, Tsam(j+r)$ be r consecutive samples included in t (these exist by Lemma 3). Let A be an alignment of p and t with $\delta(A) \leq k$ and let p_c be the subword of p which is aligned with $Tsam(j+c)$ for $c \in \{1, \ldots, r\}$. Then there is a $d \geq 0$ such that p_c is a subword of B_{d+c} for all $c \in \{1, \ldots, r\}$.*

Proof. The proof is divided into three parts.

1. We define the shift of characters within the alignment.
2. We determine the block B_{d+1} in P that corresponds to $Tsam(j + 1)$.
3. We show that p_c occurs in B_{d+c} for all $c \in \{1, \ldots, r\}$.

(1) For all i, $0 \leq i \leq |t|$, let A^i be the subalignment of A which aligns the prefix $T[b \ldots b+i-1]$ of t and the prefix $P[b' \ldots b'+i'-1]$ of p (note that i' is determined by A and i). Let A^i_{del} and A^i_{ins} be the number of deletions and insertions, respectively, in A^i. Observe that for $i = 0$, $T[b \ldots b - 1]$ and $P[b' \ldots b' - 1]$ are empty, so A^0 aligns two empty words. Hence $A^0_{del} = A^0_{ins} = 0$. We define $S(i) = A^i_{del} - A^i_{ins}$. According to Lemma 1, $|P[b' \ldots b'+i'-1]| - |T[b \ldots b+i-1]| = A^i_{del} - A^i_{ins} = S(i)$. Thus $i' = i + S(i)$. So $S(i)$ denotes a shift within A. In A, these shifts are bounded by

$$S_{min} = \min\{S(i) \mid 0 \leq i \leq |t|\} \text{ and } S_{max} = \max\{S(i) \mid 0 \leq i \leq |t|\}.$$

So S_{min} is the maximum left-shift in A, whereas S_{max} is the maximum right-shift in A. Obviously, we have

(i) $-max_{ins} \leq S_{min} \leq 0 \leq S_{max} \leq max_{del}$ and
(ii) $S_{min} \leq S(i) \leq S_{max} \leq max_{indel} + S_{min}$ because $S_{max} - S_{min} \leq max_{indel}$.

(2) For ease of presentation, we assume that $Tsam(j+1)$ is the first q-sample in t, i.e., $(j-1)h+1 < b \leq jh+1$. The offset σ of $Tsam(j+1)$ w.r.t. the beginning of t (position b) is $\sigma = jh + 1 - b$. Clearly, $0 \leq \sigma \leq h - 1$. For instance, if b coincides with the beginning of $Tsam(j + 1)$, then $\sigma = 0$. Figure 3 depicts the maximum offset $\sigma = h - 1$. We choose block B_{d+1} such that position $b' + \sigma$ is at least $|S_{min}|$ characters right to its beginning and at least $max_{indel} + q - 1 + S_{min}$ characters left to its end. To this end, let

$$d' = \lfloor \tfrac{b'+\sigma+S_{min}-1}{h} \rfloor$$

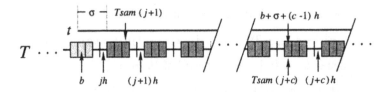

Fig. 3. Maximum offset $\sigma = h - 1$

In particular, $d'h \leq b' + \sigma + S_{min} - 1 \leq (d'+1)h - 1$. Hence $d'h + 1 + |S_{min}| \leq b' + \sigma$. Furthermore, $b' + \sigma \leq (d'+1)h - S_{min} = (d'+1)h + max_{indel} + q - 1 - (max_{indel} + q - 1 + S_{min})$. All in all, $P[d'h + 1 \ldots (d'+1)h + max_{indel} + q - 1]$ is the block we are looking for. This is block B_{d+1} for $d = d' + i_{min} + 1$.

(3) For all non-empty p_c, $c \in \{1, \ldots, r\}$, let $p_c = P[b_c \ldots e_c]$. We have to show

$$b_c \geq (d' + c - 1)h + 1$$
$$e_c \leq (d' + c)h + max_{indel} + q - 1.$$

Note that (see Figure 3) $Tsam(j + c) = T[(j + c - 1)h + 1 \ldots (j + c - 1)h + q] = T[b + \sigma + (c - 1)h \ldots b + \sigma + (c - 1)h + q - 1]$. Now since p_c is aligned with $Tsam(j + c)$ in A, it follows from (1) with $i = \sigma + (c - 1)h + 1$,

$$\begin{aligned}
b_c &= b' + i' - 1 = b' + i + S(i) - 1 \\
&= b' + \sigma + (c - 1)h + S(i) \\
&\geq b' + \sigma + (c - 1)h + S_{min} \\
&= (b' + \sigma + S_{min} - 1) + (c - 1)h + 1 \\
&\geq d'h + (c - 1)h + 1 = (d' + c - 1)h + 1.
\end{aligned}$$

Analogously, we derive with $i = \sigma + (c - 1)h + q$

$$\begin{aligned}
e_c &= b' + i' - 1 = b' + i + S(i) - 1 \\
&= b' + \sigma + (c - 1)h + q + S(i) - 1 \\
&\leq b' + \sigma + (c - 1)h + q + max_{indel} + S_{min} - 1 \\
&= (b' + \sigma + S_{min} - 1) + (c - 1)h + max_{indel} + q \\
&\leq (d' + 1)h - 1 + (c - 1)h + max_{indel} + q \\
&= (d' + c)h + max_{indel} + q - 1.
\end{aligned}$$

Now we are in a position to formulate the algorithm Local Similarity Search 1. The algorithm is divided into three phases: In a preprocessing phase, it computes the values $Best(B_i, x) = \min\{edist_\delta(y, x) \mid y$ is a subword of $B_i\}$ for every q-gram x and every block B_i, $1 \leq i \leq m' = i_{min} + i_{max} + 2$. In the subsequent filtering phase, for every j, $1 \leq j \leq n' = \lfloor \frac{n-q}{h} \rfloor + 1$, it incrementally computes the value $\sum_{c=0}^{r-1} Best(B_{i-c}, Tsam(j - c))$ and stores it in the i-th element of array $M[1 \ldots m']$. This is illustrated in Figure 4. If $M[i] \leq k$, where $j \geq r$ and $i \geq r$, then in the third phase a dynamic programming computation is applied to neighborhoods of the blocks and the q-samples to determine if there really is a local similarity.

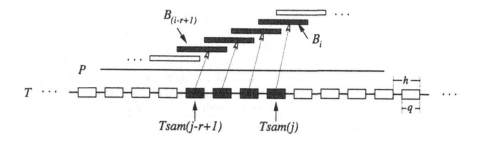

Fig. 4. Filtering phase of LSS1

Algorithm [LSS1]
1   choose appropriate values q, h, r and preprocess P
2   **for** $i := 1$ **to** m' **do** $M[i] := 0$
3   **for** $j := 1$ **to** n' **do begin**
4        $x := T[(j-1)h+1 \ldots (j-1)h+q]$
5   **if** $j > r$ **then** $x' := T[(j-r-1)h+1 \ldots (j-r-1)h+q]$
6   **else** $x' := \varepsilon$ (* empty word *)
7   **for** $i := m'$ **downto** 2 **do** $M[i] := M[i-1]$
8        $M[1] := 0$
9   **for** $i := 1$ **to** m' **do** $M[i] := M[i] + Best(B_i, x)$
10   **for** $i := r+1$ **to** m' **do** $M[i] := M[i] - Best(B_{i-r}, x')$
11   **if** $j \geq r$ **then begin**
12   **for** $i := r$ **to** m' **do begin**
13   **if** $M[i] \leq k$ **then begin**
14                   $b'_{DP} := \max\{1, (i - i_{min} - 2)h + q - l - max_{del} + 1\}$
15                   $e'_{DP} := \min\{m, (i - i_{min})h + max_{indel} + max_{del} + q - 2\}$
16                   $b_{DP} := \max\{1, (j-1)h + q - l + 1\}$
17                   $e_{DP} := \min\{n, jh + q - 1\}$
18                   $DP_{bf}(P[b'_{DP} \ldots e'_{DP}], T[b_{DP} \ldots e_{DP}])$
19   **end** (* if *)
20   **end** (* for *)
21   **end** (* if *)
22   **end** (* for *)

On line 1, once q is chosen, h may vary from q to $l - q + 1$ provided r varies accordingly. If $h = q$ is minimal, then $r = \lfloor \frac{l-q+1}{q} \rfloor$ is maximal. The empty word has been introduced on line 6 to avoid further case analysis (by definition $Best(B_i, \varepsilon) = 0$ for any i). On line 7, the values in M are shifted right (with zero extension on line 8). Then the minimum cost of aligning sample $Tsam(j)$ with a subword of block B_i is added to $M[i]$ on line 9. Since we are interested in the last r samples only, the minimum cost of aligning sample $Tsam(j - r)$ with a subword of block B_{i-r} is subtracted from $M[i]$ on line 10. To prove that $M[i] = \sum_{c=0}^{r-1} Best(B_{i-c}, Tsam(j - c))$ is an invariant of the **for**-loop on lines

3–22 is relatively easy. Finally, $M[i] \leq k$ is checked on line 13 and a dynamic programming computation is applied if necessary. Note that DP_{bf} on line 18 is a shorthand for the brute force dynamic programming approach described in the introduction.

Before proving correctness, we explain LSS1 by means of a simple example. For simplicity, let δ correspond to the Levenshtein distance. Let $l = 11$, $k = 1$, $h = 3$, $q = 2$, and $r = 3$. Let P be the string "five languages". P is covered by six blocks: $\boxed{\text{fi}}$, $\boxed{\text{five}}$, $\boxed{\text{e lan}}$, $\boxed{\text{angua}}$, $\boxed{\text{uages}}$, and $\boxed{\text{es}}$. Consider the text below, where the samples are framed.

$\boxed{\text{The}}$ l $\boxed{\text{angu}}$ ag$\boxed{\text{e}}$ h$\boxed{\text{a}}$s a$\boxed{\text{n}}$ e$\boxed{\text{ne}}$r$\boxed{\text{gy}}$ $\boxed{\text{th}}$a t $\boxed{\text{k}}$ee$\boxed{\text{p}}$s p$\boxed{\text{ac}}$e w$\boxed{\text{i}}$th ...

Table 1 shows the values of array M for $0 \leq j \leq 5$. Since $M[5] = 0 \leq 1$, where $j = 4$, DP_{bf} is applied to $P[1\ldots14]$ and $T[2\ldots13]$. Consequently, algorithm LSS1 detects that "ve language" is similar to "he language" and that "e languages" is similar to "e language ". Since P is a very short string in this small example, DP_{bf} has to be applied to the whole pattern $P = P[1\ldots14]$.

j=0	j=1	j=2	j=3	j=4	j=5
[0,0,0,0,0,0]	[2,2,2,2,2,2]	[2,3,2,4,4,4]	[2,4,4,2,5,6]	[2,4,5,3,0,5]	[2,3,5,5,4,2]

Table 1. The values of array M

Theorem 6. *LSS1 correctly solves the weighted local similarity search problem.*

Proof. Let $p = P[b'\ldots e']$ and $t = T[b\ldots e]$ be given with $edist_\delta(p,t) \leq k$ and $|t| \geq l$. In order to show correctness of LSS1, we have to show that it reports the end-points e' and e. W.l.o.g. we may assume $|t| = l$ (every string of length $\geq l$ has a suffix of length l). Let $Tsam(j - r + 1), \ldots, Tsam(j - 1), Tsam(j)$ be the last r consecutive q-samples in t. Let A be an alignment of p and t with $\delta(A) \leq k$ and let p_c be the subword of p which is aligned with $Tsam(j - c)$ for $c \in \{0, \ldots, r - 1\}$. Let A^c denote the alignment of p_c and $Tsam(j - c)$. By Theorem 5, there is an integer $d \geq r$ such that p_c is a subword of B_{d-c} for all $c \in \{0, \ldots, r - 1\}$. It follows that

$$Best(B_{d-c}, Tsam(j - c)) \leq edist_\delta(p_c, Tsam(j - c)) \leq \delta(A^c).$$

Therefore, $\displaystyle\sum_{c=0}^{r-1} Best(B_{d-c}, Tsam(j-c)) \leq \sum_{c=0}^{r-1} \delta(A^c) \leq \delta(A) \leq k.$

So DP_{bf} is applied to $P[b'_{DP}\ldots e'_{DP}]$ and $T[b_{DP}\ldots e_{DP}]$ and it remains to show:

(i) $T[b\ldots e]$ is a subword of $T[b_{DP}\ldots e_{DP}]$ and
(ii) $P[b'\ldots e']$ is a subword of $P[b'_{DP}\ldots e'_{DP}]$.

Since $T sam(j)$ is the last sample in t, we have

$$(j-1)h + q \le e \le \min\{n, jh + q - 1\}. \tag{1}$$

It is an easy consequence of (1) and $|t| = l$ that

$$b \ge \max\{1, (j-1)h + q - l + 1\}. \tag{2}$$

Hence (i) is true. In order to prove (ii), we have to show

(a) $b' \ge \max\{1, (d - i_{min} - 2)h + q - l - max_{del} + 1\}$ and
(b) $e' \le \min\{m, (d - i_{min})h + max_{indel} + max_{del} + q - 2\}$.

Choose c such that $p_c = P[b_c \ldots e_c]$ is non-empty. Since p_c is a subword of B_{d-c},

$$(d - c - i_{min} - 2)h + 1 \le b_c \le e_c \le (d - c - i_{min} - 1)h + max_{indel} + q - 1. \tag{3}$$

(a) Let A' be the subalignment of A that aligns the prefixes $P[b' \ldots b_c]$ and $T[b \ldots (j-c-1)h+1]$. Let A'_{del} and A'_{ins} be the number of deletions and insertions in A'. According to Lemma 1, $b_c - b' + 1 + A'_{ins} = (j-c-1)h + 1 - b + 1 + A'_{del}$. Thus, we infer

$$
\begin{aligned}
b' &= b_c - (j-c-1)h - 1 + b - A'_{del} + A'_{ins} \\
&\ge b_c - (j-c-1)h - 1 + (j-1)h + q - l + 1 - A'_{del} + A'_{ins} \quad (2) \\
&\ge (d - c - i_{min} - 2)h + 1 + ch + q - l - A'_{del} + A'_{ins} \quad (3) \\
&= (d - i_{min} - 2)h + 1 + q - l - A'_{del} + A'_{ins} \\
&\ge (d - i_{min} - 2)h + 1 + q - l - max_{del}.
\end{aligned}
$$

(b) Let A'' be the subalignment of A that aligns the suffixes $P[e_c \ldots e']$ and $T[(j-c-1)h + q \ldots e]$. Since $e' - e_c + 1 + A''_{ins} = e - (j-c-1)h - q + 1 + A''_{del}$, we conclude that

$$
\begin{aligned}
e' &= e - (j-c-1)h - q + 1 + e_c - 1 + A''_{del} - A''_{ins} \\
&\le jh + q - 1 - (j-c-1)h - q + e_c + A''_{del} - A''_{ins} \quad (1) \\
&\le (c+1)h - 1 + (d - c - i_{min} - 1)h + max_{indel} + q - 1 + A''_{del} - A''_{ins} \quad (3) \\
&= (d - i_{min})h + max_{indel} + q - 2 + A''_{del} - A''_{ins} \\
&\le (d - i_{min})h + max_{indel} + q - 2 + max_{del}.
\end{aligned}
$$

We next present LSS2, a variant of LSS1, which avoids the computation of $Best(B_i, x)$. Its correctness relies on Corollary 7 which again is a direct consequence of Theorem 5.

Corollary 7. *Let* $p = P[b' \ldots e']$ *and* $t = T[b \ldots e]$ *be given such that* $|t| \ge l$ *and* $edist_\delta(p, t) \le k$. *Let* $r = max_{eop} + s$ *for some* $s \in \mathbb{N}_+$. *If the sampling step* $h \le \lfloor \frac{l - q + 1}{r} \rfloor$ *satisfies* $h \ge q$, *then for any sequence of* r *consecutive* q-*samples in* t, *say* $T sam(j+1), T sam(j+2), \ldots, T sam(j+r)$, *there are numbers* $1 \le j_1 < j_2 < \ldots < j_s \le r$ *and an integer* $d \ge 0$ *such that* $T sam(j + j_c) \in Q_{d+j_c}$ *for all* $c \in \{1, \ldots, s\}$.

$$A = \begin{pmatrix} \texttt{gh-ijklmnopqrstuvwx} \\ \texttt{ghxijklxno-qrstu-wx} \end{pmatrix}$$

Fig. 5. δ = Levenshtein distance and $k = 4$

Proof. By Theorem 5, there is an integer $d \geq 0$ such that p_c occurs in B_{d+c} for all $c \in \{1, \ldots, r\}$, Now in any alignment A of p and t, $p_c \neq Tsam(j + c)$ can hold for at most max_{eop} indices, see Figure 5.

Algorithm [LSS2] Let $q, s \in \mathbb{N}_+$ be chosen such that $h \leq \lfloor \frac{l-q+1}{r} \rfloor$ satisfies $h \geq q$, where $r = max_{eop} + s$. For every q-sample $Tsam(j)$ and every block B_i, check whether $M_j[i] = \sum_{c=0}^{r-1} \varphi(Tsam(j-c) \in Q_{i-c}) \geq s$, where

$$\varphi(Tsam(j-c) \in Q_{i-c}) = \begin{cases} 1 & \text{if } Tsam(j-c) \in Q_{i-c} \\ 0 & \text{otherwise.} \end{cases}$$

If so, apply DP_{bf} to $P[b'_{DP} \ldots e'_{DP}]$ and $T[b_{DP} \ldots e_{DP}]$, where $b'_{DP}, e'_{DP}, b_{DP}, e_{DP}$ are defined as in LSS1.

In other words, LSS2 checks whether at least s out of the $r = max_{eop} + s$ consecutive samples $Tsam(j-r+1), Tsam(j-r+2), \ldots, Tsam(j)$ occur (in the same order) in the corresponding blocks $B_{i-r+1}, B_{i-r+2}, \ldots, B_i$; see Figure 4. LSS2 thus resembles algorithm LEQ [ST95].

In our previous example, LSS2 checks at every sample whether or not $s = 2$ out of $k + s = 3$ samples occur in consecutive blocks of the pattern. For instance, sample $\boxed{\text{Th}}$ does not appear in block $\boxed{\text{fi}}$, $\boxed{\text{l}}$ does not occur in $\boxed{\text{five}}$ and $\boxed{\text{ng}}$ does not appear in $\boxed{\text{o lan}}$. Thus there is no need to check whether there might be a local similarity between $\boxed{\text{five lang}}$ and $\boxed{\text{The langua}}$ because it can be concluded that there is none. On the other hand, since $\boxed{\text{l}}$ appears in $\boxed{\text{e lan}}$, $\boxed{\text{ng}}$ occurs in $\boxed{\text{angua}}$ and $\boxed{\text{ag}}$ appears in $\boxed{\text{uages}}$, LSS2 applies the same dynamic programming computation as LSS1.

3 Analysis

In order to estimate the time complexity of the algorithms LSS1 and LSS2, we first analyse (1) the preprocessing and (2) the filtering phase. To this end, let n' be the number of q-samples in T, let m' be the number of blocks in P, and let $b = h + max_{indel} + q - 1$ be the (maximum) size of a block. Note that $n' \in O(\frac{n}{h})$ and $m' \in O(\frac{m}{h})$.

(1) Table $Best(B_i, x)$ can be computed in $O(|\mathcal{A}|^q \cdot |B_i|)$ time, where $|\mathcal{A}|$ is the alphabet size, by incrementally computing values over the complete trie of q-grams. Thus the preprocessing phase of LSS1 takes $O(|\mathcal{A}|^q \cdot \frac{m}{h} \cdot b)$ time and $O(|\mathcal{A}|^q \cdot \frac{m}{h})$ space. Using standard techniques like hashing and conversion of q-grams into numbers (see for instance [ST95, Tak94]) the preprocessing phase of LSS2 takes $O(\frac{m}{h} \cdot max_{indel})$ time and space.

(2) Evaluation of one q-sample takes $O(q)$ time. Updating and checking array $M[1 \ldots m']$ for that sample requires $O(\frac{m}{h})$ time and space. It follows from $q < \frac{m}{h}$ in conjunction with the fact that T contains $O(\frac{n}{h})$ samples that the filtering phase of both LSS1 and LSS2 takes $O(\frac{mn}{h^2})$ time and $O(\frac{m}{h})$ space.

If $b\,|\mathcal{A}|^q > \frac{n}{h}$, then the preprocessing phase of LSS1 dominates the two phases, so the tables $Best(B_i, x)$ should better be computed on demand. This ensures that at most n' rather than all the $|\mathcal{A}|^q$ q-samples have to be considered. For simplicity, we assume that q is chosen so small that the filtering phase dominates the two phases.

We next show that under certain conditions, the dynamic programming phase (hence the algorithm itself) requires no more than $O(\frac{mn}{h^2})$ time on average. First, the average time complexity of the dynamic programming phase of LSS2 is analysed. To this end, suppose T is a uniformly random string over the finite alphabet $\mathcal{A}$.

Theorem 8. If $e\,l^{1+\frac{3}{s}}\,b^{1+\frac{1}{s}} \leq s\,|\mathcal{A}|^q$, then[3]

1. $Pr[M_j[i] \geq s] < b^{-1}l^{-2}$, and
2. the dynamic programming phase of LSS2 runs in $O(\frac{mn}{h^2})$ expected time.

Proof. (1) Let X_c be the random variable $\varphi(Tsam(j-c) \in Q_{i-c})$, where $0 \leq c \leq r-1$ and $r = max_{eop} + s$. So X_c takes values 0 and 1. The X_c's are independent and identically distributed because the position at which $Tsam(j-c)$ begins is beyond the last character of $Tsam(j-c-1)$. Let X be the random variable $\sum_{c=0}^{r-1} X_c$. Obviously, $X = M_j[i]$. Since there are at most $h + max_{indel}$ different q-grams in Q_{i-c}, it follows $Pr[X_c = 1] \leq \frac{h + max_{indel}}{|\mathcal{A}|^q}$. Our goal is to derive an upper bound for $Pr[X \geq s]$. Clearly,

$$Pr[X \geq s] = \sum_{t=s}^{max_{eop}+s} Pr[X = t] \leq (max_{eop} + 1)\,Pr[X = s].$$

With $(1 - Pr[X_c = 1]) \leq 1$, we first derive an upper bound for $Pr[X = s]$.

$$Pr[X = s] = \binom{max_{eop} + s}{s}\,Pr[X_c = 1]^s\,(1 - Pr[X_c = 1])^{max_{eop}}$$
$$\leq \binom{max_{eop} + s}{s}\left(\frac{h + max_{indel}}{|\mathcal{A}|^q}\right)^s.$$

It can be shown by using Stirling's formula[4] that $\binom{max_{eop}+s}{s} \leq \left(\frac{e\,(max_{eop}+s)}{s}\right)^s$. Thus,

$$Pr[X = s] \leq \left(\frac{e\,(max_{eop} + s)}{s}\right)^s\left(\frac{h + max_{indel}}{|\mathcal{A}|^q}\right)^s$$
$$= \left(\frac{e\,(max_{eop} + s)\,(h + max_{indel})}{s\,|\mathcal{A}|^q}\right)^s.$$

---

[3] e is the base of the natural logarithm.
[4] $\sqrt{2\pi n} \cdot \left(\frac{n}{e}\right)^n \leq n! \leq \sqrt{2\pi n} \cdot \left(\frac{n}{e}\right)^{n+(1/12^n)}$

Observe that $r = max_{eop} + s \leq l$ because $r \leq \frac{l-q+1}{h}$. Hence

$$Pr[X \geq s] \leq (max_{eop} + 1) \left(\frac{e\,(max_{eop} + s)\,(h + max_{indel})}{s\,|\mathcal{A}|^q} \right)^s$$

$$\leq l \left(\frac{e\,l\,b}{s\,|\mathcal{A}|^q} \right)^s.$$

In order to show $Pr[X \geq s] \leq b^{-1}l^{-2}$, it is sufficient to show $(e\,l\,b\,s^{-1}\,|\mathcal{A}|^{-q})^s \leq b^{-1}l^{-3}$. This is equivalent to $e\,l^{1+\frac{3}{s}}\,b^{1+\frac{1}{s}} \leq s\,|\mathcal{A}|^q$.

(2) We next estimate the work at $M_j[i]$ (the terminology "work at" is borrowed from [CL94]). If $M_j[i] \geq s$, then DP_{bf} is applied to $p = P[b'_{DP} \ldots e'_{DP}]$ and $t = T[b_{DP} \ldots e_{DP}]$. We have seen in the introduction that this requires $O(l \cdot |p| \cdot |t - l + 1|)$ time. Note that $|p| = l + 2h + max_{indel} + 2max_{del} - 2 \in O(l)$ and $|t - l + 1| = h \in O(b)$. Therefore, the required time is in $O(b\,l^2)$. If $M_j[i] < s$, no dynamic programming computation is needed. Hence

$$E[\text{work at } M_j[i]] = E[\text{work at } M_j[i] \mid M_j[i] \geq s] \cdot Pr[M_j[i] \geq s] \in O(1).$$

The fact that the expected total work for the dynamic programming phase is in $O(\frac{mn}{h^2})$ finally follows from $m' \in O(\frac{m}{h})$, $n' \in O(\frac{n}{h})$, and, for some constant c,

$$E[\text{total work}] \leq E[\sum_{\substack{1 \leq i \leq m' \\ 1 \leq j \leq n'}} \text{work at } M_j[i]] \leq \sum_{\substack{1 \leq i \leq m' \\ 1 \leq j \leq n'}} E[\text{work at } M_j[i]] = m' \cdot n' \cdot c$$

The condition $e\,l^{1+\frac{3}{s}}\,b^{1+\frac{1}{s}} \leq s\,|\mathcal{A}|^q$ can be interpreted as follows. The larger the block size b, the more likely an arbitrary q-sample occurs in it. Thus b must be relatively small in comparison to $|\mathcal{A}|^q$. Since b depends on h which again depends on l, l also must be relatively small compared to $|\mathcal{A}|^q$. Finally, s plays a role here as well because the larger it is, the more likely s out of $max_{eop} + s$ q-samples do not occur (in the same order) in $max_{eop} + s$ blocks (hence the permitted size of b and l also hinges on s).

So, if the parameters satisfy the above theorem, then LSS2 runs in $O(\frac{mn}{h^2})$ time. The same is true for LSS1 because it uses a stronger filter than LSS2. More precisely, if $r = max_{eop} + s$, then $\sum_{c=0}^{r-1} Best(B_{i-c}, Tsam(j - c)) \leq k$ implies $M_j[i] = \sum_{c=0}^{r-1} \varphi(Tsam(j - c) \in Q_{i-c}) \geq s$. This is because there are at most max_{eop} differences—consequently for at most max_{eop} out of $r = max_{eop} + s$ indices $Tsam(j - c) \notin Q_{i-c}$ can hold (cf. Corollary 7). The converse does not hold. Hence $Pr[\sum_{c=0}^{r-1} Best(B_{i-c}, Tsam(j - c)) \leq k] < Pr[M_j[i] \geq s]$.

There is a very important special case: If $h \in \Omega(\sqrt{m})$, then the filtering phase runs in $O(mn/\sqrt{m}^2) = O(n)$ time and the dynamic programming phase requires $O(n)$ time on the average, provided that $e\,l^{1+\frac{3}{s}}\,b^{1+\frac{1}{s}} \leq s\,|\mathcal{A}|^q$. However, $h \in \Omega(\sqrt{m})$ implies $r = max_{eop} + s \in O(l/\sqrt{m})$. So the algorithms run in linear expected time only for stringent threshold values (infrequent errors, that is).

4 Experimental results

In order to compare our algorithms with Myers' algorithm, their respective filtration efficiencies have been evaluated. The filtration efficiency quantifies the number of entries in matrices that have to be computed in the dynamic programming phase. A filtration efficiency of 90% means that only 10% of the $l \cdot m \cdot (n - l + 1)$ entries have to be computed (cf. introduction). In the experiments, both P and T were uniformly random strings. In the first experiment we used the Levenshtein distance and an alphabet of size 20. The length of the text T is 1 000 000 and the minimum permitted length of a local similarity is $l = 40$. In our first test series pattern P is 400 characters long and the mismatch ratio k/l varies from 0% to 50%. Since the time and space complexity of the preprocessing phase of both Myers' algorithm and LSS1 depends on $|\mathcal{A}|^q$, q should be relatively small. Here, $q = 3$ seems to be reasonable. Table 2 shows the results of the first test series. Myers algorithm, LSS1, and LSS2 achieve a filtration efficiency $\geq$ 95% for mismatch ratios up to 25%, 42.5%, and 35%, respectively.

		MYERS	LSS1	LSS2				
	k	$q = 3$	$q = 3$		q	h	r	s
	≤ 7	100.0	100.0	100.0	3	4	9	2
20 %	8	100.0	100.0	100.0	3	3	12	4
	9	99.8	100.0	100.0	3	3	12	3
	10	98.2	100.0	100.0	3	3	12	2
	11	89.7	100.0	100.0	2	3	13	2
30 %	12	61.6	100.0	100.0	2	2	19	7
	13	15.9	100.0	99.9	2	2	19	6
	14	2.2	100.0	98.8	2	2	19	5
	15	0.2	100.0	90.7	2	2	19	4
40 %	16	0.0	99.5	59.2	2	2	19	3
	17	0.0	96.4	14.9	2	2	19	2
	18	0.0	82.6	0.7	1	2	20	2
	19	0.0	50.0	0.0	1	1	40	21
50 %	20	0.0	15.5	0.0	1	1	40	20

Table 2. Filtration efficiencies

Finding good values for h and r (resp. s) is a rather delicate matter because large h result in small r which might lead to a bad filtration efficiency. On the other hand, large r result in small h, so the filtration phase might run slower than necessary. The values of h and r (resp. s) were determined according to the following heuristic: First set[5] $r' = k + 2$ and compute $h' = \lfloor \frac{l - q + 1}{r'} \rfloor$. If $h' \geq q$, then set $h = h'$ and $r = \lfloor \frac{l - q + 1}{h} \rfloor$ (resp. $s = r - k$). For instance, for $k = 8$, this yields $h = 3$ and $r = 12$ (resp. $s = 4$). In case of LSS1, if $h' < q$, then we may

---

[5] Intuitively, the more errors are allowed, the more q-samples should be considered.

set $h = q$ and $r = \lfloor \frac{l-q+1}{q} \rfloor$ (so for $k = 20$, this still yields $h = 3$ and $r = 12$). For LSS2, the requirement $h \geq q$ can only be met by decreasing q. This is the reason why our experiments show that applying LSS2 with $q = 2$ results in the same filtration efficiency as for higher values of q. For LSS2, the values of q, h, r, and s are tabulated above.

In the second test series of the first experiment, pattern length m varies. Table 3 shows the maximum k for which the filtration efficiency exceeds 95%. In Myers' algorithm, k decreases as m increases, whereas LSS1 and LSS2 show a much better behavior. This result confirms our theoretical considerations.

m	100	200	300	400	500	600	1000	2000	3000	4000
Myers ($q = 3$)	14	12	11	10	10	9	8	7	5	4
LSS1 ($q = 3$)	17	17	17	17	17	17	17	17	17	17
LSS2 ($q = 2$)	14	14	14	14	14	14	14	14	14	14

Table 3. Maximum k with filtration efficiency $\geq 95\%$

In the second experiment, we made a similar test series for the DNA alphabet $\mathcal{A} = \{A, C, G, T\}$ and the symmetric cost function δ defined in Table 4. δ is taken from [Kur96] and it is called transversion/transition weight function there.

δ	ε	A	C	G	T
ε					
A	3	0			
C	3	2	0		
G	3	1	2	0	
T	3	2	1	2	0

Table 4. Cost function δ

Bases A and G are called purine, and bases C and T are called pyrimidine. Kurtz [Kur96] writes: "The transversion/transition weight function reflects the biological fact that a purine/purine and a pyrimidine/pyrimidine replacement is much more likely to occur than a purine/pyrimidine replacement. Moreover, it takes into account that a deletion or an insertion of a base occurs more seldom." Table 5 shows the results for $|T| = 1\,000\,000$ and $l = 40$.

m	60	80	100	120	140	160	180	200	220	240
Myers ($q = 4$)	6	5	4	3	2	2	1	1	1	0
LSS1 ($q = 4$)	14	14	14	14	14	14	14	14	14	14
LSS2 ($q = 2$)	8	8	8	8	8	8	8	8	8	8

Table 5. Maximum k with filtration efficiency $\geq 95\%$

Acknowledgements: I am indebted to Michael Jandrey who provided the C implementation of the algorithms and suggested many improvements. Thanks go to Stefan Kurtz for useful comments on a previous version of the paper and to Gene Myers for fruitful discussions on approximate string searching. Furthermore, I am grateful to Peter Bruns for a discussion on the probability of the event $X_c = 1$.

References

[CL92]   W.I. Chang and J. Lampe. Theoretical and Empirical Comparisons of Approximate String Matching Algorithms. In *Proc. CPM*, pages 175–184. LNCS **644**, 1992.

[CL94]   W.I. Chang and E.L. Lawler. Sublinear Approximate String Matching and Biological Applications. *Algorithmica*, 12(4/5):327–344, 1994.

[CM94]   W.I. Chang and T.G. Marr. Approximate String Matching and Local Similarity. In *Proc. CPM*, pages 259–273. LNCS **807**, 1994.

[GG88]   Z. Galil and R. Giancarlo. Data Structures and Algorithms for Approximate String Matching. *Journal of Complexity*, 4:33–72, 1988.

[GP90]   Z. Galil and K. Park. An Improved Algorithm for Approximate String Matching. *SIAM Journal on Computing*, 19(6):989–999, 1990.

[Kur96]   S. Kurtz. Approximate String Searching under Weighted Edit Distance. In *Proc. Third South American Workshop on String Processing*, pages 156–170. Carlton University Press, 1996.

[LV88]   G.M. Landau and U. Vishkin. Fast String Matching with k Differences. *Journal of Computer and Systems Sciences*, 37:63–78, 1988.

[Mye94a] E.W. Myers. A Sublinear Algorithm for Approximate Keyword Searching. *Algorithmica*, 12(4/5):345–374, 1994.

[Mye94b] E.W. Myers. Algorithmic Advances for Searching Biosequence Databases. In *Computational Methods in Genome Research*, pages 121–135. Plenum Press, 1994.

[Sel80]   P.H. Sellers. The Theory and Computation of Evolutionary Distances: Pattern Recognition. *Journal of Algorithms*, 1:359–373, 1980.

[ST95]   E. Sutinen and J. Tarhio. On Using q-Gram Locations in Approximate String Matching. In *Proc. ESA*, pages 327–340. LNCS **979**, 1995.

[ST96]   E. Sutinen and J. Tarhio. Filtration with q-Samples in Approximate Matching. In *Proc. CPM*, pages 50–63. LNCS **1075**, 1996.

[Tak94]   T. Takaoka. Approximate Pattern Matching with Samples. In *Proc. ISAAC*, pages 234–242. LNCS **834**, 1994.

[TU93]   J. Tarhio and E. Ukkonen. Approximate Boyer-Moore String Matching. *SIAM Journal on Computing*, 22(2):243–260, 1993.

[Ukk85]   E. Ukkonen. Finding Approximate Patterns in Strings. *Journal of Algorithms*, 6:132–137, 1985.

[Ukk92]   E. Ukkonen. Approximate String-Matching with q-Grams and Maximal Matches. *Theoretical Computer Science*, 92(1):191–211, 1992.

Trie-Based Data Structures for Sequence Assembly

Ting Chen
Steven S. Skiena*
Department of Computer Science
State University of New York
Stony Brook, NY 11794-4400
{tichen—skiena}@cs.sunysb.edu

Abstract. We investigate the application of trie-based data structures, *suffix trees* and *suffix arrays* in the problem of overlap detection in fragment assembly. Both data structures are theoretically and experimentally analyzed on speed and space. By using heuristics, we can greatly reduce the calls to the time-consuming dynamic programming, and have improved the speed of overlap detection up to 1,000 times with high accuracy in our collaborative DNA sequencing with Brookhaven National Laboratory. We also studied the problem of approximating maximum space savings in tries structures for *unification factoring* in logic programming, which is proved to be hard.

1 Introduction

Trie-based data structures for strings have proven themselves in a wide variety of text-searching and biological applications. Several distinct string data structures have been developed, including suffix trees and suffix arrays, motivated by tradeoffs between construction time and space, search time, virtual memory performance, and the complexity of implementation.

In this paper, we consider two different applications for string data structures – (1) fragment assembly for DNA sequences and (2) unification factoring for optimizing logic programs. For the biological application, we present experimental results on the performance of three different data structures, dynamic suffix trees, static suffix trees and suffix arrays, for fast fragment assembly. These are a product of our work with Dr. William Studier's genome sequencing group at Brookhaven National Laboratory to build an assembler for sequencing the one-megabase genome of *Borrelia Burgdorferi*, the bacterium which causes Lyme disease. For the logic programming application, we demonstrate that, surprisingly, the complexity of optimizing unification factoring for datalog programs is considerably higher than that for more general Prolog programs. This is a product of our work with the XSB logic programming group at Stony Brook.

Although the applications are quite different, a common theme runs through them. Both require space-efficient data structures for strings. Sequence assembly requires fast substring search on large sets of strings. Unification factoring requires minimum size tries capable of representing a fixed set of strings. Indeed, our work

---

* Supported by ONR award 400x116yip01 and NSF Grant CCR-9625669.

with unification factoring suggests a new direction for reducing the size of trie and suffix-tree data structures for other string applications.

Our specific results on fragment assembly include:

- A fast sequence assembler based on exact-match overlap detection, which proves capable of detecting the overlaps of 3,800 fragments of the *Borrelia Burgdorferi* sequence within ten minutes on a Sparc1000. By comparison, the Brookhaven group's previous assembler program takes approximately fifteen hours on the same data. Indeed, our assembler proved fast enough to eliminate the need for incremental fragment assembly, the original problem with which we began this work. Our times are certainly comparable with such state-of-the-art assemblers as the TIGR assembler [15], which assembled the 24,304 fragments of the bacterium *Haemophilus influenzae* in 30 hours.
- A careful experimental study of the impact of exact-match length on the accuracy of overlap detection, which demonstrates that over a wide range of sequencing error rates (including those realized by the Brookhaven group) exact-match suffices for accurate assembly. This study compares the overlap of unprocessed ABI reads (fragments) generated our program with the overlap graph induced in the final assembly on clean data by the biologists. Using a transitive overlap recovery strategy, our exact-matching program misses only *nine* of 4,320 edges on 35kb of *Borrelia* data compared with an exhaustive pairwise Smith-Waterman computation, taking over 1,000 times as long.
- An experimental comparison between three data structures for fast overlap detection – static suffix trees, dynamic suffix trees, and suffix arrays, which shows suffix arrays to be a clear winner over all data structures in both construction and traversal time, and space.

Unification in logic programs is performed by constructing a trie from the arguments of the rule heads and performing a depth-first search of this trie against a specific goal. The time complexity of this operation is a function of the number of edges in the trie. An *open goal* contains all distinct variables in its arguments, like $p(X, Y, Z)$ and hence matches everything. In unifying an open goal against a set of clause heads, each symbol in all the clause heads will be bound to some variable. By doing a depth-first traversal of this minimum edge trie, we minimize the number of operations performed in unifying the goal with all of the clause heads.

For a set of m rule heads, each with n arguments, this trie can range in size from $n + m$ to $n \cdot m$, depending upon the length of common prefixes among the rule heads. Since we have all the rule heads at compilation time, we can exploit our freedom to reorder the character positions in the trie in order to reduce its size. Instead of the root node always representing the first argument in the trie, we can choose to have it represent the third argument.

Consider the following example consisting of four strings: $s_1 = aaa$, $s_2 = baa$, $s_3 = cbb$, and $s_4 = dbb$. A trie constructed according to the original string position order $(1, 2, 3)$ uses a total of 12 edges, as shown in Figure 1. However, by permuting the character order to $(2, 3, 1)$, we obtain a trie with only 8 edges. Note that different permutations may be used along various paths of the same trie, since we assume that each internal node (and not level) contains a specification of the next probe position.

The problem of *unification factoring* is to use this freedom to build a minimize size trie for the set of rule heads. Beyond the context of unification, this suggests a new idea for reducing the space needed for trie-like data structures for many off-line string search applications, since there is no reason to compare characters in left-to-right order when the entire string sits in random-access memory. Any root-to-leaf path down the trie defines the full string with some permutation of its characters. The encoding becomes unambiguous when each internal node contains the index of the next character position to be probed.

Note that there is a potential for enormous space reduction even for tries of highly structured strings. For example, consider the trie of the following n binary strings, each m characters long. The first $\lg n$ characters of each of the ith string will represent i written in binary, while the last $n - \lg n$ characters will be all the same character. Building a conventional search trie on these strings will use

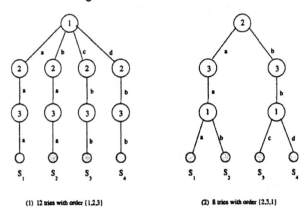

(1) 12 tries with order {1,2,3}   (2) 8 tries with order {2,3,1}

Fig. 1. Two different tries for the same set of strings.

$(m-\lg n)n+2n-1$ edges and quadratic space. However, by permuting the character order to as to probe first from position $\lg n + 1$ yields a linear size trie with only $m - \lg n + 2n - 1$ edges.

We investigate approximation algorithms for minimizing the size of such permuted order tries. Unfortunately, our main result is a negative one, that it is hard to approximate the savings of such a construction to within a polynomial factor. Still, we believe that heuristics for constructing such data structures can have interesting behavior in practice.

Our paper is organized as follow. Section 2 provides an introduction to the problem of fragment assembly for DNA sequencing. Section 3 presents the ideas behind our assembler, and experimental results of its performance. Section 4 provides an introduction to unification factoring for logic programming, while Section 5 presents our applicable inapproximatibility results.

2 Overlap Detection for Fragment Assembly

The fragment assembly problem has been characterized as finding a shortest common super-string of a set of fragments within a certain assumed error rate ϵ. The common approach to this problem is to divide assembly into three stages: *overlap detection, layout generation,* and *consensus sequence construction* [12, 8]. In overlap detection, every fragment has to be compared against all other fragments for similar substrings which define possible overlaps between them. There may exist conflicts among these overlap relations, to be resolved in the layout stage, which determines the *orientation* of each fragment (i.e. whether it belongs to the upper or lower strand of the molecule) selects the subset of overlaps which most reasonably reflect the relations of their physical maps in the genome to determine their *ordering* and also approximate locations. Finally, the consensus stage builds the multiple-alignment of all regions where two or more fragments overlap with each other, and generates a single consensus DNA sequence.

Figure 2 presents the flow of control of Brookhaven's directed primer walking approach to DNA sequencing, being used in the *Borrelia Burgdorferi* project. Although their technique differs somewhat from conventional shotgun sequencing, the overlap detection problem is identical, and forms the focus of our work.

The result of overlap detection is to generate an *overlap graph,* where vertices correspond to fragments, with edges between pairs of fragments which overlap. Edge weights can be used to measure the confidence of detected overlaps. Overlap

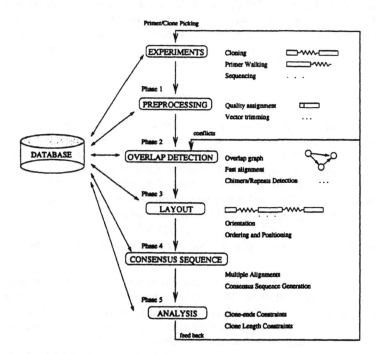

Fig. 2. The fragment assembly process for directed primer walking.

detection is the most time-consuming part of typical assembly programs. Given n fragments, there are $n(n+1)/2$ overlap candidates to consider. The Smith-Waterman dynamic programming [16] algorithm takes $O(l^2)$ time to detect whether two l-length fragments truly overlap.

For small-size sequencing projects (say cosmid level), where $n \approx 500$ and $L \approx 400$, a program with $O(n^2 l^2)$ complexity is feasible, taking several hours. However, for megabase genome-level sequencing projects, such as $Borrelia$, where $n \approx 20,000$, faster algorithms are needed.

The problem of fragment assembly has received considerable attention – see [12] for a detailed survey. Kececiolglu and Myers [8] have shown that given an assumption of maximum error rate ϵ, the edit distance of two fragments can be solved in $O(\epsilon l^2)$ time by aligning suffixes incrementally. For fragments of total length nl, the overlap detection can be done in time $O(\epsilon n^2 l^2)$. If the error rate ϵ is low, i.e. 2%, this algorithm will be 50 times faster than the complete dynamic programming. In a large shotgun sequencing project of a complete genome (1,830,137 base pairs) from the bacterium $Haemophilus$ $influenzae$, the TIGR assembler [15], assembles 24,304 fragments in 30 hours. It builds a table of all 10 base pairs oligonucleotide subsequences to generate a list of potential fragment overlaps. Then it uses a fast initial comparison of fragments (similar to BLAST [1]) to eliminate false overlaps in the list before applying the Smith-Waterman dynamic programming algorithm. Phrap [7] compares pairwise fragments by an efficient implementation of the Smith-Waterman algorithm called SWAT, which is claimed to be ten times faster than BLAST. SWAT uses recursion and word-packing to search similarities between two fragments and stores the alignment information for the significant matches. Then based on one or more matching words found, it scores two fragments within a constrained bands of the Smith-Waterman matrix.

2.1 Exact Matches and Overlap Reconstruction

In fact, for real sequencing projects, there are only $O(n)$ true overlaps, because the typical shotgun sequencing strategy uses roughly six-times genome coverage. Thus on average, each fragment will physically overlap only a constant number of other fragments. One approach to avoiding quadratic behavior is to filter the comparisons by search for all exact matches of a certain length k (say $k = 14$) as a threshold to quickly reject many non-overlapping pairs of fragments. For two random l-length fragments, the probability that they do not share any k exact matches is $(1 - 1/4^k)^l \approx e^{-l^2/4^k}$, which is 99.905% for $k = 14$ and $l = 500$, so false matches are rare. This idea implies a speedup more than 1,000 times using 14 exact matches.

The potential danger is that sequencing errors may render exact matching too unreliable for overlap detection. We suggest a new strategy to reconstruct some missed overlaps by the transitive relations between three fragments f_i, f_j, f_t: if both f_i and f_j are overlapped with f_t and their alignments on f_t are overlapped, then f_i and f_j are likely overlapped with each other. There are two cases shown in Figure 3. The first case is obvious. The second one can greatly reduce the conflicts happened

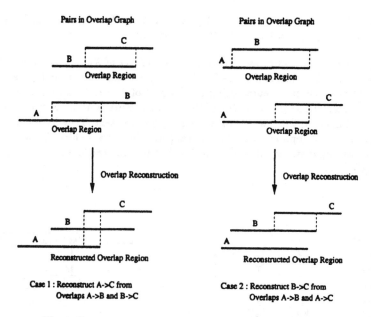

Fig. 3. Reconstruct missed overlaps from transitive relation.

in the *Layout* phase; layout algorithms such as finding maximum weight path may have trouble in linking A, B and C into a single path.

Another main contribution of this paper is a careful experimental analysis of the accuracy of exact-match filtering for both real and simulated data. We investigate the tradeoff between the exact-match length k and the accuracy to be achieved. The larger the value of k, the more likely we will miss some real overlaps, although it is faster and more likely the candidates we find will be true overlaps.

The sequencing error rate various from laboratory to laboratory, with about 2% per base error rates being typical. The end of fragments (beyond 350 base pairs) have higher error rate than in the middle (between 50 and 350 bps). For two fragments whose physical locations on the genome overlap in a region of 60 nucleotides, a 2% error rate yields an average of less than 3 errors of this region. Thus there is a very high probability that their must exist an exact match of length ≥ 14. We find that exact matching works well even at higher error rates: most of the missed overlaps can be reconstructed using a third fragment with a transitive relation, because of the redundancy in sequencing coverage. Moreover, the error rates have been steadily reduced with the development of sequencing techniques, and the exact matches strategy will be more and more powerful in the sequencing.

2.2 Suffix Trees and Suffix Arrays

In this paper, we evaluated both suffix trees [3] and suffix arrays [11] data structures for overlap detection. Suffix arrays show a significant advantage over suffix trees, not only on space but also on speed. We review these data structures. For both, we are given a text string $X = x_1 x_2 x_3 ... x_n$, where each x_i is a member of an alphabet

Σ, and seek to preprocess X such that given a pattern $P = p_1 p_2 p_3 ... p_m$, $(p_i \in \Sigma)$, the set $\{i : x_i ... x_{i+m-1} = P\}$ can be found efficiently.

- *Suffix Trees* – the compressed trie for all the suffixes of $X_1, ..., X_n$, where $X_i = x_i x_{i+1} ... x_n$. A sample suffix tree for the string *aabbaab* is shown in Figure 4. All the edges of the tree, representing a substring of X, can be implemented by a pair of pointers, so the total size of the data structure is linear. Suffix trees can be built in $O(m)$ time, where m is the length of the text. They have a memory requirement of approximately $17m$ bytes [11], which is a problem for large texts (such as megabase sequencing projects), but permit searching for a string p in $O(|p|)$ time.

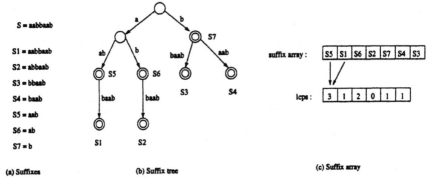

Fig. 4. Suffix tree and suffix array for *aabbaab*.

- *Suffix Array* – basically a sorted list of all the suffixes of X. If it is coupled with an array of the *longest common prefixes (lcps)* of adjacent elements in the suffix array, string searches can be answered quickly using binary search. Suffix arrays can be built in expected $O(m)$ time, where m is the length of the text. They have a memory requirement of only $6m$ bytes, and support searching for a string p in $O(|p| + \lg m)$ time.

There is no significant complexity difference in building these two data structures. Although the computation of a suffix array consists of sorting M strings, there exist $O(M)$ construction algorithms. One idea is to construct a suffix tree, then build the array with a traversal of the tree. In practice, we can do it in expected linear time via bucket-sort. The best known encoding of technique for the suffix tree requires approximately 17 bytes per index (symbol) while suffix array needs only 4 bytes for identifying each suffix and 2 bytes for each *longest common prefix*. For a very long string or multiple strings, those numbers will be doubled. Both data structures need only constant space for each index, but in the application of long strings, suffix array will be much more efficient than suffix tree: the whole suffix array is more likely to stay in the memory instead of swapping into the secondary storage. Despite the advantage of suffix trees to search a p length pattern in $O(p)$ time, experiments done by Myers [11] show that for 100,000 characters text files, suffix arrays are at least as fast as suffix trees.

A second idea to avoid the $O(l^2)$ expense of a full Smith-Waterman computation is to use a linear or super-linear heuristic to compare two fragments, such as FASTA [13] and BLAST [1]. We use a FASTA-like approach in our implementation.

3 Exact Matching for Fast Sequence Assembly

3.1 Implementation Details for Data Structures

We built two different programs for finding overlaps among DNA sequences, which are strings consisting of symbols a, g, c, t, n, where a, g, c, t are nucleotides and n means unknown.

One suffix tree is constructed by a linear time algorithm and contains each fragment and its reverse complement. To search all the common substrings at least length k, we traverse the tree, and for any node whose distance to the root (as the number of text symbols) is larger than and equal to k, we return the subtrees whose leaf-suffixes share a common prefix at least k.

There are two distinct ways to implement suffix trees, which differ on how the out-edges of a node are encoded. One option is to fully encode the out-edges as a vector which has 5 pointers for the DNA fragments. Let M be the number of suffixes and I as the number of nodes in the suffix tree. This structure requires $4M + 32I$ bytes, consisting of 20 bytes for each out-edges vector, 6 bytes for an arc(or an in-edge), 4 bytes for each suffix, 2 bytes for the distance from a node to the root, and 4 bytes for each suffix link. A second encoding technique dynamically allocates space for nonempty out-edges only. It requires only $4M + 16I$ bytes.

The *suffix array (sa)* and the *longest common prefixes (lcps)* array can be constructed in expected linear time. To find all the fragments sharing common substrings at least length k, we can traverse the *lcps* array and return all the intervals $[l, r]$ satisfying $lcps[l-1] < k$, $lcps[r+1] < k$, and for any $l \le i \le r$, $lcps[i] \ge k$. It is obvious that for any such an interval $[l, r]$ in *lcps*, the suffixes of $sa[l]$, $sa[l+1]$, ..., $sa[r]$, $sa[r+1]$ in suffix array sa share a common prefix at least length k. In Figure 4, for $k = 1$, the interval $[1, 3]$ in *lcps* satisfies that suffixes S_5, S_1, S_6, and S_2 share the common longest prefix a.

The suffix array can be implemented as follows. We first encode the fragments into long integers. For example, every 6 indexes can be encoded into a number between 0 and $5^6 - 1$, and this simple coding technique can speed up the whole construction of suffix array. Then we bucket-sort all the suffixes and quick-sort each bucket. The *longest common prefixes* array can be quickly computed within each bucket by comparing the encoded integers. The structure requires only 6 bytes for each suffix in the sorted array and *lcps* and 4 bytes for auxiliary encoding, for a total of $10M$ bytes, where M is the number of suffixes.

We evaluated our overlap detection on the following data sets:

1. Edited shotgun sequencing data from a 35kb cosmid of *Borrelia* from Brookhaven, consisting of 448 fragments, with a total length of 187,105 base pairs.
2. Raw shotgun sequencing data from the same cosmid, with total length of 189,286 base pairs and about 5% errors.
3. Simulated shotgun fragments from the cosmid of (1), containing 610 fragments and 246,479 base pairs, with 2% errors.

4. Same fragments as (3) but with 5% errors.

5. Same fragments as (3) but with 7% errors.

6. Same fragments as (3) but with 10% errors.

7. Shotgun sequencing data from the full *Borrelia Burgdorferi* sequencing project at Brookhaven National Laboratory. It consists of 4,612 fragments totaling 2,032,740 base pairs with an unknown error rate estimated at 2-5%.

Test sets (2) (3) (4) and (5) were generated by the shotgun simulation program Genfrag [6] which randomly splices the 35k cosmid DNA sequence into 610 shot gun fragments, totaling 7 times coverage and randomly generate 2%, 5%, 7% and 10% errors (insertion, deletion and mutation) into the fragments.

| | | | Space (Bytes/Base Pair) | | |
| | | | Suffix Trees | | |
Data set	Fragments	Base pairs	Dynamic	Vector	Suffix Arrays
(1)Edited 35k	448	187,105	47.5	68.5	20
(2)Raw 35k	448	189,296	67.0	113.8	20
(3)Simulated (2% errors)	610	246,318	65.6	112.7	20
(4)Simulated (5% errors)	610	246,476	69.3	117.3	20
(5)Simulated (7% errors)	610	246,388	68.5	117.1	20
(6)Simulated (10% error)	610	246,373	67.2	109.3	20
(7)BNL Project	4,612	2,032,738	62.6	105.4	20

| | | | Construction Time (Seconds) | | |
| | | | Suffix Trees | | |
Data set	Fragments	Base pairs	Dynamic	Vector	Suffix Arrays
(1)Edited 35k	448	187,105	40	31	51
(2)Raw 35k	448	189,296	40	36	25
(3)Simulated (2% errors)	610	246,318	38	38	24
(4)Simulated (5% errors)	610	246,476	50	50	34
(5)Simulated (7% errors)	610	246,388	39	41	25
(6)Simulated (10% errors)	610	246,373	38	39	24
(7)BNL Project	4,612	2,032,738	443	481	334

| | | | Traversal Time (Seconds) | | |
| | | | Suffix Trees | | |
Data set	Fragments	Base pairs	Dynamic	Vector	Suffix Arrays
(1)Edited 35k	448	187,105	7	2.8	0.22
(2)Raw 35k	448	189,296	12	4.4	0.22
(3)Simulated (2% errors)	610	246,318	14	5.8	0.28
(4)Simulated (5% errors)	610	246,476	15	6.0	0.31
(5)Simulated (7% errors)	610	246,388	16	6.0	0.29
(6)Simulated (10% errors)	610	246,373	15	5.9	0.29
(7)BNL Project	4,612	2,032,738	118	47.1	2.47

Table 1. Comparing Suffix Tree and Suffix Array time and space performance on sequence data sets

Table 1 summarizes timing and space experiments on each of the data sets, running on Sparc100 Workstation with 512 Mb RAM. The length of each DNA fragment ranges from 100 to 1000 base pairs. Each program was run twice and the timings averaged to compensate for system load. The traversal time for each data structure is important because it is proportional to the time for exhaustive search for exact matches. Each fragment actually represents two DNA sequences: both itself and its reverse complement. Thus the actual number of text symbols are the double of the *base pairs (bps)*.

Table 1 shows that the dynamic suffix trees save about half the space of the vector suffix trees, and its constructions time is competitive. The reason it is slower to traverse is because the out-edges of a node have to be decoded before its children are visited. The suffix arrays are much more efficient in space, and they will beat the suffix trees in big sequencing projects. The construction time of suffix arrays is better than the suffix trees. The only exception on data (1) is because the edited fragments share very long common substrings and make the computation of *lcps* harder. Another advantage of the suffix arrays is that it is fast to traverse. The traversing time measures the expense to perform the exact matching. Its linear structure makes this work much easier and faster.

3.2 Accuracy

The accuracy of exact matching strategy is measured by *Sensitivity* and *Specificity*, defined as follows.

$$Sensitivity = \frac{\#TrueOverlapsDetected}{\#TrueOverlaps}$$

$$Specificity = \frac{\#TrueOverlapsDetected}{\#DetectedOverlaps}$$

A high *Sensitivity* program will rarely miss any true overlaps, and a high *Specificity* program will rarely claim non-overlaps pairs to be overlaps. Generally, a good strategy must be high in both numbers.

Figure 5 shows the *Sensitivity* under different error rates, which calculates how many actual overlaps are detected by k length exact matches. The first curve in each plot gives the number of actual overlaps with different length thresholds, from 10 to 40. The four curves under it from top to bottom correspond to the number of overlaps found in different data sets with errors 2%, 5%, 7% and 10% respectively. Whenever the sequencing error rate is less than 5%, we rarely miss true overlaps, even for $k = 20$, but with the errors of 10%, $k = 15$ misses 15% of the overlaps with overlap length at least 40. This figure suggests that with high error rate fragments (> 7%), smaller match lengths perform better. With low error rates ($\approx 2\%$), there is no significant difference between a range of ks, so we can select the length which gives us the best running time. Fortunately, most of the sequencing data we have seen falls within the low error rates, so we have the freedom to make the tradeoff.

Table 2 shows how many of the candidate overlaps are detected, and how many of them represent true overlaps (Specificity). A small match-length k will preserve most of the true overlaps but contain many false overlaps, while a large k identifies few false overlaps but misses some true overlaps. In general, the number of true overlaps in the candidates reflects the potential accuracy of k-match search, while the number of false overlaps in candidates measures the amount of work to identify them. Ideally, we seek a value of k with the most candidates true and few true overlaps missed.

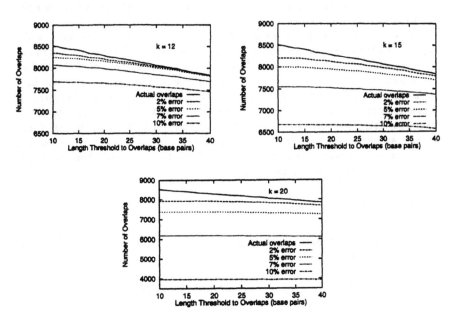

Fig. 5. Number of Detected Overlaps when k=12, 15, 20 under errors 2%, 5%, 7%, 10%.

	2% Error		5% Error		7% Error		10% Error	
	True ovlps	Spec	True ovlps	Spec	True overlaps	Spec	True ovlps	Spec
k	/Candidates	%	/Candidates	%	/Candidates	%	/Candidates	%
12	8,346/50,968	16.4	8,252/41,984	19.7	8,074/38,386	21.0	7,698/32,744	23.5
13	8,296/22,304	37.2	8,172/18,886	43.3	7,902/17,332	45.6	7,386/14,920	49.5
14	8,250/12,830	64.3	8,084/11,686	69.2	7,732/10,672	72.5	7,084/9,538	74.3
15	8,202/9,628	85.2	8,000/9,108	87.8	7,552/8,496	88.9	6,672/7,424	89.9
16	8,150/8,602	94.7	7,876/8,166	96.4	7,344/7,630	96.3	6,198/6,414	96.6
17	8,094/8,352	96.9	7,750/7,860	98.6	7,092/7,204	98.4	5,750/5,812	98.9
18	8,032/8,036	100.0	7,642/7,650	99.9	6,816/6,846	99.6	5,200/5,222	99.6
19	7,994/7,994	100.0	7,502/7,508	100.0	6,536/6,538	100.0	4,572/4,578	99.9
20	7,926/7,926	100.0	7,380/7,380	100.0	6,190/6,190	100.0	3,956/3,956	100.0

Table 2. Specificity of overlap detection on Borrelia sequence with simulated errors.

3.3 Performance

In addition to the simulated fragments, we have run our program on raw ABI machine sequencing data, after only vector trimming. Our program picks the candidates with at least one length-k exact-match, and then evaluates the quality of overlaps by FASTA-like dynamic programming and retain them with at least 25 matches and at most 20% errors. Thus we use a procedure to reconstruct overlaps without a k length common substrings. Later, we apply the strategy of overlap reconstruction described before to detected missed overlaps. We compared the sensitivity, the detected overlaps over all overlaps, the specificity, and the speed of our reconstructed data with that of performing a full Smith-Waterman on all pairs of sequences in Table 3.

Table 3 shows that the overwhelming majority of the overlaps can reconstructed using large exact-matches. Further, all the overlaps we claim are correct. This is very promising because we can reconstruct almost all the overlaps we could if we

	kmer+DP			Reconstruct			Time
k	Ovlps	Sen(%)	Spe(%)	Ovlps	Sen(%)	Spe(%)	Sec.
12	3,963	91.6	100.0	4,031	93.3	100.0	567
13	3,935	91.1	100.0	4,031	93.3	100.0	146
14	3,862	89.9	100.0	4,025	93.3	100.0	98
15	3,834	88.7	100.0	4,008	92.8	100.0	79
16	3,794	87.8	100.0	4,008	92.8	100.0	70
17	3,746	86.7	100.0	4,008	92.8	100.0	67
18	3,683	85.3	100.0	4,003	92.7	100.0	65
19	3,601	83.8	100.0	3,998	92.5	100.0	64
20	3,577	82.8	100.0	3,998	92.5	100.0	63
DP	4,040	93.5	100.0	4,040	93.5	100.0	151,200

Table 3. Sensitivity, Specificity and Speed of k-mer overlap detection on raw 35kb Borrelia sequence, comparing to 4,320 true overlaps (length $\geq$ 25).

had run Smith-Waterman on all pairs of fragments. For example, for $k = 12$, we can reconstruct 68 overlaps and miss only 9 (0.2%) overlaps compared to the overlaps found by fully dynamic programming (DP). Even for $k = 20$, we successfully reconstruct 521 overlaps with only 42 (1.0%) missed. Considering the dynamic programming requiring more than 40 hours to do the job, that we can achieve 99% accuracy in about a minute by a simple transitive relation strategy is significant. We can reconstruct 143 overlaps and miss only 15 (0.4%) overlaps in 2 minutes (in Sparc1000) compared to the overlaps found by fully dynamic programming (DP) run for 40 hours.

The gap of 280 overlaps between dynamic programming and true overlaps, is because the overlap regions between fragments are either having low scores or contain more than 20% errors.

Among all those missed overlaps when $k = 14$, only one overlap proves significant, without which we break the 35k cosmid into two disjoint blocks. This overlap has high errors, even so we are able to reconstruct it by performing block ends comparison.

4 Introduction to Unification Factoring

Unification is the basic computational mechanism in Prolog, and other logic programming languages. A Prolog program consists of an ordered list of rules, where each rule consists of a head with an associated action whenever that rule head matches or unifies with the current computation.

An execution of a Prolog program starts by specifying a goal, say $p(a, X, Y)$, where a is a constant and X and Y are variables. The system then systematically matches the head of the goal with the head of all rules which can be *unified* with the goal. Unification means binding the variables with the constants if it is possible to match them. For example, consider the set of rule heads $p(a, b, c)$, $p(a, b, d)$, $p(a, c, c)$, and $p(b, a, c)$. The goal $p(a, X, Y)$ would match all of the first three rules, since X and Y can be bound to match the extra characters. The goal $p(a, X, X)$ would only match the third rule, since the variable bound to the second and third position must be the same.

Unification factoring for logic programming was first considered in by Dawson, et.al. [4, 5] who give a dynamic programming algorithm for optimizing the trie size when the strings have an imposed left-right order, as is the case in Prolog programs. Experimental results showed that unification factoring substantially sped up typical Prolog programs. For datalog programs, i.e. Prolog programs without variables, the problem of minimizing trie size was shown to be NP-complete. Lin [9] showed that an augmented version of the trie minimization problem was even harder.

Below, we consider the question of approximation algorithms for unification factoring, i.e. producing a small size trie for a given set of strings. We prove a surprising but negative result, that it is impossible to approximate minimum size trie to within a polynomial factor unless $P = NP$. Along the way, we prove the inapproximatibility of a new variant of subgraph isomorphism.

5 Inapproximatibility Results for Unification Factoring

We will relate the problem of unification factoring to the *edge-maximum complete bipartite subgraph* problem. A complete bipartite subgraph defines two disjoint sets of vertices V_1 and V_2, $V_1, V_2 \subset V$, such that $(v_1, v_2) \in E$ for any $v_1 \in V_1$ and $v_2 \in V_2$. The edge-maximum complete bipartite subgraph of G contains the the maximum number of bipartite edges. i.e. the largest product of $|V_1| \cdot |V_2|$. Edges are permitted to be incident on two vertices either V_1 or V_2, but they do not contribute to the number of bipartite edges.

The vertex-maximum induced complete bipartite subgraph problem has been shown to be hard to approximate to a polynomial factor by Lund and Yannakakis [10] and Simon [14]. However, these do not resolve the problem of approximating vertex-maximum complete bipartite subgraphs. Note that the vertex-maximum complete bipartite subgraph is easy to approximate to within a factor of two by simply selecting the highest degree vertex of the graph and its neighborhood. In Section 5.1, we prove the inapproximatibility results for this subgraph problem. In Section 5.2, we use this to demonstrate the hardness of unification factoring.

Proofs will be omitted for space reasons, to appear in the full version of the paper.

5.1 Edge-Maximum Complete Bipartite Subgraph

Consider the following transformation from an arbitrary graph $G = (V, E)$ to a bipartite graph H. H will contain the pair of vertices v_i, v_i' for each vertex v_i of G. For each edge (v_i, v_j) of G, H will contain edges (v_i, v_j') and (v_j, v_i'). Finally, H will contain edges (v_i, v_i') for $1 \leq i \leq n$.

Lemma 1. *If there is a clique C with n_a vertices in G, there must be a complete bipartite subgraph of H with n_a^2 edges.*

Proof. The subgraph formed by inducing $\cup_{i \in C}(v_{i1}, v_{i2})$ for a given clique $C \in G$ is $K_{|C|,|C|}$. $\qquad\square$

Lemma 2. *If there is a complete bipartite subgraph S of n^{1+b} edges in H, then there must exist a clique of n^b vertices in G.*

Proof. The vertices of S can be two-colored, and let b' be the cardinality of the smaller set. Since the cardinality of the larger set is $\leq n$, $b' \geq b$ to realize n^{1+b} edges. By extending S to be a maximal complete bipartite graph, the vertices of b' must define a clique in G. $\qquad\square$

This reduction demonstrates that clique is intimately related to the complete bipartite subgraph problem (CBS), however it does not suffice to show the hardness of approximation of CBS. Note that for small cliques ($\leq n^{1/2}$ vertices) the resulting complete bipartite subgraph may be too small be the largest in H. We must demonstrate that is it hard to find cliques of size $\geq n^{1/2}$.

Lemma 3. *The problem of finding a clique of size $n^{1/2}$ in a graph G containing a clique of size $n^{5/6}$ is NP-hard.*

Proof. The result of [2] demonstrates that it is hard to find a clique within a factor of $n^{1/3}$ times optimal. Thus there exists a d such that finding a clique of size n^d in a graph containing a clique of size $n^{1/3+d}$ is NP-hard. If $d > 1/2$, we are done. Otherwise, consider the graph product $G' = G \times K_n$. This implies that finding a clique in G' of size n^{1+d} in a graph containing a clique of size $n^{4/3+d}$ is NP-hard, where the number of vertices in $G' = n^2$, giving the result. $\qquad\square$

5.2 Minimum-Size Trie

Since the trivial trie for a set of m strings each of length n uses mn edges, we define the *savings* SV of a trie T to be the number of edges saved over the trivial trie, i.e. $SV = mn - |T|$. Thus the optimal trie maximizes the amount of savings.

Theorem 4. *If finding a maximum complete bipartite subgraph from undirected graph G cannot be approximated to within an N^c factor, where N is the number of total vertices in G, then the maximum savings trie cannot be approximated to within a $M^c / \log M$ factor, where M is the number of strings in the trie instance.*

Proof. Consider the following reduction from an input graph $G = (V, E)$ to a set of strings. For each vertex $v_i \in V$, we construct a string s_i of length $n = |V|$ such that for all j, $1 \leq j \leq n$, the jth character of s, $s_i[j] = 1$ if $(v_i, v_j) \in E$; and a unique symbol $\alpha_{i,j}$ otherwise. The set of strings S consist of $\{s_1, s_2, \ldots, s_N\}$.

Consider any complete bipartite subgraph of G, defined by disjoint sets of vertices $V_1 = \{v_{j_1}, v_{j_2}, \ldots, v_{j_l}\}$ and $V_2 = \{v_{i_1}, v_{i_2}, \ldots, v_{i_k}\}$. This subgraph contains bipartite edges $BE = kl$. This subgraph defines a trie with at least $kl/2$ saves, by using the character positions $i_1, \ldots, i_k$ as a path from root of the trie. Since all strings $s_{j_1}, s_{j_2}, \ldots, s_{j_l}$ share the same value of the probed characters in common the strings are clustered together through a height of $k+1$ so the total saves of this suffix tree SV is at least:

$$SV = \sum_{h=1}^{k} h \cdot (n_{h+1} - 1) \geq k(n_{k+1} - 1) \geq k(l-1) \geq kl/2$$

where we assume $l > 1$ without loss of generality. Since this holds for any complete bipartite subgraph of G, it holds for the maximum CBS. Let SV_{opt} denote the savings of the optimal trie of strings S, and let BE_{max} denote the number of bipartite edges in the maximum complete bipartite subgraph of G. Then

$$BE_{max} \leq 2SV_{opt}$$

Now consider any trie of S with SV saves. We claim that we can construct a complete bipartite subgraph for G containing at least $SV/\log N$ edges.

In any trie for the set S, all savings in the trie must result from a single path from the root, as in Figure 6, because at each probe position the set of strings is broken into singletons except for those containing a 1 at the given position. Once a string belongs to a singleton set, no further saves can be credited to it. Thus the total amount of savings is

$$SV = \sum_{j=1}^{k} (j-1) n_j$$

For any j, $1 \leq j \leq k$, the vertex sets $\{v_{i_0}, v_{i_1}, \ldots, v_{i_j}\}$ and $n_{j+1}, \ldots, n_k$ defines a complete bipartite subgraph G_j, since each string under these branches has symbol

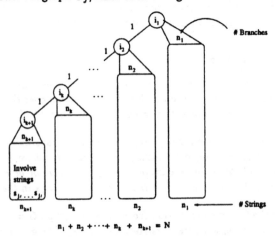

Fig. 6. Constructing a suffix tree from complete bipartite subgraph.

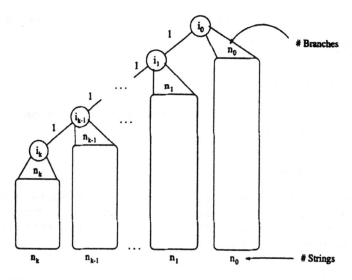

Fig. 7. Constructing a complete bipartite subgraph from a suffix tree.

1 at i_0th, i_2th, ..., i_jth positions; on the other hand, it means each vertex under these branches is incident to vertices $v_{i_0}, v_{i_1}, \ldots, v_{i_j}$. The number of bipartite edges for a given j is

$$BE_j = j \sum_{h=j+1}^{k} n_h$$

Let r be the value which maximises this, so $BE_r = \max_j BE_j$. Further

$$j \sum_{h=j+1}^{k} n_h \leq BE_r$$

and

$$\sum_{h=j+1}^{k} n_h \leq BE_r/j$$

Thus,

$$SV = \sum_{j=1}^{k} (j-1)n_j = \sum_{j=2}^{k} \sum_{i=j}^{k} n_i \leq \sum_{j=2}^{k} (BE_r/(j-1)) \leq BE_r \cdot \log N$$

which means there exists a complete bipartite subgraph with at least $SV/\log N$ edges for any trie of S with SV saves.

Since such a subgraph (or r) can be found in linear time, there is an algorithm can approximate the maximum saves trie within $N^c/\log N$ factor, we can approximate the maximum complete bipartite subgraph in N^c factor:

$$BE_{max} \leq 2SV_{opt} \leq SV \cdot (2N^c/\log N) \leq (BE_r \cdot \log N) \cdot (2N^c/\log N) \leq BE_r \cdot 2N^c$$

giving the result. □

6   Open Problem

Does a constant-factor approximation algorithm exist for the minimum-size trie problem on binary strings?

Acknowledgments

We thank Bill Studier and the rest of the Brookhaven group for interesting discussions on primer walking and sequencing. We thank IV Ramakrishnan for introducing us to unification factoring, and Steve Dawson, Keri Ko, C.R. Ramakrishnan, and Terry Swift for useful discussions.

References

1. S.F. Altschul, W. Gish, W. Miller, E.W. Myers, and D.J. Lipman. Basic local alignment search tool. *J. Mol. Biol.*, 215:403–410, 1990.
2. M Bellare, O. Goldreich, and M. Sudan. Free bits, PCPs, and non-approximability – towards tight results. In *Proc. IEEE 36th Symp. Foundations of Computer Science*, pages 422–431, 1995.
3. D.R. Clark and J.I. Munro. Efficient suffix trees on secondary storage. In *Proc. Seventh ACM Symp. on Discrete Algorithms (SODA)*, pages 383–391, 1996.
4. S. Dawson, C.R. Ramakrishnan, I.V. Ramakrishnan, K. Sagonas, T. Swift, and D.S. Warren. Unification factoring for efficient execution of logic programs. In *2nd ACM Symposium on Principles of Programming Languages (POPL '95)*, pages 247–258, 1995.
5. S. Dawson, C.R. Ramakrishnan, and T. Swift. Principles and practice of unification factoring. In *ACM Trans. on Programming Languages (TOPLAS)*, pages 528–563, 1996.
6. M.L. Engle and C. Burks. Artificially generated data sets for testing DNA fragment assembly algorithms. *Genomics*, 16:286–288, 1993.
7. P. Green. Documentation for phrap. Genome Center, University of Washington, http://bozeman.mbt.washington.edu, 1996.
8. J. Kececioglu and E.W. Myers. Exact and approximate algorithms for the sequence reconstruction problem. *Algorithmica*, 13:5–51, 1995.
9. C.-L. Lin. Optimizing tries for ordered pattern matching is π_2^p-complete. In *Proc. 10th IEEE Structures in Complexity Theory Conference*, pages 238–244, 1995.
10. C. Lund and M. Yannakakis. The approximation of maximum subgraph problems. In *Proc. 20th ICALP*, pages 40–51, 1992.
11. U. Manber and E.W. Myers. Suffix arrays : A new method for on-line string searches. *SIAM J. Computing*, 22:935–948, 1993.
12. E. W. Myers. Towards simplifying and accurately formulating fragment assembly. *J. Comp. Biol.*, 2(2):275–290, 1995.
13. W.R. Pearson and D.J. Lipman. Improved tools for biological sequence comparison. In *Proc. Natl. Acad. Sci.*, pages 2444–2448, 1988.

14. H. Simon. On approximate solutions for combinatorial optimization problems. *SIAM J. Discrete Math.*, 3:294–310, 1990.
15. G.G. Sutton, O. White, M.D. Admas, and A.R. Kerlavage. TIGR assembler: a new tool for assembling large shotgun sequencing projects. *Genome Science and Technology*, 1:9–19, 1995.
16. M. S. Waterman. *Introduction to Computational Biology.* Chapman & Hall, London, UK, 1995.

Flexible Identification of Structural Objects in Nucleic Acid Sequences: Palindromes, Mirror Repeats, Pseudoknots and Triple Helices

Marie-France Sagot[1,2] and Alain Viari[2]

[1] Institut Gaspard Monge, Université de Marne-la-Vallée, 2, rue de la Butte Verte, 93160 - Noisy-le-Grand
[2] Atelier de BioInformatique, Université de Paris 6, 12, rue Cuvier, 75005 - Paris

Abstract. This paper presents algorithms for flexibly identifying structural objects in nucleic acid sequences. These objects are palindromes, mirror repeats, pseudoknots and triple helices. We further explore here the idea of a model against which the words in a sequence are compared for finding these structural objects [17]. In the present case, models are words defined over the alphabet of nucleotides that have both direct and inverse occurrences in the sequence. Moreover, errors (substitutions, deletions and insertions) are allowed between a model and its inverse occurrences. Helix stems may therefore present bulges or interior loops, and mirror repeats need not be exact. Reasonably efficient performance comes from the fact that the parts composing the structures are kept separated until the end and that filtering for valid occurrences (occurrences that may form part of such a structure) can be done in $O(n)$ time where n is the length of the sequence. The time complexity for the searching phase (that is, before the structural parts are put together at the end) of both algorithms presented here (one for palindromes and mirror repeats, the other for pseudoknots and triple helices) is then $O(nk(e+1)(1 + min\{d_{max} - d_{min} + 1 + e, k^e \mid \Sigma \mid^e\}))$ where n is the length of the sequence, d_{max} and d_{min} are, respectively, the maximal and minimal length of a hairpin loop, k is either the maximum length k_{max} of a model, is a fixed length or represents the maximum value of a range of lengths, e is the maximum number of errors allowed (substitutions, deletions and insertions) and $\mid \Sigma \mid$ is the size of the alphabet of nucleotides.
keywords : nucleic acid sequence, nucleic structural object, palindrome, mirror repeat, pseudoknot, triple helix, approximate comparison, model, direct occurrence, (complementary) inverse occurrence.

1 Introduction

We present in this paper algorithms for flexibly identifying structural objects in nucleic acid sequences. These objects are palindromes, mirror repeats, pseudoknots and triple helices. The reason why these structural units have been chosen rather than other ones is that they are among the simplest ones present in DNA or RNA and that the algorithms for locating them are very similar.

Palindromes and mirror repeats are the simplest structures of the four, and represent the basic elements composing a helix in RNA (palindromes) or DNA (mirror repeats). The palindromes considered here are of course not lexical palindromes but biological ones and correspond to two segments of RNA that complement one another in inverse order [11]. The accepted complementary base pairs are the Watson-Crick A-T(U), C-G pairs, although the G-T(U) pair could also be considered. Mirror repeats are inverse exact (i.e. not complementary) repeats and may form part of triple helices in DNA [14]. The third part of the triple helix belongs to the other strand and links itself with both repeats. Palindromes represent secondary structures that are not always local as the distance between the two halves of the palindrome may be very big. Mirror repeats that may belong to a triple helix in DNA are less well known objects.

Pseudoknots and triple pairs on the other hand form part of the tertiary structure of a nucleic acid sequence. A pseudoknot in a folded RNA molecule has a hairpin loop plus a single strand folded back to form base pairs with the bases in the loop [7] [8] [16]. A base triple is an approximately planar group of three bases involving at least one hydrogen bond joining each pair. A triple helix is a contiguous series of such base triples.

Algorithms for finding palindromes follow in general a naïve approach [23] or restrict themselves to exact comparisons [9] [12]. When only mismatches are allowed, the first method has a time complexity that is $O(nk(d_{max} - d_{min} + 1))$ where n is the length of the sequence, d_{max} and d_{min} are, respectively, the maximal and minimal length of a hairpin loop and k is the length of the helix stems composing the structure. This last value has to be established beforehand. The exact method has a complexity in $O(n \log k + n(d_{max} - d_{min} + 1))$ where k is, in this case, the maximum length k_{max} of the helices or is a fixed length. The first part of the formula corresponds to the search for the halves of potential helices, the second to the process of putting the halves together. Algorithms for locating pseudoknots on the other hand are usually thermodynamic simulation or grammar-based approaches, either general [5] [10] [21] [22] [25] [26] or specifically designed for finding pseudoknots [1] [4] [13]. When an analysis of the complexity of such searches is given, it usually appears to be high even though these methods are heuristics ($O(n^3)$ for [4]). Slightly better time complexities seem to be obtained by the approaches that do not allow for deletions and insertions. Indels are not believed to exist in pseudoknots, but this may simply be because pseudoknots with them have seldom been looked for. Finally, locating mirror repeats or triple helices seems to have been a far less studied problem.

Our main aim in this paper is to try to fill some of these gaps by proposing algorithms that are reasonably efficient and flexible in their definition of what constitutes structures such as these. The algorithms presented are not predictive in the sense that we cannot say which of the potentially structural objects found will actually be part of the final, global structure of the molecule, we just find all those that may do so. These algorithms, in particular the pseudoknots and triple helices finder, may be used independently but they are meant to form part of a more general program of global structure prediction or identification such as the one given in [2] or [3]. The first one is a database scanning tool based on a descriptive grammar of more complex structural objects. The second is a structure prediction algorithm.

This paper is a continuation of some of our previous ones [17] [18] [19] [20] in the sense that the problem of errors will be dealt with the help of an object called a model against which the comparisons are made. In this case, a model is either a word or a pair of words over the same alphabet as that of the sequences that corresponds exactly to one half of the structures we search for. This half is the second part of either one (palindromes, mirror repeats) or two (pseudoknots, triple helices) helix stems. It corresponds only approximately to the other half of the structures.

However, we present two innovations in relation to the previous work. The first one is the fact that a model may now be "read" in a sequence in both directions. We therefore introduce the idea of a direct and of an inverse occurrence. The second innovation is that we manipulate here representatives of equivalence classes of the occurrences of a model m instead of all occurrences as was done before. This allows us to obtain a better performance in theory as well as in practice. It also means we may find a solution space that is slightly larger than the one actually searched for, but the latter one may be efficiently recovered afterwards.

Importantly also, by not putting together the diverse parts of a structural object until the end, we are able, during the whole phase of the search for the structures, to depend not on the number of such objects there is in a sequence, but instead on the number of parts that compose them. In the case of exact palindromes for instance, an upper bound for their number is $O(n^2)$ where n is the length of the sequence, while the total number of the parts composing them is bounded over by $O(n)$.

Where errors are allowed, we are then able to propose algorithms that find all the parts composing the structural objects mentioned above in $O(nk(e + 1)(1 + min\{d_{max} - d_{min} + 1 + e, k^e \mid \Sigma \mid^e\}))$ where n, d_{max} and d_{min} are as before, k is in this case either the maximum length k_{max} of a model, is a fixed length or represents the maximum value of a range of lengths, e is the maximum number of errors allowed (substitutions, deletions and insertions) and $\mid \Sigma \mid$ is the size of the alphabet of nucleotides. Putting the parts together requires $O(N)$ time, where N is the number of possible structural objects and is majored by $O(n(d_{max} - d_{min} + 1))$ in the case of palindromes and mirror repeats or by $O(n(d_{max} - d_{min} + 1)^3)$ in the case of pseudoknots and triple helices. This majoration is reached only for very degenerated sequences (as, for instance, a sequence composed of n As in the case of mirror repeats and for $d_{max} - d_{min} + 1 = n$). An $O(n \log k(e + 1)(1 + min\{d_{max} - d_{min} + 1 + e, k^e \mid \Sigma \mid^e\}))$ version of the algorithm for the first searching phase is possible but may not be very interesting in practice when errors are permitted and so has not been implemented.

This paper is organized as follows. In section 2, we present the problem of finding palindromes and mirror repeats. Basic definitions are given, then we introduce the algorithm and a sketch of its complexity. In section 3, we show how to extend the algorithm to deal with the more complicated structures composing a pseudoknot or a triple helix. We end by a discussion on the limitations of the approach given here for treating nucleic structural problems of a more general nature and give some indications on how we propose to extend this work in the future.

2 Palindromes and Mirror Repeats

2.1 Basic Definitions

Let Σ be the alphabet of nucleotides, that is, $\Sigma = \{A, C, G, T \text{ or } U\}$ and let a sequence s be an element of Σ^*. A word u of length k is an element of Σ^k for $k \geq 1$ and u is said to be a word in s if $s = xuy$ with $x, y \in \Sigma^*$.

Definition 2.1 *Given $u = s_i...s_j$ a word in s, we call i the start position of u in s and j its end position.*

Definition 2.2 *Given two words $u = u_1...u_k$ and $v = v_1...v_l$, we say that u is equal to v (denoted by $u = v$) if $k = l$ and $u_i = v_i$ for $1 \leq i \leq k$.*

Definition 2.3 *Given a word $u = u_1...u_k \in \Sigma^k$, we call inverse of u the word $\bar{u} = u_k...u_1$.*

Definition 2.4 *Given $u = s_i...s_j$ a word in s, we call j the start_inverse position and i the end_inverse position of its inverse $\bar{u}$ in s.*

Notation 2.1 *We denote by $\mathcal{M}_c$ the 4×2 matrix of the nucleotides complementary base pairs. We have:*

$$\mathcal{M}_c = \begin{array}{|c|c|} \hline A & T\ (U) \\ \hline C & G \\ \hline G & C \\ \hline T\ (U) & A \\ \hline \end{array}$$

where $(A,\ T)$ and (C,G) are the Watson-Crick base pairs.

As mentioned in the introduction, we could add to this table the (G, T or U) pair. Of course, this would increase the number of potential structures.

Definition 2.5 *Given a word $u = u_1...u_k \in \Sigma^k$ and a matrix $\mathcal{M}_c$ of complementary base pairs, we call complementary inverse of u the word $\overline{u_c} = \mathcal{M}_c(u_k)... \mathcal{M}_c(u_1)$.*

Observe that if the pair (G, T or U) is added to the definition of $\mathcal{M}_c$, a word could have more than one complementary inverse.

Definition 2.6 *Given non-negative integers e, d_{min} and d_{max}, we call $(u, v, e, d_{min}, d_{max})$ an approximate palindrome (resp. mirror repeat) in s if:*

- u and v are words in s;
- $dist_L(u, \overline{v_c})$ (resp. $dist_L(u, \overline{v})$) is no more than e where $dist_L(x, y)$ is the Levenshtein (or edit) distance between x and y (it is the minimum number of substitutions, deletions and insertions necessary to convert x into y);
- $d_{min} \leq d \leq d_{max}$ where d is the distance between the end position f_v of v and the start position i_u of u in s $(d = i_u - f_v + 1)$.

If $e = 0$, we have of course an exact palindrome (resp. mirror repeat).

Mismatches in a palindrome or repeat correspond to interior loops, and deletions and insertions to bulges. The segment of DNA or RNA sequence between the two halves of a palindrome or mirror repeat represent the hairpin loops (see figure 1).

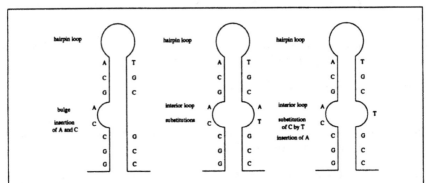

Fig. 1. Mismatches and deletions/insertions corresponding to bulges and interior loops.

Definition 2.7 A model m of length k is an element of Σ^k.

Definition 2.8 Given a sequence s and non-negative integers e, d_{min} and d_{max}, we say that m is an approximate palindromic (mirror) model present in s and satisfying the constraints (e, d_{min}, d_{max}) if there exist words u and v in s that verify:

- $u = m$;
- $(u, v, e, d_{min}, d_{max})$ is an approximate palindrome (resp. mirror repeat) in s.

From $dist_L(u, \overline{v_c}) \leq e$ (resp. $dist_L(u, \overline{v}) \leq e$), we deduce $dist_L(m, \overline{v_c}) \leq e$ (resp. $dist_L(m, \overline{v}) \leq e$).

We say that u is an (exact) direct occurrence of m in s and v an approximate complementary inverse (resp. inverse) occurrence of m.

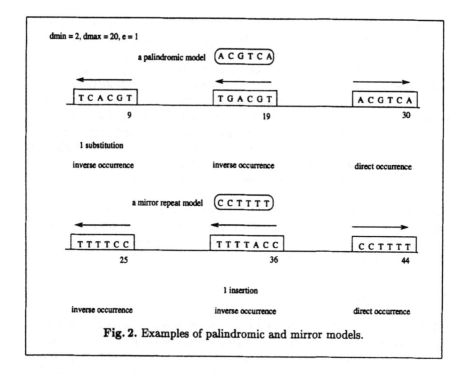

Fig. 2. Examples of palindromic and mirror models.

Example 2.1 *See figure 2.*

Observe that the fact that direct occurrences are always exact and start after the complementary inverse (resp. inverse) occurrences with which they form a palindrome (resp. mirror repeat) does not add any further constraint on the structural objects that the algorithm is able to find than the ones introduced by e, d_{min} and d_{max}. It just allows for a simpler definition of the objects searched for and, subsidiarily, to an easier analysis of the algorithm's complexity.

Definition 2.9 *We note $OD(m)$ the set of all direct occurrences of a palindromic (resp. mirror) model m present in s and $OI(m)$ its set of complementary inverse (resp. inverse) occurrences.*

With both palindromic and mirror models, we need not have $OD(m) \cap OI(m) = \emptyset$. For instance, mirror model $m =$ AAA is present in sequence s = AAAAAAAAA four times with the constraints ($e = 0, d_{min} = 0, d_{max} = 0$) (at positions 4, 5, 6 and 7) and the word AAA starting at position 4 is both a direct and an inverse occurrence of m. In the same way, palindromic model $m =$ AGCT which is twice present in $s =$ AGCTAGCTAGCT with the same constraints (at positions 5 and 10) has the word AGCT starting at position 5 as both a direct and a complementary inverse occurrence.

We may note already that, as announced in the introduction, the parts of the structural objects searched for are not kept "assembled" in the sets of occurrences of a model, but are instead kept separated until the end. In the exact case for instance ($e = 0$), keeping them assembled would require $O(n^2)$ space in the worst case while keeping them separated requires at most $2n$ space. Note however that, although the parts of a structure are kept apart, only valid ones are preserved. If a word u in s is such that $u = m$ for m a model, u will be stocked in $OD(m)$ only if there exists v in s such that $(u, v, e, d_{min}, d_{max})$ is a palindrome or a mirror repeat, and vice-versa. What is even more important to observe is that checking for this validity takes only $O(n)$ time as we show later on.

We can then now state the problem we wish to solve in the following way:

The Palindrome or Mirror Repeat Problem *Given a sequence s and non-negative integers e, d_{min} and d_{max}, the problem we propose to solve is to find all approximate palindromic models (resp. mirror models) present in s that satisfy the constraints (e, d_{min}, d_{max}).*

2.2 Algorithm

2.2.1 Main idea The main idea of the algorithm is to progressively construct models having at least one direct occurrence and one approximate complementary inverse (resp. inverse) occurrence that form a palindrome (resp. mirror repeat) verifying the constraints (e, d_{min}, d_{max}). To simplify matters, we shall henceforward treat only the case of mirror repeats with the understanding that palindromes are dealt with in exactly the same way except for the introduction of a matrix of complementary bases.

The construction of models is done by increasing length in a way that is similar to what was done in [17] [18] [19] [20]. There are now two differences though. The first one is that, as we saw, a word that is an occurrence of a model is related not just to the model, but also to at least one other occurrence of the same model. The relation of an occurrence to a model is either an exact match or an upper bounded Levenshtein distance while the relation to another occurrence is positional (it is the distance between the end position of an occurrence and the start position of the other). A second difference in relation to the construction of models as done in our previous papers is that we introduce here the idea of an equivalence relation between inverse occurrences. Before we do that however, we formally introduce the rules of model construction.

2.2.2 Rules of Model Construction We start by introducing a new definition and two notations.

Definition 2.10 *We say that a position i in a sequence s represents a possible occurrence u (resp. $\bar{u}$) of a model m if u (resp. $\bar{u}$) is an exact match of m (resp. is at a distance at most e from m). We note $POD(m)$ and $POI(m)$ the sets of possible occurrences (direct or inverse respectively) of m.*

The rules of construction are then given by the following lemmas. As usual, let s be a sequence of length n and m a model of length k.

For now an occurrence $s_i...s_f$ (direct or inverse) of a model m is identified by:

- its end position f in s if it is a direct occurrence;
- its start and end positions (i, f) and its number of errors num_errors against the model otherwise (notice that we must have $num\_errors \leq e$ for any valid occurrence).

An example is given in figure 3.

Lemma 2.1 $f \in POD(m' = m\alpha)$ with $\alpha \in \Sigma$ if, and only if:

- $f - 1 \in OD(m)$
- $f \leq n$ and $s_f = \alpha$

Lemma 2.2 $(i, f, num\_errors) \in POI(m' = m\alpha)$ with $\alpha \in \Sigma$ if, and only if, at least one of the following conditions is verified:

(match)	$(i + 1, f, num\_errors) \in OI(m)$, $i \geq 1$ and $s_i = \alpha$
(substitution)	$(i + 1, f, num\_errors - 1) \in OI(m)$, $i \geq 1$ and $s_i \neq \alpha$
(deletion)	$(i, f, num\_errors - 1) \in OI(m)$
(insertion)	$\exists g > 0$ (g is the minimal value for which this is true)

$$such\ that$$
$$(i + g + 1, f, num\_errors - g - 1) \in OI(m), i \geq 1$$
$$and\ s_i \neq \alpha$$
$$or$$
$$(i + g + 1, f, num\_errors - g) \in OI(m), i \geq 1\ and$$
$$s_i = \alpha$$

Lemma 2.3 $f_d \in OD(m' = m\alpha)$ if, and only if, $f_d \in POD(m')$ and $\exists\ (i, f_i, num\_errors) \in POI(m')$ such that $d_{min} \leq f_d - k - f_i \leq d_{max}$
   Conversely:
   $(i, f_i, num\_errors) \in OI(m' = m\alpha)$ if, and only if, $(i, f_i, num\_errors) \in POI(m')$ and $\exists\ f_d \in POD(m')$ such that $d_{min} \leq f_d - k - f_i \leq d_{max}$

An illustration is given in figure 3.

2.2.3 Equivalence Classes of Occurrences

The idea of an equivalence relation on the occurrences of a model will concern here only inverse occurrences because this relation is trivially true for direct ones.

Let us consider the following model $m =$ ACCAGTG and let us suppose that $e = 2$ and that GTGAACGA is a word starting at position i and ending at position f in s. Its inverse is AGCAAGTG. If having constructed model m up to length 4, that is we have constructed the model $m' =$ ACCA, we identify the inverse occurrences by their end positions, that is, their start_inverse positions

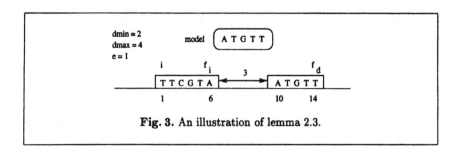

Fig. 3. An illustration of lemma 2.3.

(see definition 2.4), we need to keep all those whose start_inverse positions are the same but which have different lengths because we do not know which ones will be extendable later on. If instead of that, we identify the inverse occurrences by their end_inverse positions, we need to keep only one occurrence per position even when several have the same end_inverse position. The one kept is the one having less errors against the model and the reason why preserving this one is enough is because if it cannot be extended, none of the other occurrences whose end_inverse position is the same may be extended either. Figure 4 gives an example of that. We may therefore establish the following equivalence relation E on the set of inverse occurrences of a model m. Given two words $u = u_i...u_f$ and $v = v_k...v_l$ in s that belong to $OI(m)$, we say that u is equivalent to v, noted $u \equiv v$, if, and only if, $i = k$. We identify an equivalence class of E by the common end_inverse position i of its elements and we represent it by the occurrence with end_inverse position i for which $dist_L(\overline{u}, m)$ is minimal among all $v \in OI(m)$ having same end_inverse position. We note $\overline{OI(m)}$ the quotient $OI(m)/\equiv$ and $\overline{POI(m)}$ the quotient $POI(m)/\equiv$.

Occurrences are therefore now identified by:

- their end positions f in s if they are direct occurrences;
- the end_inverse position i of any element of a class of inverse occurrences and the start_inverse position f and number of errors num_errors of the representative of the class against the model otherwise (once again, we have $num\_errors \leq e$ for valid occurrences).

An example is given in figure 5.

In this case, Lemmas 2.2 and 2.3 above become:

Lemma 2.4 $(i, f, num\_errors) \in \overline{POI(m' = m\alpha)}$ *with* $\alpha \in \Sigma$ *if, and only if:*

- *at least one of the following conditions is verified:*

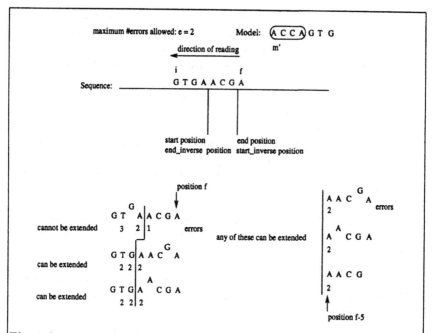

Fig. 4. An example that shows why keeping end positions instead of start ones leads to no loss of occurrences.

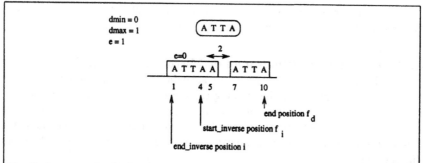

Fig. 5. An example of why the representative of a class of inverse occurrences may not be the one that keeps verifying the constraint on the length of the hairpin loops.

(match)	$(i+1, f, num\_errors) \in \overline{OI(m)}$, $i \geq 1$ and $s_i = \alpha$
(substitution)	$(i+1, f, num\_errors - 1) \in \overline{OI(m)}$, $i \geq 1$ and $s_i \neq \alpha$
(deletion)	$(i, f, num\_errors - 1) \in \overline{OI(m)}$
(insertion)	$\exists g > 0$ *(g is the minimal value for which this is true)*

such that

$$(i+g+1, f, num\_errors - g - 1) \in OI(m), \ i \geq 1$$
and $s_i \neq \alpha$

or

$$(i+g+1, f, num\_errors - g) \in OI(m), \ i \geq 1 \text{ and}$$
$s_i = \alpha$

- (i, f, num_errors) *is the occurrence with end_inverse position i whose Levenshtein distance from m is minimal*

Lemma 2.5 $f_d \in OD(m' = m\alpha)$ *if, and only if,* $f_d \in POD(m')$ *and* $\exists \ (i, f_i, num\_errors) \in \overline{POI(m')}$ *such that* $d_{min} - num\_errors \leq f_d - k - f_i \leq d_{max} + num\_errors$

Conversely:

$\overline{(i, f_i, num\_errors)} \in \overline{OI(m' = m\alpha)}$ *if, and only if,* $(i, f_i, num\_errors) \in \overline{POI(m')} \ \exists \ f_d \in POD(m')$ *such that* $d_{min} - num\_errors \leq f_d - k - f_i \leq d_{max} + num\_errors$

The need to consider a larger interval (its length is increased by $2 \times num\_errors$ for each occurrence) for the length of the hairpin loops comes from the fact that the representative chosen for a class of inverse occurrences may not be the one that keeps verifying the constraint on the length of these loops. An example is given in figure 5. Taking as representative the one presenting less errors against the model and enlarging the interval allows us to miss no valid occurrence. It may preserve a few that are not. These may be eliminated at the end of the algorithm as will be explained when we discuss the overall algorithm's complexity.

To simplify, we call from now on the double inequality $d_{min} - num\_errors \leq f_d - k - f_i \leq d_{max} + num\_errors$ the (d_{min}, d_{max}) constraint.

2.2.4 Filtering for Palindromes or Mirror Repeats

Applying the rules given by lemmas 2.1 and 2.4 is relatively straightforward and will not be further detailed here except to say that finding the representative (i, f, num_errors) of a class of inverse occurrences can be done in $O(e + 1)$ time, or $O(1)$ time if bit vectors are employed to represent the $(e + 1)$ occurrences of a same class. We shall now say a word about how we can apply the rule given by lemma 2.5 in an efficient way.

The point to notice is that the sets $OD(m)$ and $\overline{OI(m)}$ are kept permanently ordered (this follows naturally from the way models are constructed). The sets $POD(m')$ and $\overline{POI(m')}$ obtained by an application of lemmas 2.1 and 2.4 are therefore ordered too. Filtering them to eliminate direct or inverse occurrences that do no longer form part of a structure can then be done in time linear with the sum of the cardinality of both sets. An idea of the algorithm that performs such filtering is given in figure 6. One can compare this to the operation of emptying two bottles of water in synchrony. When the level of the water in one bottle stops at a certain point, it is because the level in the other bottle is going down, and vice versa. The total time needed to empty both bottles is then proportional to the quantity of water they contain. Such a simple operation is enough because we just need to flag those occurrences (direct or inverse) that form part of at least one mirror repeat or palindrome, we do not need to establish with what other occurrences it does so. The only delicate point arises when the level of water in the second bottle (corresponding to $POD(m')$) goes down a little too far. This happens when the last occurrence of $POD(m')$ is successful (i.e. it is flagged) while the current one is not (because $(f_d - k - f_i) > d_{max} + num\_errors$). In this case, one must temporarily stop emptying $POD(m')$ and go back to emptying $\overline{POI}(m')$. However, the next occurrence of $\overline{POI}(m')$ must be tested for validity not against the next or the current one in $POD(m')$ but against the last occurrence considered before the current one if it was successful or against the current one otherwise. The function RestoreLevel($POD(m')$) shown in figure 6 serves therefore the purpose of getting the level in $POD(m')$ back to the "right" place. It is not further described here. In terms of complexity, this means an occurrence of $POD(m')$ may be considered more than once during the filtering operation. However, for each element of $\overline{POI}(m')$ we do not get back more than two elements in $POD(m')$, the delay is therefore of at most 2 for each element of $\overline{POI}(m')$ and the final time complexity remains proportional to n.

2.3 Sketch of Complexity

We sketch here the complexity of the algorithm. The recursive form used for constructing the models leads to a formulation of its time complexity that is also recursive. The question we must therefore ask in order to find it is, how many operations do we have to do to construct the models of length $k+1$ from those of length k?

Since a palindromic or mirror model m of length k must have at least one direct occurrence, there can be at most $O(n)$ models of length k. Conversely, these $O(n)$ models have at most a total of $O(n)$ direct occurrences between them. Inverse occurrences on the other hand may belong to more than one model. Given one such occurrence, the number of models it may belong to is bounded over by $min\{d_{max} - d_{min} + 1 + e, k^e \mid \Sigma \mid^e\}$. The first term $(d_{max} - d_{min} + 1 + e)$ comes from the fact that a model must have an exact direct occurrence "close" enough to the inverse one. The second term $(k^e \mid \Sigma \mid^e)$ represents the models in the neighboorhood [15] of the inverse occurrence. The time needed to build the models of length $k+1$ from those of length k is therefore proportional to:

$$\underbrace{n}_{1} + \underbrace{n \, min\{d_{max} - d_{min} + 1 + e, k^e \mid \Sigma \mid^e\}}_{2} \underbrace{(e+1))}_{3} \times \underbrace{3}_{4}$$

where

- 1 is the total number of direct occurrences of all models of length k
- 2 is the total number of inverse occurrences of all models of length k

Let $POD(m' = m\alpha)$ and $\overline{POI(m' = m\alpha)}$ be the sets of direct and inverse occurrences of the model m' we are in the process of constructing. These sets are ordered by increasing position in s.

```
1:      get the first element (i, f_i, num_errors) of POI(m')
2:      do
3:          get the next element f_d of POD(m')
4:          if the corresponding occurrences satisfy the (d_min, d_max) constraint
5:              flag them as valid occurrences
6:      until (f_d − k − f_i > d_max + num_errors or no element f_d is available)
7:      RestoreLevel(POD(m'))
8:      get the next element (i, f_i, num_errors) of POI(m')
9:      if none is available
10:         stop
11:     else goto 2
```

The sets $\overline{OI(m')}$ and $OD(m')$ are then the elements of $\overline{POI(m')}$ and $POD(m')$ flagged as valid occurrences.

Fig. 6. Filtering to keep only the parts that can form a structure.

- 3 is the number of possible ways of extending an inverse occurrence (a direct occurrence can be extended in only one way)
- 4 comes from the fact that occurrences have to be extended (one time the first term of the product) and then to be filtered to determine which ones can be part of a structure (2 times that term)

The total construction takes then $O(nk(e+1)(1+min\{d_{max}-d_{min}+1+e, k^e \mid \Sigma \mid^e\}))$ time. Since a single model is being constructed at each instant, the space requirement is $O(n)$. We can make the length of a model double at each step instead of incrementing it by just one unit at a time. This leads to an $O(n \log k(e+1)(1+min\{d_{max}-d_{min}+1+e, k^e \mid \Sigma \mid^e\}))$ algorithm. In practice though, since k is a small value in general, such an approach does not allow us to gain much time when errors are allowed. When errors are forbidden, it results in an $O(n \log k)$ algorithm for the search phase.

As mentioned in section 2.2.3, the solution space obtained by applying the rule given in lemma 2.5 may be slightly bigger than the one corresponding to the solution of the problem as stated in section 2.1. Filtering that space means considering each element of $\overline{OI(m')}$, that is, each representative of a class of indirect occurrences (i, f_i, num_errors) for each model m and verifying if $d_{min} \leq f_d - k - f_i \leq d_{max}$. If that is not the case, a pattern matching algorithm such as the one given in [24] allows us to recover in $O((k + e)e)$ time all f_i' such that $d_{min} \leq f_d - k - f_i' \leq d_{max}$ and $\overline{m}$ occurs in s with at most num_errors starting at position i and ending at position f_i'. The whole operation takes time proportional to $n(k + e)e \, min\{d_{max} - d_{min} + 1 + e, k^e \mid \Sigma \mid^e\}$ which does not increase the algorithm's overall theoretical complexity when $e^2 \leq k$.

Finally, printing the resulting structures takes $O(N)$ time where N is the number of palindromes or mirror repeats that satisfy the constraints. This number is bounded over by $O(n(d_{max} - d_{min} + 1))$. Where $d_{max} = n$ and $d_{min} = 0$, this gives an $O(n^2)$ majoration.

2.4 Discussion

We wish to call attention to two points here. The first one is the fact that the time complexity of the algorithm in practice becomes independent of the length L of the hairpin loops for values of L above a certain threshold L_0. This is not true of the naïve approaches such as Waterman's [23] for instance. This is shown experimentally in the curves given in figure 7(a), (b) and (c). The sequence used for the test is a randomly generated one of length $n = 25000$, the x-axis plots the lengths of the hairpin loops (varying between 1000 and 25000 with a step of 1000) and the y-axis the execution time (in seconds). The structures searched for were palindromes, with stem lengths (i.e. models lengths) of 5 or more nucleotides. This experiment was run on a Silicon Graphics Indigo WorkStation (R4000). Figures 8(a), (b) and (c) show also that the time complexity of the search phase of the algorithm is in practice linear with the length of the sequence (each sequence is a randomly generated one of the corresponding length, stem lengths are superior or equal to 5 and the lengths of hairpin loops may vary between 0 and 600). In both figures (7 and 8), plot (a) corresponds to 0 error, (b) to 1 substitution at most authorized, and (c) to 1 error at most (substitution, deletion or insertion).

The second point we wish to call attention to is the fact that allowing for errors in the heads and feet of helices is not very meaningful in general. It is however easy to add a constraint into the algorithm forbidding errors at a given distance p from either end of an helix without increasing its time and space complexity. For errors at the head, we just need to disallow them while $\mid m \mid < p$ where m is the model being constructed. Coloring the complementary inverse (resp. inverse) occurrences that have an error in the last p symbols allows us to determine in $O(1)$ time which ones should be discarded in the final printing phase.

3   PseudoKnots and Triple Helices

3.1 Basic Definitions

Definition 3.1 *Given non-negative integers e, $d[i]_{min}$, $d[i]_{max}$ for $0 \leq i \leq 3$, we call $((u_1, v_1), (u_2, v_2), e, d[i]_{min}, d[i]_{max})$ an approximate pseudoknot in s if:*

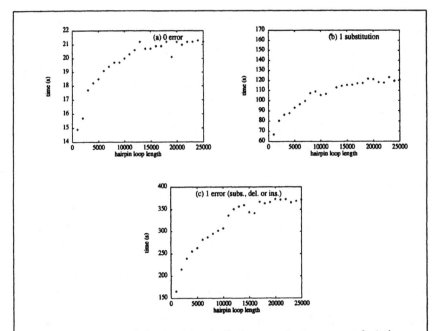

Fig. 7. Execution time of the algorithm (palindrome and mirror repeat finder) versus length of hairpin loop for 0 error, 1 substitution at most authorized, and 1 error (substitution, deletion or insertion).

- $(u_1, v_1, e, d[0]_{min}, d[0]_{max})$ and $(u_2, v_2, e, d[0]_{min}, d[0]_{max})$ are approximate palindromes in s
- $d[i]_{min} \leq d[i] \leq d[i]_{max}$ where $d[i]$ is the distance between:
  for i=1 the end position of v_1 and the start position of v_2
  for i=2 the end position of u_1 and the start position of u_2
  for i=3 the end position of v_2 and the start position of u_1

Triple helices are elements $((u_1, v_1), (u_2, v_2), e, d[i]_{min}, d[i]_{max})$ for which i takes only the values 0, 1 and 2 but which in addition verify that the start positions of v_2 and u_1 are the same.

In practice, we set $d[0]_{min} = 0$ and $d[0]_{max} = \infty$ and verify only the $(d[i]_{min}, d[i]_{max})$ constraints for $1 \leq i \leq 3$.

Example 3.1 *See figure 9.*

When working with pseudoknots or triple helices, models are no longer words as with palindromes or mirror repeats but pairs of words.

Definition 3.2 *A model (m_1, m_2) is an element of $\Sigma^{k_1} \times \Sigma^{k_2}$.*

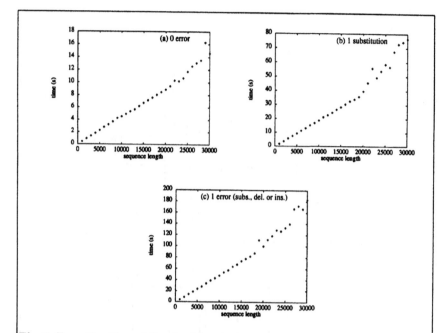

Fig. 8. Execution time of the algorithm (palindrome and mirror repeat finder) versus length of sequence for 0 error, 1 substitution at most authorized, and 1 error (substitution, deletion or insertion).

We shall for now consider only models for which $k_1 = k_2$. If any one of the two may be further extended, this can be done at the end, when printing the structural objects. Note that in the case of triple helices, models m_1 and m_2 are not completely independent.

Definition 3.3 *Given a sequence s and non-negative integers e, $d[i]_{min}$ and $d[i]_{max}$ for $0 \le i \le 3$ (resp. for $0 \le i \le 2$), we say that (m_1, m_2) is an approximate pseudoknot (resp. triple helix) model present in s and satisfying the constraints $(e, d[i]_{min}, d[i]_{max})$ if there exist pairs of words (u_1, v_1) and (u_2, v_2) in s that verify:*

- $u_1 = m_1$ *and* $u_2 = m_2$;
- $((u_1, v_1), (u_2, v_2), e, d[i]_{min}, d[i]_{max})$ *is an approximate pseudoknot (resp. triple helix) in s.*

Definition 3.4 *We note $OD_1(m_1)$ (resp. $OD_2(m_2)$) the set of all direct occurrences of the first (resp. second) palindrome in the pseudoknot or triple helix model (m_1, m_2) present in s and $OI_1(m_1)$ (resp. $OI_2(m_2)$) the sets of complementary inverse occurrences.*

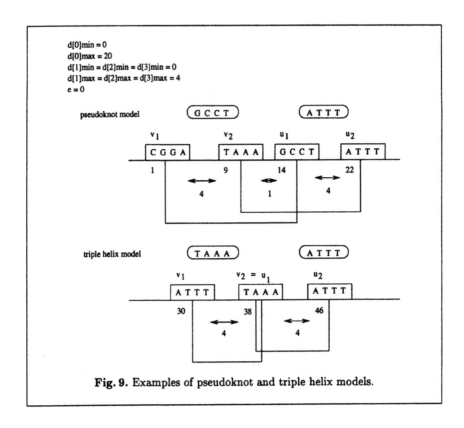

Fig. 9. Examples of pseudoknot and triple helix models.

$POD_1(m_1)$, $POD_2(m_2)$, $POI_1(m_1)$, $POI_2(m_2)$ are the corresponding sets of possible occurrences.

3.2 Constructing the Models and Filtering for Pseudoknots and Triple Helices

Lemmas 2.1 and 2.4 remain valid for the initial stages of the construction of the models (m_1, m_2) (construction of the palindromic models m_1 and m_2) but lemma 2.5 is now different. We give it here for pseudoknots. In the case of triple helices, the last pair of inequalities is not verified and the equality of the start positions of v_2 and u_1 is checked only at the end of the algorithm.

Lemma 3.1 $f1_d \in OD_1(m'_1 = m_1\alpha)$ if, and only if, $\underline{f1_d \in POD_1(m'_1)}$ and $[\exists\, f2_d \in POD_1(m'_2 = m_2\alpha)$, $(i1, f1_i, num\_errors1) \in \overline{POI_1(m'_1)}$ and $(i2, f2_i, num\_errors2) \in \overline{POI_2(m'_2)}]$ such that:

- $d[1]_{min} - num\_errors1 \le i2 - f1_i - 1 \le d[1]_{max} + num\_errors1$
- $d[2]_{min} \le f2_d - f1_d - 1 \le d[2]_{max}$
- $d[3]_{min} - num\_errors2 \le f1_d - k - f2_i \le d[3]_{max} + num\_errors2$

Similar conditions hold for $f2_d$, $(i1, f1_i, num\_errors1)$ *and* $(i2, f2_i, num\_errors2)$ *and are not given here.*

Example 3.2 *See again figure 9.*

To simplify, we call from now on the double inequalities of the preceding lemma for $1 \leq i \leq 3$ the $(d[i]_{min}, d[i]_{max})$ constraints.

The algorithm that filters for pseudoknots (or triple helices) by verifying the rules given in the preceding lemma is now slightly more complex than the previous one. An idea of it is given in figure 11. It uses the function RestoreLevel(SetofOccurrences) described previously (see section 2.2.4). Once again, we can compare this to the operation of emptying this time four bottles of water in synchrony. When the level in three out of the four bottles stops at a given point, it is because the level in the fourth one is going down. At each moment, one bottle is filling out while the water in the other ones is staying still. Furthermore, if one considers the level of water in a bottle to be directly related to the position of the occurrences in the sequence, then the bottles must be emptied in a given order and the level of water in one bottle must never be higher than the level in the bottles that precede it in the ordering or lower than the level in the bottles that follow it (see figure 10). This equilibrium must be maintained all the time while the bottles are being emptied so as not to lose any valid occurrence. Emptying the four bottles takes then time proportional to the total quantity of water they initially contained.

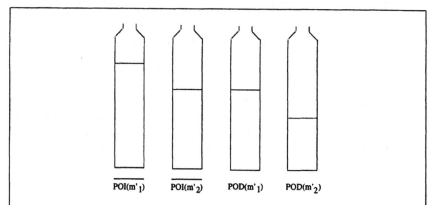

POI(m'₁)   POI(m'₂)   POD(m'₁)   POD(m'₂)

Fig. 10. Filtering operation regarded as emptying the contents of four bottles in synchrony.

3.3 Complexity

The overall complexity of the algorithm for constructing pseudoknot or triple helix models is just a constant time bigger than that for palindromic or mirror models given in section 2.3. This comes from the fact that the occurrences of pseudoknots or triple helices are kept separated as before and that filtering for the structures remains linear in the total number of such occurrences there is at any given step in the algorithm. Eliminating the false occurrences at the end (see the end of section 2.3) takes the same time. However putting the parts together for printing, that is, actually printing the occurrences as structures, may now be a more complex task. The time needed is still $O(N)$ where N is the number of pseudoknots or triple helices there is in a sequence, but N is now bounded over by $O(n.(d_{max} - d_{min} + 1)^3)$ where $d_{max} - d_{min} = max(d[i]_{max} - d[i]_{min})$ for $1 \leq i \leq 3$. This is of course a worst case situation that is observed only in very degenerated sequences. In the exact case, it may be reached if, for instance, $k << n$, $d_{max} - d_{min} + 1 = n$ and $s = A^{\frac{n}{4}} C^{\frac{n}{4}} T^{\frac{n}{4}} G^{\frac{n}{4}}$. Finally, extending the length of one of the two palindromes when maximal pseudoknots or triple helices are searched for takes $O(N.(k_{max} - k'_{max}))$ where k_{max} is the maximal length of the longest palindromes found in a structure and k'_{max} the maximal length of the smallest. We may of course have $k_{max} = k'_{max}$ (none of the palindromes can be extended).

3.4 Discussion

We give in figures 12(a), (b) and (c) plots showing that the time complexity of the search phase of the algorithm for identifying pseudoknots and triple helices is in practice linear in the length of the sequence as was the case for the palindromes and mirror repeats finder. The experiment was run on the same machine as previously, with the following parameters :

- length of models searched for ranging between 4 and 20;
- $d[1]_{min} = 0$ for $1 < i < 3$;
- $d[1]_{max} = 20$ for $1 < i < 3$.

The sequences were randomly generated, the x-axis plots their lengths (varying between 1000 and 20000 with a step of 1000) and the y-axis the execution time (in seconds). Plot (a) corresponds to 0 error, (b) to 1 substitution at most authorized, and (c) to 1 error at most (substitution, deletion or insertion).

4 Perspectives

As has been shown in [2], procedural algorithms are not ideally suited for looking for structures more complex than the ones we have dealt with here. The pseudoknot and triple helix finders are an example of that. In the procedural approach, the rules governing such structures have to be written directly into the algorithms's code. This soon becomes an awkward business for more elaborated

Let $POD(m'_j = m_j\alpha)$ and $\overline{POI(m'_j = m_j\alpha)}$ for $j = 1, 2$ be the sets of direct and inverse occurrences of the model (m'_1, m'_2) we are in the process of constructing. As before, these sets are ordered by increasing position in s.

```
 1:    get the first element (i1, f1_i, num_errors1) of POI(m'_1)
 2:    do
 3:        get the next element (i2, f2_i, num_errors2) of POI(m'_2)
 4:        do
 5:            get the next element f1_d of POD(m'_1)
 6:            do
 7:                get the next element f2_d of POD(m'_2)
 8:                if the corresponding occurrences satisfy the (d_min, d_max)
                      constraint
 9:                    flag them as valid occurrences
10:            until (f2_d - f1_d - 1 > d[2]_max or no element f2_d is available)
11:            RestoreLevel(POD(m'_2))
12:            get the next element f1_d of POD(m'_1)
13:            if none is available
14:                RestoreLevel(POD(m'_1))
15:                    goto 2
16:            else goto 6
17:            until (f1_d - k - f2_i > d[3]_max + num_errors2 or no element f1_d
                 is available)
18:            RestoreLevel(POD(m'_1))
19:            get the next element (i2, f2_i, num_errors2) of POI(m'_2)
20:            if none is available
21:                RestoreLevel(POI(m'_2))
22:                goto 1 replacing "first" by "next"
23:            else goto 4
24:        until ((i2 - f1_i - 1) > d[1]_max + num_errors1)
25:        RestoreLevel(POI(m'_2))
26:        get the next element (i1, f1_i, num_errors1) of POI(m'_1)
27:        if none is available
28:            stop
29:        else goto 2
```

The sets $\overline{OI(m'_j)}$ and $OD(m'_j)$ for $j = 1, 2$ are then the elements of $\overline{POI(m'_j)}$ and $POD(m'_j)$ flagged as valid occurrences.

Fig. 11. Filtering to keep only the parts that can form a pseudoknot.

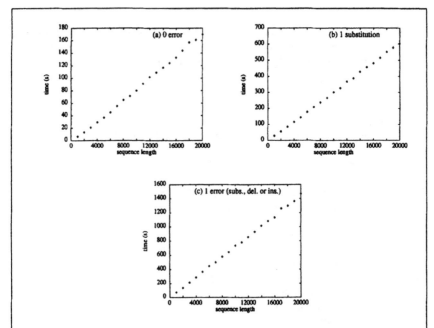

Fig. 12. Execution time of the algorithm (pseudoknot and triple helix finder) versus length of sequence for 0 error, 1 substitution at most authorized, and 1 error (substitution, deletion or insertion).

structures. However, it is an approach that is quite justified for palindromes which form the basic units of such more complex structures, as well as for mirror repeats, pseudoknots and triple helices that are either considered important objects in themselves, or thought to be frequent in DNA or RNA molecules. Looking for these simpler structural objects in a quick yet flexible way becomes then necessary and the algorithms we propose here try to fill this need.

Another idea concerning the structure of nucleic acids that is absent in this paper is the fact that molecules having a same function often have the same structure. It is therefore interesting when trying to determine the structure of an RNA for instance, to proceed by analogy with other molecules having same structure. However, this requires again having a global vision of such structures. It also requires that we know how to define the similarity between two, or more, RNA structures. This is a difficult problem, and one that does not seem to have been dealt with in a completely satisfying way. This is what we propose to do in our future work.

Finally, the idea of "reading" a sequence in both directions, and of allowing for indirect as well as direct occurrences of a model, is interesting also when one is looking for sequence motifs instead of structural objects as in this paper. In such cases, one may be confronted with the problem of having to identify a motif common to a set of sequences read from a database without knowing whether the sequence annotated in the database corresponds to the strand that contains the signal searched for. This problem has been illustrated by Waterman [6] in the case of the promotor signals implicated in *Escherichia coli* transcription process.

Acknowledgements

The authors would like to thank Maxime Crochemore for a careful reading of the manuscript and for his advice and E. W. Myers for an exchange of ideas that do not bear directly on this work but greatly influenced it.
The first author would also like to express her gratitude to Giuseppina Rindone from the Institut Gaspard Monge for having encouraged her to finish this work. Both wish to thank the referees for correcting some mistakes in the definitions. Finally, we wish to thank the Institut Pasteur (Project PowerGene) for partial financing of this work, in particular Antoine Danchin for his support of it.

References

1. J. P. Abrahams, M. v. d. Berg, E. v. Batenburg, and C. Pleij. Prediction of RNA secondary structure, including pseudoknotting, by computer simulation. *Comput. Appli. Biosci.*, 8:243–248, 1992.
2. B. Billoud, M. Kontic, and A. Viari. Palingol: a declarative programming language to describe nucleic acids' secondary structures and to scan sequence databases. *Nucleic Acids Res.*, 24:1395–1403, 1996.
3. D. Bouthinon, H. Soldano, and B. Billoud. Apprentissage d'un concept commun à un ensemble d'objets dont la description est hypothétique : application à la découverte de structures secondaires d'ARN. In *11èmes Journées Françaises d'Apprentissage*, 1996.
4. M. Brown and C. Wilson. RNA pseudoknot modeling using intersections of stochastic context free grammars with applications to database search. manuscript - University of California, Santa Cruz, Oct. 1995, 1995.
5. J.-H. Chen, S.-Y. Le, and J. V. Maizel. A procedure for RNA pseudoknot prediction. *Comput. Appli. Biosci.*, 8:243–248, 1992.
6. D. J. Galas, M. Eggert, and M. S. Waterman. Rigorous pattern-recognition methods for DNA sequences. Analysis of promoter sequences from *Escherichia coli. J. Mol. Biol.*, 186:117–128, 1985.
7. I. Tinoco Jr., P. W. Davis, C. C. Hardin, J. D. Puglisi, G. T. Walker, and J. Wyatt. RNA structures from A to Z. In *Cold Spring Harbor Symposia on Quantitative Biology*, volume LII, pages 135–146. Cold Spring Harbor Laboratory, 1987.
8. N. A. Kolchanov, I. I. Titov, I. E. Vlassova, and V. V. Vlassov. Chemical and computer probing of RNA structure. In W. E. Cohn and K. Moldave, editors, *Progress in Nucleic Acid Research and Molecular Biology*, pages 131–196. Academic Press, 1996.
9. M. Kontic. Palingol. Langage pour la description et la recherche de structures secondaires dans les séquences nucléotidiques, 1993. DEA d'Intelligence Artificielle, Université de Paris Nord.

10. F. Lefebvre. An optimized parsing algorithm well suited for RNA folding. In *Proceedings First International Conference on Intelligent Systems for Molecular Biology*, Cambridge, England, 1995.

11. B. Lewin. *Genes V.* Oxford University Press, 1994.

12. H. M. Martinez. An efficient method for finding repeats in molecular sequences. *Nucleic Acids Res.*, 11:4629–4634, 1983.

13. H. M. Martinez. Detecting pseudoknots and other local base-pairing structures in RNA sequences. 183:306–317, 1990.

14. S. M. Murkin, V. I. Lyamichev, K. N. Druhlyak, V. N. Dobrynin, S. A. Filipov, and M. D. Frank-Kamenetskii. DNA H form requires a homopurine-homopyrimidine mirror repeat. *Nature*, 330:495–497, 1987.

15. E. W. Myers. A sublinear algorithm for approximate keyword searching. *Algorithmica*, 12:345–374, 1994.

16. C. W. A. Pleij and L. Bosch. RNA pseudoknots: structure, detection, and prediction. 180:289–303, 1989.

17. M.-F. Sagot, V. Escalier, A. Viari, and H. Soldano. Searching for repeated words in a text allowing for mismatches and gaps. pages 87–100, Viñas del Mar, Chili, 1995. Second South American Workshop on String Processing.

18. M.-F. Sagot and A. Viari. A double combinatorial approach to discovering patterns in biological sequences. In D. Hirschberg and G. Myers, editors, *Combinatorial Pattern Matching*, volume 1075 of *Lecture Notes in Computer Science*, pages 186–208. Springer-Verlag, 1996.

19. M.-F. Sagot, A. Viari, and H. Soldano. A distance-based block searching algorithm. pages 322–331, Cambridge, England, 1995. Third International Symposium on Intelligent Systems for Molecular Biology.

20. M.-F. Sagot, A. Viari, and H. Soldano. Multiple comparison: a peptide matching approach. In Z. Galil and E. Ukkonen, editors, *Combinatorial Pattern Matching*, volume 937 of *Lecture Notes in Computer Science*, pages 366–385. Springer-Verlag, 1995. to appear in *Theoret. Comput. Sci.*

21. Y. Sakakibara, M. Brown, R. Hughey, I. S. Mian, K. Sjolander, R. C. Underwood, and D. Haussler. Stochastic context-free grammars for tRNA modeling. *Nucleic Acids Res.*, 22:5112–5120, 1994.

22. D. Searls. The linguistics of DNA. *American Scientist*, 80:579–591, 1992.

23. M. S. Waterman. Consensus methods for folding single-stranded nucleic acids. In M. S. Waterman, editor, *Mathematical Methods for DNA Sequences*, pages 185–224. CRC Press, 1989.

24. S. Wu, U. Manber, and E. W. Myers. An O(NP) sequence comparison algorithm. *Inf. Proc. Letters*, 35:317–323, 1990.

25. M. Zuker and D. Sankoff. RNA secondary structures and their prediction. *Bull. Math. Biol.*, 46:591–621, 1984.

26. M. Zuker and P. Stiegler. Optimal computer folding of large RNA sequences using thermodynamics and auxiliary information. *Nucleic Acids Res.*, 9:133–148, 1981.

Banishing Bias from Consensus Sequences

Amir Ben-Dor[1], Giuseppe Lancia[2], Jennifer Perone[3], and R. Ravi[2]

[1] Dept. of Computer Science, Technion, Haifa 32000, Israel.
[2] GSIA, Carnegie Mellon University, Pittsburgh, PA 15213.
[3] New York University School of Medicine, New York.

Abstract. With the exploding size of genome databases, it is becoming increasingly important to devise search procedures that extract relevant information from them. One such procedure is particularly effective in finding new, distant members of a given family of related sequences: start with a multiple alignment of the given members of the family and use an integral or fractional consensus sequence derived from the alignment to further probe the database. However, the multiple alignment constructed to begin with may be biased due to skew in the sample of sequences used to construct it.

We suggest strategies to overcome the problem of bias in building consensus sequences. When the intention is to build a fractional consensus sequence (often termed a profile), we propose assigning weights to the sequences such that the resulting fractional sequence has roughly the same similarity score against each of the sequences in the family. We call such fractional consensus sequences *balanced profiles*. On the other hand, when only regular sequences can be used in the search, we propose that the consensus sequence have minimum maximum distance from any sequence in the family to avoid bias. Such sequences are NP-hard to compute exactly, so we present an approximation algorithm with very good performance ratio based on randomized rounding of an integer programming formulation of the problem. We also mention applications of the rounding method to selection of probes for disease detection and to construction of consensus maps.

1 Introduction

Efforts in genome projects have led to wide availability of genetic information in the form of nucleic acid and protein sequences. This is reflected by the exponentially increasing sizes of several sequence databases such as SwissProt [BB92].

Proteins are comprised of sequences of amino acids, or residues, which determine their structure and thereby function. Many proteins exist in many different organisms, or in several different forms in the same organism. The sets of these proteins are called families. These families exhibit structural, and therefore presumably sequential, similarities. The careful study of one protein in a family can provide information concerning the function, or predicted function of other proteins in that family. Likewise, the classification of a relatively unstudied protein into a well-defined family can offer insight toward its structure and function. In order to study families of proteins, the technique of multiple alignment is used.

Multiple alignments allow the simultaneous comparison of several sequences. Using blocks of conserved regions identified from multiple alignment data, a database can be probed for sequences with similarity with all of the sequences in the alignment. In order to do so, it is first necessary to derive from the multiple alignment a single sequence called the *consensus* which best represents all the aligned sequences, that can be used to search the database. Alternatively, a *profile* can be derived: this is a numerical representation of the multiple alignment [GME87], which, for each position and each residue, scores the likelihood that the given residue will appear at the indicated position in the protein alignment. Intuitively, we can think of the profile as a "fractional" consensus in which at any position, some fraction of each residue is present instead of just one.

Sequence collections are seldom a fair representation of the diversity of sequences consistent with a given protein structure conserved in a family. An example is the set of all currently available globin sequences, of which more than half are vertebrate α- and β-globins, while the remaining subfamilies are represented by much fewer sequences. A reason for this bias is that experimental sequence collections are not accurate representations of the diversity associated with the structure of a given protein in nature. This is partly by necessity since there are a select few organisms that are suitable for scientific research. Hence, biases in sequence databases tend to exist toward common experimental model organisms which are intensively sequenced.

A problem resulting from such biased database collections is that multiple alignments and consensus sequences that are built from such collections tend to be biased as well. For example, consider a multiple alignment constructed from a group of closely related proteins and one distant family member. The close proteins will dominate the consensus sequence and preclude retrieval of sequences which may bear more resemblance to the outlyer. Thus in the globin example, a profile or consensus built from a multiple alignment of all currently known globins would effectively recognize vertebrate globin sequences, while invertebrate globins would be poorly recognized.

The more intrinsic problem here is that the consensus built from a multiple alignment of a skewed sample from a family may not reflect the sequence homology of the family. This is what renders it ineffective in identifying distant members of the family that may be present in the database. In this paper, we describe two approaches to banish bias from consensus sequences for the two cases of constructing fractional and integral consensus sequences.

1.1 Weights for Unbiased Profile Construction

The traditional approach to correcting bias in constructing profiles from a multiple alignment is to weight the different sequences in the alignment differently in constructing the profile. A plethora of weighting schemes have been proposed in the literature [ACL89, GSC94, HH94, THG94, SA90, LXB94, EMD95, KM95]. The basic idea is to emphasize under-represented sequences by giving them high weights, while de-emphasizing over-represented sequences by giving them low weights. It is an open problem to determine a system of weights that results in the profile that can be used to search the database most effectively for biologically relevant signals. In Section 2 we discuss some of the existing weighting schemes.

We propose a new method for sequence weighting whose goal is to yield a profile that has roughly the same similarity score to each of the sequences in the alignment. We call a profile of this type a *balanced profile* and the problem of determining the corresponding weights will be called the Balanced Profile Weight Assignment Problem. For this problem we outline a simple iterative algorithm which converges to the desired weights. Preliminary experiments with an implementation of this algorithm indicate that this method may be more effective than an unweighted profile construction, especially when the alignment is composed of several divergent sequences. We elaborate on this in Section 3.

1.2 Unbiased Consensus Sequences

A similar fairness problem arises also when building the (integral) consensus sequence. One way to define a consensus sequence is to require that it minimizes its total distance from the sequences of the alignment (sum-of-pairs criterion), but this objective is biased toward overrepresented sequences. To overcome this bias, we define the consensus as the sequence whose maximal distance from any of the sequences in the alignment is minimum. Under this definition, the problem of determining the consensus turns out to be NP-hard. In this paper we model this consensus problem as an integer programming problem and give an approximation algorithm based on randomized rounding applied to its fractional relaxation.

1.3 Two applications of randomized rounding

Consider a problem arising in the design of probes for disease detection. The probes work by hybridizing with complementary strands of sufficient similarity. To design such probes to be specific for a particular strain of bacteria, we would like the probe sequence to be as close as possible to the genetic sequence from this strain while staying as far away as possible (in Hamming distance) from the sequences of all the other strains. We propose approximation algorithms for finding near-optimal probes by applying the rounding mentioned earlier.

Another application is to the construction of a consensus map from a variety of physical maps, all of which identify the location of the same set of markers linearly along the same fragment of DNA. The construction of a consensus map that is unbiased against any skew in the input data can be formulated as an integer program. Applying randomized rounding gives good approximate solutions to this problem as well.

The rest of the paper is organized as follows. In section 2 we describe some of the existing schemes for weighting sequences in an alignment and address the Balanced Profile Weight Assignment Problem. Section 3 reports computational results of database search with our method as opposed to the unweighted one. Section 4 describes the randomized rounding approximation algorithm for the consensus problem. Finally in section 5 we outline extensions of the randomized rounding technique to designing probes for disease detection and to physical map construction.

2 Profiles and Weighting Schemes

Let $\Sigma = \{\sigma_1, \ldots, \sigma_{|\Sigma|}\}$ be a finite alphabet (in particular, we can take Σ to be the set of 20 amino acids). We will consider elements of Σ^n, called *sequences*. Let $d : \Sigma \times \Sigma \longrightarrow \mathcal{R}$ be a distance function (e.g., for amino-acids $d = \text{PAM-250}$). We generalize d to sequences in Σ^n in the natural way, i.e., $d(x, y) = \sum_{i=1}^{n} d(x_i, y_i)$. A *weighted multiple alignment* is a vector of k sequences, i.e. a $k \times n$ matrix over $\Sigma \cup \{-\}$ (where $-$ denotes the blank character), together with a set $w_1, \ldots, w_k$ of weights for the sequences.

Given a multiple alignment, a *profile* is an $n \times |\Sigma|$ real–valued matrix P. Each entry $P[i, j]$ of P scores the likelihood with which σ_j is the i-th symbol of the sequences in the alignment (blanks are traditionally excluded from the profile). A profile can be constructed for a group of k aligned sequences each of equal length n using PROFILEMAKE, Genetic Computer Group's profile building tool [GCG94]. Profiles can also reflect values for any gaps which may appear in the alignment; however, gaps will not be discussed here (see [GME87] for further information). A database can then be searched with a profile by using the GCG's procedure PROFILESEARCH [GCG94], which implements an alignment algorithm by Smith and Waterman, based on dynamic programming [SW81].

When using PROFILEMAKE, the scores at position r, character σ, are defined as follows:

$$P[r, \sigma] = \sum_{j=1}^{|\Sigma|} W_j s(\sigma_j, \sigma) \tag{1}$$

where s is the Dayhoff similarity score [DBH83]. The weight W_j depends on the number of occurrences of each type of residue at each alignment position and the sum of the weighted number of sequences. PROFILEMAKE can use either linear weighting:

$$W_j = \frac{\sum_{i=1}^{k} w_i \delta_{i,j}}{\sum_{i=1}^{k} w_i}$$

or logarithmic weighting:

$$W_j = \frac{\ln\left[1 - \sum_{i=1}^{k} w_i \delta_{i,j} \Big/ \left(1 + \sum_{i=1}^{k} w_i\right)\right]}{\ln\left[1 \Big/ \left(1 + \sum_{i=1}^{k} w_i\right)\right]}$$

where $\delta_{i,j}$ is equal to 1 if sequence i has residue σ_j at the current position, and is equal to 0 otherwise. As expected with either of these methods the weight will be zero if the given amino acid does not occur in the alignment position, and it will be one if it is the only amino acid which appears at that position.

Weights assigned to each sequence affect these values. In the unweighted system, all weights w_i are equal, so the contribution to W_j is the same for all sequences. In this case, the W_j value calculated for a residue σ_j is simply the fraction of input sequences that contain residue σ_j at the current position. When different weights are assigned to sequences, each sequences contribution to W_j is scaled by its weight. Sequences can be accentuated or de-emphasized to reflect closely related sequences by varying these weights. Intuitively, weights on distant sequences should be larger than the weights of the closely related sequences. Strategies to determine appropriate weights are discussed next.

While there is a consensus about the necessity of weighting sequences when searching with multiple alignments, there is considerable debate concerning what weighting method should be used. Many of the differences in these weighting systems are based on differing opinions on the problem as well as defining correct behavior. Some of the existing weighting techniques are the following (see also [VS93]):

- *Pairwise Distance:* A sequence weight is set to be equal to the sum of the distances from this sequence to all other sequences in the alignment. The idea is that a far-off sequence which is under-represented in the population, will have higher weight than a single sequence in a cluster of closely related sequences, thus correcting the bias. If D is the square matrix of pairwise distances between the sequences, we have $w = D\mathbf{1}$ for this method, where $\mathbf{1}$ is a vector with 1 in each component.
- *Voronoi Weighting:* This method, introduced by Sibbald and Argos [SA90], relies on constructing a Voronoi diagram from the sequences, based on pairwise distances [SA90]. A hypothetical population of sequences is built using the information from the real sequence alignment. A Voronoi diagram for this population is constructed using the input sequences as Voronoi centers. The weight for each input sequence is proportional to its Voronoi volume, i.e. the volume of the Voronoi polygon occupied by the sequence. Instead of generating all sequences, estimates of this volume can be obtained by random sampling.
- *Weighting by Phylogeny:* Altschul, Carroll, and Lipman describe a method for assigning weights based on an implied evolutionary relationship among the given sequence set [ACL89]. Using a tree constructed with all sequences in an alignment, weights can be determined by inverting a matrix of variances and covariances between pairs of sequences that is inferred from the tree.
- *Balanced Profiles:* The notion of balanced profiles appears in the work of Vingron and Sibbald [VS93], where they also draw a parallel between this scheme and that of Altschul et al. mentioned above [ACL89]. Even though the sequence weights are defined in exactly the same way as we do here, the way in which the weights are used in their method to compute the profile

is different from ours. As mentioned earlier, in our method, the profile at each position is computed using a combination of the weights as well as the underlying distance function (see Equation 1), while in Vingron and Sibbald the profile is defined simply as the weighted sum of the sequences in the alignment (i.e. $P[i, \sigma]$ is the sum of the weights for the sequences having residue σ in column i). With their definition, the weights yield a balanced profile if $Dw = \lambda 1$ so that, if D is invertible, w can be found by solving $w = \lambda D^{-1} 1$. This objective is pursued in Vingron and Argos [VA89] where the weights of sequences far from the (unweighted) profile are increased thus moving the weighted profile away from nearby sequences and toward distant ones.

- *Maximum Discrimination:* This probabilistic method introduced by Eddy, Mitchison and Durbin [EMD95], uses Hidden Markov Model to model the protein family (based on the given multiple alignment). The objective is to find a HMM that maximizes the probability that all of the sequences participating in the multiple alignment will be produced by the model (as opposed to being produced by a random model). The following gradient descent training method is proposed: Find the sequence (or sequences) with the lowest score, and change the model probabilities so as to better recognize those sequences (this method is called the Maxmin algorithm in [EMD95]). We remark that in our algorithm, a very similar training method is used; However, no probabilistic modeling of the multiple alignment is employed and hence our algorithm is much simpler and faster.

2.1 The Balanced Profile Weight Assignment Problem

The main utility of a sequence weighting scheme in building profiles is its potential to correct for bias in the input sequences. Motivated by this intended application of weighting schemes, we define a *balanced profile* to be one with almost identical similarity to all sequences participating in its construction. In other words, all sequences are chosen with the same intensity by this profile. The problem of weighting the sequences to build a profile then becomes one of setting the weights so that the resulting profile is balanced.

A profile can be thought of as a sequence of weighted residuals. A natural way to define the similarity score of a profile versus a target sequence is, by linearity, to compute the weighted sum, for all the positions in the sequence, of the similarities of residues in the profile and that in the target. When this scoring function is used, computing the weights which yield a balanced profile would simply amount to solving a system of linear equations in the variables $\{w_i\}$ [4]. However, this method can not be applied to the scoring function used by PROFILESEARCH, since it is not additive over the columns (e.g. the score for a run of k gaps is not the same as k times the score of one gap).

We propose an algorithm which is independent from the details of the scoring function adopted in PROFILESEARCH, or in any other scoring procedure to be

---

[4] This remark also appears in the article by Vingron and Sibbald[VS93]

used. Our algorithm is iterative and starts by giving equal weights to all the sequences. At each iteration the profile is computed using the current weights. If the profile is not balanced, the weights are updated and a new iteration is executed. Preliminary computational experiments with the scoring function used by PROFILESEARCH have shown that the algorithm convergences on the average very fast (under 20 iterations) to the final weights. A formal statement of the procedure follows.

Repeat

1. Build a profile from the multiple alignment (initially, all weights are equal).
2. Score this profile against the sequences in the alignment.
3. **If** the profile is not balanced
   then change the weight of either the sequences with the highest or lowest score.

until the profile is balanced.

In order to choose which weight to change (Step 3), the following procedure is followed. Define $score_{mid} = (score_{max} + score_{min})/2$. Determine how many scores lie on either side of $score_{mid}$. If there are fewer scores below $score_{mid}$, raise the weight of all sequences with the lowest score by δ, and if there are fewer weights above $score_{mid}$, then lower the scores of the sequences of the highest score by δ, where $\delta = (score_{max} - score_{min})/score_{max}$. The choice of δ reflects the dependence on the step size on the current spread of scores, moving rapidly when the spread is high and more carefully otherwise.

In order to prevent negative weights, if the weight of the sequences with the scores to be lowered is below a threshold, simply increase the lowest scoring sequence's weight instead. The iterations are stopped when all scores are within some specified error bound of each other, i.e. roughly similar. This method of weight changing assumes that increasing the weight of one sequence while holding everything else constant, will increase the amount by which the profile reflects that sequence compared to the others. This is true for both the linear and logarithmic weighting schemes used by PROFILEMAKE.

3 Search Results with Balanced Profiles

Balanced profiles were constructed for protein blocks and searched against SwissProt. The results of these searches are compared to searches with profiles of the same blocks constructed using equal weights. The data for these experiments were provided by Henikoff and Henikoff, which they used to calibrate their weight setting scheme against an equal-weights model [HH94]. Known protein families were extracted from PROSITE, and subsets of these families were used to construct blocks with PROTOMAT [Bai92]. Weights were assigned to sequences with the balanced profile scheme, and profiles were constructed from these blocks.

In order to compare the balanced profile search results to the ones with equal-weights, the PROFILESEARCH results are compared to the sequences known to be in the protein family of the block subset. Thirty-seven blocks (corresponding to as many families) were arbitrarily chosen from the complete data set of Henikoff and Henikoff. Thirty seven pairs of profiles were constructed from these blocks using the balanced weights and equal weights for each case, and these were used to search the database for members of the corresponding blocks. The overview of the computational results is presented in Table 1.

Number of Sequences	Equal Weights	Balanced Profile	Ties
0-50	6	4	27
51-100	2	4	31
101-200	4	3	30
201-300	3	1	33
301-400	0	1	36
401-500	0	1	36
501-1000	4	3	30
1001-5000	3	2	32
over 5000	0	1	36
overall	4	4	29

Table 1. Overview of results - the number of wins for each scheme and the ties are shown for each of the ranges. E.g., among the 37 tests, Equal weight profiles recovered more sequences from the family 6 times than the balanced profile and the reverse happened 4 times, while in the remaining 27 cases, they were tied in the recovery rate in this range. The final row tabulates overall wins considered over all the ranges for the 37 tests. Note that every row sums to the total number of tests - 37.

The small sample size does not allow conjecture toward significant conclusions, however it appears that the balanced profile will perform no worse on average than the equal-weights profile. Simply by inspection, it appears that the balanced profile performed better when the blocks which composed the multiple alignment were more divergent. This is to be expected since this is exactly the type of data that exploits the benefits of using balanced profiles. As seen in Table 1, the overwhelming number of results are ties. This is likely a result of the nature of the experimental dataset. The majority of blocks used in these experiments were composed of sequences with a high degree of similarity. When data sets are constructed from very similar sequences, the weights generated by the balanced profile method should not significantly impact the database search. Those blocks which resulted in ties often had more residues which were completely conserved over all sequences than those for which the balanced profile won. These preliminary experiments have helped to provide some insight into potential situations where balanced profiles would be a helpful search tool. Further experiments with datasets with a wide variety of skew will enable the specific usefulness of balanced profiles to be further defined.

4 Approximation Algorithms for the Consensus Problem

The *sphere* in Σ^n of *radius* r with center $a \in \Sigma^n$ is the set of all sequences $v \in \Sigma^n$ such that $d(v, a) \leq r$, and is denoted by $S(a, r)$. Given a sequence x and a set $V \subseteq \Sigma^n$, the *radius* of V with respect to x, denoted by $R(V, x)$ is the the smallest integer r such that $V \subseteq S(x, r)$. We define the *radius* of V, denoted by $R(V)$, as $R(V) = \min_{x \in \Sigma^n} R(V, x)$.

The consensus problem is then the following: given a set of sequences $V \subseteq \Sigma^n$, representing the rows of some multiple alignment, find a sequence $c \in \Sigma^n$ (the consensus) such that $R(V, c) = R(V)$ (in other words, find the sequence c which minimizes $\max_{a \in V} d(c, a)$).

It has been shown by Frances and Litman ([FL94]) that determining the consensus is NP-complete (using a reduction from 3-SAT), in the special case where the alphabet is binary and the distance measure is the Hamming distance. A slight modification of their reduction generalizes the hardness result to arbitrary finite alphabet Σ. Assuming the distance function d takes only rational values, and using the fact that scaling d does not change the problem, allows us to generalize the hardness results to arbitrary distance functions d. Because of the computational complexity of the problem, exact methods for finding the consensus sequence may require too much computing time and heuristic procedures should be sought instead. In particular, we are interested in fast performance-guarantee algorithms.

Our version of the consensus problem arises as a natural alternate objective for tree alignment in the special case when the tree is a star with the given sequences at the leaves and the internal node has to be computed so as to minimize the bottleneck cost of the tree – namely, minimize the maximum distance from the internal sequence to the other input sequences. This is termed bottleneck tree alignments by Ravi and Kececioglu [RK95]. Thus our method for finding unbiased consensus sequences also finds near-optimal bottleneck tree alignments for the star. However, our method is applicable only for the case when the distance function does not allow gaps, and is therefore not applicable to the general version of the tree problem allowing arbitrary edit distances.

A trivial 2-approximation algorithm for the consensus problem consists simply in picking any of the given sequences as the consensus [RK95]. In fact, assuming that the distance between sequences satisfies the triangle inequality, we have the following.

Claim 1. *Let V be a set of sequences; then any sequence $v \in V$ gives a 2-approximation of the radius of V.*

Proof. Let v^* be an optimal solution, and let r denote the minimal radius of a sphere around v^* that contains V, that is $d(v^*, x) \leq r$ for every $x \in V$. Let v be an arbitrary sequence in V. We need to show that for every word $w \in V$, the distance between w and v is at most $2r$. Using the inequalities $d(v^*, v) \leq r$ and

$d(v^\star, w) \leq r$ together with the triangle inequality $d(v, w) \leq d(v, v^\star) + d(v^\star, w)$, completes the proof.

□

4.1 Near-optimal approximation using randomized rounding

In this section we use the method of randomized rounding [RT87, Rag88, MR95] to achieve a near-optimal solution to the consensus problem. The method can be roughly described as follows: First we formulate our problem as an integer-programming problem, using zero-one variables. Then, we relax the integrality constrains, so that the variables are allowed to have fractional values. An optimal solution to this linear programming problem can be found in polynomial time [Kar84]. To obtain a solution to the original, integer program, we "round" the solution of the relaxed problem, using the fractional values in each column as probabilities. We then show that with high probability, the value of the rounded solution is close to the value of the non-integral (optimal) solution.

The consensus problem can be cast as a zero-one linear program as follows. Let c be the consensus sequence to be determined. For every symbol $\sigma \in \Sigma$, and every column i ($1 \leq i \leq n$), we use a zero-one variable $x_{i,\sigma}$ to indicate whether $c_i = \sigma$. Note that we do not allow the blank character to occur as part of the consensus sequence. Our integer program can be expressed as follows.

$$\begin{aligned}
Minimize \qquad & r \\
s.t. \qquad & \sum_\sigma x_{i,\sigma} = 1 \qquad \forall i \in \{1, \ldots n\} & (2) \\
& \sum_{i,\sigma} x_{i,\sigma} d(\sigma, v_i) \leq r \; \forall v \in V & (3) \\
where \qquad & x_{i,\sigma} \in \{0, 1\} & (4)
\end{aligned}$$

The constraint (2) ensures that a unique symbol is chosen for each position of c. The constraint (3) specifies that the total distance between any member of V and c is at most r (i.e, $R(V, c) \leq r$). The objective function seeks a solution of minimum radius, r, with the zero-one constraint imposed. Let r_0 denote the value of the objective function in the optimum solution to the program above. Since this problem is NP-Hard, we do not hope to compute r_0 efficiently. Instead, we solve its *linear programming relaxation*.

We replace the integrality constraint (4) with $x_{i,\sigma} \geq 0$. In other words, we allow $x_{i,\sigma}$ to assume real values between 0 and 1 (the constraint $x_{i,\sigma} \leq 1$ is implicit in the constraint (2)). Let $\hat{r}$ be the value of the objective function for this problem and $\hat{x}_{i,\sigma}$ its solution. Since the linear program is a relaxation of the integer program, it is clear that $r_0 \geq \hat{r}$. The $\hat{x}_{i,\sigma}$'s may be fractional values, and therefore may not constitute a feasible solution to the integer program. We must therefore "round" these fractional values to 0's and 1's to obtain a feasible solution x.

Note that the fractional solution $\hat{x}_{i,\sigma}$ still satisfies the constraints of the original linear program, in particular, for each i, the values $\{\hat{x}_{i,\sigma}\}_{\sigma \in \Sigma}$ are all non-negative, and satisfy $\sum_{\sigma \in \Sigma} \hat{x}_{i,\sigma} = 1$. Therefore, they define a probability distribution over Σ. The rounding process goes as follows. Independently for each i, choose a symbol for c_i according to the probability distribution defined by $\{\hat{x}_{i,\sigma}\}_{\sigma \in \Sigma}$. That is, $Pr(c_i = \sigma) = \hat{x}_{i,\sigma}$.

Consider some fixed sequence $v \in V$; The expected value of the distance between c and v satisfies the following.

$$
\begin{aligned}
E[d(c,v)] &= E[\sum_{i=1}^{n} d(c_i, v_i)] \\
&= \sum_{i=1}^{n} E[d(c_i, v_i)] \quad \text{(by linearity of expectation)} \\
&= \sum_{i=1}^{n} \sum_{\sigma \in \Sigma} Pr(c_i = \sigma) d(\sigma, v_i) \\
&= \sum_{i=1}^{n} \sum_{\sigma \in \Sigma} \hat{x}_{i,\sigma} d(\sigma, v_i) \\
&\leq \hat{r} \\
&\leq r_0
\end{aligned}
$$

Following [RT87], we use Hoeffding's bound [Hoe63] (See, e.g., [MR95]).

Lemma 2. *Let $X_1, X_2, \ldots, X_n$ be n independent random variables, each ranging over the real interval $[a, b]$. Let S be a random variable denoting the sum of the X_i's, and δ any nonnegative real number. Then*

$$ Pr(S \geq (1+\delta)E[S]) \leq e^{-\frac{E[S]\delta^2}{3(b-a)^2}} $$

Notice that each random variable $d(c_i, v_i)$ range over $[0, D]$, where D is an upper bound on the distance function, e.g. $D = \max_{\alpha, \beta \in \Sigma} \{d(\alpha, \beta)\}$. Let $\epsilon > 0$ be a positive constant. For

$$ \delta = D\sqrt{\frac{3}{E[S]} \log \frac{|V|}{\epsilon}}, $$

we get $Pr(d(c,v) \geq r_0(1+\delta)) \leq \frac{\epsilon}{|V|}$. Summing over every member of V, we have $Pr(R(V,c) \geq r_0(1+\delta)) \leq \epsilon$. We get the following theorem.

Theorem 3. *Let $V \subseteq \Sigma^n$ be a set of sequences, and let c_{opt} denote the optimal consensus. Let r_0 denote $R(V, c_{opt})$. Let c be the rounded solution to the relaxed integer program above, and let r denote $R(V, c)$. Then*

$$ Pr\left(r > r_0 + D\sqrt{3r_0 \log \frac{|V|}{\epsilon}} \right) < \epsilon $$

where $\epsilon > 0$ is a constant.

Notice that this probabilistic algorithm can be de-randomized using standard techniques of conditional probabilities, along with pessimistic estimators as in [Rag88].

Remark. Using a different version of the Hoeffding bound: $Pr(S - E[S] > \delta) \leq e^{-\frac{2\delta^2}{(b-a)^2 n}}$, we can derive the following result.

$$Pr\left(r > r_0 + D\sqrt{\frac{n}{2}\log\frac{|V|}{\epsilon}}\right) < \epsilon$$

where $\epsilon > 0$ is a constant. This version is better when $r_0 \gg D\sqrt{\frac{n}{2}\log\frac{|V|}{\epsilon}}$.

5 Extensions

5.1 Selecting Probes for Bacterial Infections

Assume a patient has a bacterial infection. Let S be the set consisting of short nucleotides sequences which are specific DNA (either chromosomal or ribosomal) of several possible bacteria that might be the cause of the infection. Let $T \subseteq S$ be a set of sequences of the target bacteria species that actually cause the infection.

DNA diagnosis for the bacterial infection ([DKK88, MM90]) is a technique using the complementary nature of DNA nucleotides to decide whether the cause of the infection is from the set T. The idea is to choose a DNA sequence, called the *DNA probe*, such that its complement is "close" to the sequences in T, and "far away" from the sequences in $S \setminus T$. In practice, even if the complementary sequence of a probe has a few mismatches with a specific substring of the target sequences, the probe forms duplexes with some of the target sequences. Thus for any given probe, the accuracy of the diagnosis using this probe is a function of two parameters:

- The maximum number of mismatches between the complementary sequence of the probe and any nucleotide sequence from the target bacteria. The smaller this number is, the smaller is the probability of a false negative.
- The minimum number of mismatches between the complementary sequence of the probe and any nucleotide sequences of a bacteria not in the target species. The larger this number is, the smaller is the probability of a false positive.

We say that a sequence t is a *k-separator* with respect to $\langle T, S \setminus T \rangle$ if

$$\min_{v \in S \setminus T} d(t, v) - \max_{v \in T} d(t, v) = k.$$

The Probe–Selection Problem can now be stated as follows: given two sets of sequences, T and $S \setminus T$, find a sequence t with maximum separation.

Note that this problem generalizes the consensus problem (which occurs when $T = S$) and is therefore NP-complete. By casting the problem as a 0-1 program and using randomized rounding as in the preceding section, we obtain the following result.

Theorem 4. *Let $T \subseteq S$ be two sets of sequences, and let t_{opt} denote the optimal $\langle T, S \setminus T \rangle$-separator. Let k_{opt} denote $\min_{v \in S \setminus T} d(t_{opt}, v) - \max_{v \in T} d(t_{opt}, v)$. Let t be the rounded solution to the relaxed integer program, and let k denote $\min_{v \in S \setminus T} d(t, v) - \max_{v \in T} d(t, v)$. Then*

$$Pr\left(k < k_{opt} - D\sqrt{4k_{opt} \log \frac{m}{\epsilon}} \right) < \epsilon$$

where $m = \max\{|T|, |S \setminus T|\}$, and $\epsilon > 0$ is a constant.

5.2 Building Consensus Maps

Assume a human chromosome (modeled by the real interval $[0,1]$) is known to contain n specific markers. Various mapping techniques are used to order those markers, and even suggest chromosomal locations of them.

Assume we are given a set $\mathcal{M}$ of physical maps of the chromosome. Each map consists of the locations of n markers. That is, each map can be represented by a vector $v \in [0,1]^n$. The distance between two maps, $v, u \in [0,1]^n$ is defined as:

$$d(v, u) = \sum_{i=1,\ldots,n} |v_i - u_i|$$

The goal is to find a map which is a good representative of all maps.

A trivial solution is to choose each marker location in the consensus map as the average location of that marker in the input maps. This solution, however, introduce bias. Alternatively, we can use our bottleneck criterion and apply randomized rounding method to choose a consensus map with almost no bias.

Notice that we can not simply introduce a real variable $y_i \in [0,1]$, for each marker location as the resulting program will not be linear. In order to write an linear program we need to approximate the infinite alphabet $\Sigma = [0,1]$ with a polynomial size alphabet $\Sigma' = \{\frac{1}{m+1}, \frac{2}{m+1}, \ldots, \frac{m}{m+1}\}$. We restrict each marker location in the consensus map to choose a position from Σ'. Naturally, we lose precision, but we can choose m large enough so the inaccuracy introduced will be not more than the inaccuracy introduced by the rounding phase.

We introduce mn zero-one variables $x_{i,j}$, where $x_{i,j} = 1$ if and only if the i-th marker location in the consensus map c is $\frac{i}{m+1}$. The resulting integer linear program is therefore:

$$
\begin{aligned}
&Minimize && r \\
&s.t. && \sum_j x_{i,j} = 1 && \forall i \in \{1, \ldots n\} \\
& && \sum_{i,j} x_{i,j} d(\tfrac{j}{m+1}, v_i) \leq r \ \forall v \in \mathcal{M} \\
&where && x_{i,j} \in \{0,1\}
\end{aligned}
$$

To get the consensus map, we relax the integrality constraints, and round the linear solution, as in the previous section.

Notice that we can bound the distance between the Σ-optimal map to the Σ'-optimal map by $\frac{n}{m}$ for $m = n\sqrt{n\,|\mathcal{M}|}$, and using the fact that the distance function d is bounded by 1 (i.e. $D = 1$), we have the following.

Theorem 5. *Let $\mathcal{M}$ be a set of maps, and let c_{opt} denote the optimal consensus map. Let r_0 denote $R(\mathcal{M}, c_{opt})$. Let c be the rounded solution to the relaxed integer program above, and let r denote $R(\mathcal{M}, c)$. Then*

$$Pr\left(r > r_0 + 3\sqrt{\frac{n}{2}\log\frac{2\,|\mathcal{M}|}{\epsilon}}\right) < \epsilon$$

where $\epsilon > 0$ is a constant.

6 Open Questions

In order to check the performance of the balanced profile method more accurately, the most divergent protein families should be selected for searching. This way, we would be testing the cases that the weighting systems are developed for, namely sequence sets which may result in bias. A fair comparison with weighting schemes other than equal weights must be done to determine its relative efficacy.

While our method for consensus sequences applies to compute bottleneck tree alignments for a star, they do not extend directly to arbitrary tree topologies. As a first step, an extension of our method to the general version of the star bottleneck problem allowing edit distances should be investigated. Then, it would be interesting to see if our technique can be further extended to the following problem: given a leaf-labeled tree, find ancestral sequence labels at the internal nodes so that the maximum cost of any edge (edit-distance or even Hamming distance between the endpoints) in the tree is minimized. A logarithmic approximation for this problem even with edit-distances is already known [RK95].

References

[ACL89] Stephen F. Altschul, Raymond J. Carroll, and David J. Lipman. Weights for Data Related by a Tree. *Journal of Molecular Biology*, 207, 647–653, 1989.

[Bai92] A. Bairoch. PROSITE: A Dictionary of Sites and patterns in Proteins. *Nucleic Acids Research*, 20, 2019–2022, 1992.

[BB92] A. Bairoch, and B. Boeckmann. The SWISSPROT Protein Sequence Data Bank. *Nucleic Acids Research*, 20, 2019–2022, 1992.

[DBH83] M.O. Dayhoff, W.C. Barker and L.T. Hunt. Establishing homologies in protein sequences. *Methods Enzymol.*, 91:524–545, 1983.

[DKK88] R. Dular, R. Kajioka, and S. Kasatiya. Comparison of gene-probe commercial kit and culture technique for the diagnosis of mycoplasma pneumoniae infection. *J. of Clinical Microbiology*, 26(5):1068–1069, May 1988.

[EMD95] S.R. Eddy, G. Mitchison, and R. Durbin. Maximum discrimination hidden Markov models of sequence consensus. *J. of Computational Biology*, 2:9-23. 1995.

[FL94]   M. Frances and A. Litman. On covering problems of codes. Technical Report 827, Technion, Israel, July 1994.

[GCG94]   *Program Manual for the Wisconsin Package*, Version 8, September 1994, Genetics Computer Group, 575 Science Drive, Madison, Wisconsin, USA 53711.

[GME87]   M. Gribskov, A. D. McLachlan, and D. Eisenberg. Profile Analysis: Detection of Distantly Related Proteins. *Proceedings of the National Academy of Science, U.S.A.*, 84, 4355–4358, 1987.

[GSC94]   M. Gerstein, E. Sonnhammer, and C. Chothia. Volume Changes in protein evolution. *J. Mol. Biol.*, 235:1067-1078, 1994.

[HH94]   Steven Henikoff and Jorja G. Henikoff. Position-based Sequence Weights. *J. Mol. Biol.*, 243, 574–578, 1994.

[Hoe63]   W. Hoeffding. Probability inequalities for sums of bound random variables. *J. Amer. Statist. Assoc.*, 58:13–30, 1963.

[ISNH94] M. Ito, K. Shimizu, M. Nakanishi, and A. Hashimoto. Polynomial-time algorithms for computing characteristic strings. *Proc. CPM 94*, LNCS 807:274–288, 1994.

[Kar84]   N. Karmarkar. A new polynomial time algorithm for linear programming, *Combinatorica*, 4:373-395, 1984.

[KM95]   A. Krogh, and G. Mitchison. Maximum entropy weighting of aligned sequences of protein or DNA, in *Proc. Third Int. Conf. on Intelligent System for Mol. Biol.*, (C. Rawlings, D. Clark, R. Altman, L. Hunter, T. Lengauer, S. Wodak, eds.) pp. 215-221, AAAI Press, Menlo Park, CA, 1995.

[LXB94]   R. Luthy, I. Xenarios, and P. Bicher. Improving the sensitivity of the sequence profile method, *Protein Science*, 3:139–146, 1994.

[MM90]   A.J.L. Macario and E.C.De. Macario. *Gene Probes for Bacteria*. Academic Press, 1990.

[MR95]   R. Motwani and P. Raghavan. *Randomized Algorithms*. Cambridge University Press, 1995.

[Rag88]   P. Raghavan. A probabilistic construction of deterministic algorithms: Approximating packing integer programs. *Journal of Computer and System Sciences*, 37:130–143, 1988.

[RK95]   R. Ravi and J. D. Kececioglu. Approximation algorithms for multiple sequence alignment under a fixed evolutionary tree, *Proc. CPM 95*, LNCS 937:330–339, 1995.

[RT87]   P. Raghavan and C.D. Thompson. Randomized rounding: a technique for provably good algorithms and algorithmic proofs, *Combinatorica*, 7:365–374, 1987.

[SA90]   Peter R. Sibbald and Patrick Argos. Weighting Aligned Protein or Nucleic Acid Sequences to Correct for Unequal Representation. *Journal of Molecular Biology*, 216, 813–818, 1990.

[SW81]   T.F. Smith and M.S. Waterman. Comparison of Biosequences. *Adv. Appl. Math.*, 482–489, 1981.

[THG94]   J.D. Thompson, D.G. Higgins and T.J. Gibson. Improved sensitivity of profile searches through the use of sequence weights and gap excision, *Comput. Applic. Biosci.*, 10:19–29, 1994.

[VA89]   M. Vingron and P. Argos. A fast and sensitive multiple sequence alignment algorithm. *Comput. Appl. Biosci.*, 5:115–121, 1989.

[VS93]   M. Vingron and P.R. Sibbald. Weighting in sequence space: A comparison of methods in terms of generalized sequences. *Proc. Natl. Acad. Sci. USA*, 90:8777-8781, 1993.

On the Nadeau-Taylor Theory of Conserved Chromosome Segments

David Sankoff * , Marie-Noelle Parent, Isabelle Marchand, Vincent Ferretti

Centre de recherches mathématiques, Université de Montréal, CP 6128 Succursale
Centre-ville, Montréal, Québec, H3C 3J7

Abstract. The quantification of comparative genomics dates from 1984
with the work of Nadeau and Taylor on estimating interchromosomal exchange rates based on the rearrangement of chromosomal segments in human
versus mouse genomes. We reformulate their analysis in terms of a probabilistic model based on spatial homogeneity and independence of breakpoints and
gene distribution. We study the marginal distribution of the number of genes
per segment and the distribution of the number of non- empty segments as a
function of the number of genes and segments. We propose a rapid algorithm
for identifying a given number of conserved segments in noisy comparative
map data. Finally, we propose a model which incorporates a degree of inhomogeneity in the distribution of genes and/or breakpoints. Comparative
maps of human and mouse genomes serve as test data throughout.

1 Introduction

During evolution, inter- and intrachromosomal exchanges such as reciprocal translocation, transposition and inversion disrupt the order of genes along the chromosome
(Figure 1).

In comparing two divergent genomes, a contiguous stretch of chromosome in
which the number and order of homologous genes is the same in both species, i.e.
has not been interrupted by any of the rearrangement processes that have occurred
in either lineage, is called a *conserved segment*. The number of conserved segments
increases as they are disrupted by new events, so that they tend to become shorter
over time. The number of chromosomal segments conserved during the divergence
of two species can be used to measure their genomic distance.

An early and influential contribution to the quantitative methodology of comparative genomics was made by Nadeau and Taylor in 1984 [3], focusing on interchromosomal exchange as the major mechanism in the rearrangement of mammalian

---
* E-mail: sankoff@ere.umontreal.ca. Research supported by grants from the Natural Sciences and Engineering Research Council of Canada and the Canadian Genome Analysis and Technology program. DS is a fellow of the Canadian Institute for Advanced
Research.

genomes. Our formulation of the Nadeau-Taylor model of genomic divergence assumes that each reciprocal translocation breaks chromosomes at random points on two randomly chosen chromosomes. As a consequence when we compare two divergent genomes, the endpoints of the conserved segments making up each chromosome are uniformly and independently distributed along its length (spatial homogeneity of breakpoints). We also assume that which genes of a genome are discovered and mapped first does not depend on their position on the chromosome (spatial homogeneity of gene distribution), nor on their proximity to each other (independence of map positions).

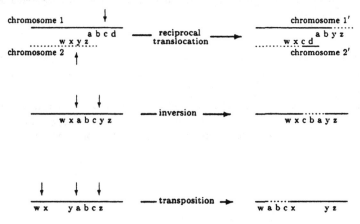

Fig. 1. Schematic view of genome rearrangement processes. Letters represent positions of genes. Vertical arrows at left indicate breakpoints introduced into original genome. Reciprocal translocation (top) exchanges end segments of two chromosomes. Inversion (center) reverses the order of genes between two breakpoints (dotted segment at right). Transposition (bottom) removes a segment defined by two breakpoints and inserts it at another breakpoint (dotted segment at right), in the same chromosome or another. Gene order conserved (possibly inverted) within segments.

2 The Marginal Probability of r-gene Segments

In trying to count the number of conserved segments for the quantification of evolution, we must deal with underestimation due to conserved segments in which genes have not yet been identified in one or both species. There are two in Figure 2: one from chromosome 4 of Genome 2 and the other from chromosome 17. This is particularly important if there are relatively few genes common to the data sets for a pair of species, so that many or most of the conserved segments are not represented in the comparison, and genomic distance may be severely underestimated. Nadeau and Taylor [3] in 1984 could only treat 13 segments out of the 130 or so now known to exist (see Section 4.3 below).

We model the genome as a single long unit broken at n random breakpoints into $n + 1$ segments, within each of which gene order has been conserved with reference to some other genome. (Little is lost in not distinguishing between breakpoints and

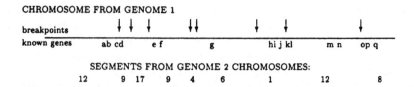

CHROMOSOME FROM GENOME 1

breakpoints

known genes   ab cd   e f   g   hi j kl   m n   op q

SEGMENTS FROM GENOME 2 CHROMOSOMES:

12   9 17   9   4   6   1   12   8

Fig. 2. Fictitious example of conserved segments indicated on a chromosome from Genome 1, with each segment labeled between its endpoints (adjacent arrows) as to which chromosome it is found on in Genome 2. Homologous genes that have been discovered to date are indicated with letters.

concatenation boundaries separating two successive chromosomes [5].) The marginal probability that a segment contain r genes is given by the following theorem [7].

Theorem 1. *Consider a linear interval of length 1, with $n > 0$ uniformly distributed breakpoints that partition the interval into $n + 1$ segments. Suppose there are m genes also distributed uniformly on the interval between 0 and 1, and independently*

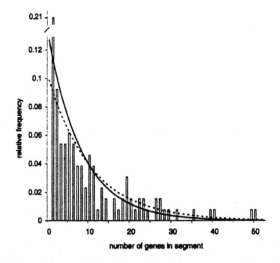

Fig. 3. Comparison of relative frequencies $n_r / \sum_{r>0} n_r$ of segments containing r genes with predictions of Nadeau-Taylor model. Value of n in formula for Q is taken to be 141 (dotted curve) or 181 (uninterrupted curve), as estimated by the maximum likelihood method of Section 3.2 or the Kolmogorov-Smirnov method of Section 5, respectively. Both curves show values for $Q(0)$, though zero is not in the range of the conditional distribution, to permit a comparison of the estimated number $K(m, n)Q(0)$ of unobserved (empty) segments with the predictions $K(m, n)Q(r)$ for positive r, where $K(m, n) = (n+1)m/(n+m)$. Three data points are off-scale, with $r = 54, 65$ and 83 and the vertical axis is interrupted to allow an expanded scale, facilitating more detailed visualization of $f(r)$ and $Q(r)$, $r > 1$.

of the breakpoints. For an arbitrary segment, the probability that it contains r genes,
$0 \le r \le m$, *is then*

$$\Pi(r) = \frac{n}{n+m} \binom{m}{r} / \binom{n+m-1}{r}.$$

We can only partially compare the theoretical distribution $\Pi(r)$ with n_r, the number of segments observed to contain r genes, since we cannot observe n_0, the number of segments containing no identified genes. We can at least compare the relative frequencies $f(r) = \frac{n_r}{\sum_{r>0} n_r}$ with the conditional probabilities $Q(r) = \Pi(r \mid r > 0)$. This is seen in Figure 3, where the largest discrepancy is the comparison between $f(1)$ and $Q(1)$. We will discuss this discrepancy, how to interpret it, and the consequences of ignoring it, in Section 5.

3 The Inference Problem

It might seem that the number of segments n_r observed to contain r genes, for $r = 1, 2...$, would be useful data for inference about the Nadeau-Taylor model, in particular about n, the unknown number of breakpoints. Though we will see in Section 5 that these data are indeed useful for generalizing the model, they are not necessary for the basic distribution given in Theorem 1.

3.1 The Sufficiency of the Number of Observed Segments

It is remarkable that to estimate n from m and the n_r, for $r = 1, 2...$, only the number of non-empty segments $a = \sum_{r>0} n_r$ is important [4].

Theorem 2. *The variable a is a sufficient statistic for the estimation of n.*

3.2 Estimating n from a

To estimate n, we study $P(a, m, n)$, the probability of observing a non-empty segments if there are m genes and n breakpoints. Combinatorial arguments give:

Theorem 3.

$$P(a, m, n) = \frac{\binom{m-1}{a-1}\binom{n+1}{a}}{\binom{n+m}{m}}$$

After observing m and a it is an easy matter to find the value of n which maximizes P, i.e. the maximum likelihood estimate.

Another approach, for extremely large values of the parameters, is to use the mean and variance of $P(a, m, n)$:

$$E(a, m, n) = \frac{(n+1)m}{(n+m)}, \quad \text{Var}(a, m, n) = \frac{(n+1)nm(m-1)}{(n+m-1)(n+m)^2}$$

A gaussian approximation allows accurate calculation for high values of m and n. To do maximum likelihood estimation, the log of the gaussian density with $\mu = E(a, m, n), \sigma^2 = \text{Var}(a, m, n)$ is differentiated with respect to n and set equal to zero. The solution is the only positive root of the following degree 6 polynomial:

$$m^3 - 2am^3 + a^2m^3 - 2m^4 + 4am^4 - 2a^2m^4 + m^5 - 2am^5 + a^2m^5 - 2am^2n$$

$$+a^2m^2n + m^3n + 2am^3n - a^2m^3n - 3m^4n + 4am^4n - 2a^2m^4n + 2m^5n$$

$$-4am^5n + 2a^2m^5n + mn^2 - a^2mn^2 - 4m^2n^2 - 4am^2n^2 + 7a^2m^2n^2$$

$$+4m^3n^2 + 10am^3n^2 - 13a^2m^3n^2 - 2m^4n^2 - 4am^4n^2 + 7a^2m^4n^2 + m^5n^2$$

$$-2am^5n^2 - a^2n^3 - mn^3 - 2amn^3 + 9a^2mn^3 - m^2n^3 + 2am^2n^3 - 17a^2m^2n^3$$

$$+3m^3n^3 + 6am^3n^3 + 8a^2m^3n^3 - 2m^4n^3 - 4am^4n^3 + 3a^2n^4 - 3mn^4 - 2amn^4$$

$$-8a^2mn^4 + 4m^2n^4 + 8am^2n^4 + 2a^2m^2n^4 - 3m^3n^4 - m^4n^4 - a^2n^5 - mn^5$$

$$+2amn^5 - 2a^2mn^5 + 4am^2n^5 - 2m^3n^5 - a^2n^6 + 2amn^6 - m^2n^6$$

The approximation is not necessary for current data levels, but both methods give, for $m = 1423, a = 130$, valid values for the man-mouse comparison in the summer of 1996 (cf. Section 4.3), an estimate of 141 for n, suggesting that less than 10% of the segments have not yet been observed.

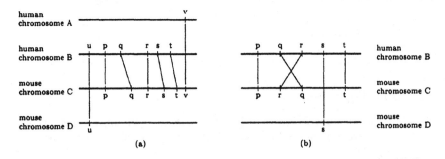

(a)   (b)

Fig. 4. (a). Schematic example of conserved segment in a human chromosome B and a mouse chromosome C. Genes u and v have homologues elsewhere in the mouse and human genomes, respectively, and thus limit the leftward and rightward extension of the segment. (b). Experimental mistake in the chromosomal assignment of s to mouse chromosome D, quantitative error in the assignment of q and/or r in the human or mouse map, or inversion of qr or transposition of q or r, results in the erroneous identification of three segments, p, qr, t, instead of just one, in human chromosome B and mouse chromosome C, and an additional one, s, in human chromosome B and mouse chromosome D.

4 The Identification of Conserved Segments

Conserved segments were defined in the Introduction to be regions of chromosomes in two related species in which both gene content and gene order are parallel (Figure 4(a)). As map data accumulate, however, it becomes increasingly difficult to find segments that satisfy the criteria of content and order perfectly. This can be attributed in part to experimental error - either gross mistakes in chromosomal assignment of genes or quantitative errors in map positions affecting apparent gene order. In addition, in the comparison of multichromosomal species such as humans and mice, we may wish to consider the segment structure to be that produced by translocation, and to consider as "noise" the effects of high rates of inversion and transpositions of small regions of chromosomes (Figure 4(b)).

Our hypothesis is that we can recover the configuration of conserved segments resulting from the evolutionary history of reciprocal translocations, and thus account for the gross differences between the genomes, by minimizing appropriately weighted mapping error plus rearrangement costs.

We do this with a variant of single link stepwise cluster analysis performed simultaneously on all conserved synteny sets (sets of genes occurring in common on one human chromosome and one mouse chromosome), with the interim results from each cluster analysis affecting the current state of all other cluster analyses [6].

4.1 The Objective Function

Let $c \leq c_1 c_2$ be the total number of conserved synteny sets, where c_1 and c_2 are the number of chromosomes in species 1 and species 2, respectively. c is also the smallest number of segments that can be produced by any analysis, grouping all genes belonging to a conserved synteny, no matter how dispersed they are along the chromosome, into a single conserved segment, not allowing for a single conserved synteny to be the result of two or more translocation events. At the other extreme, if we assume that each gene defines a different conserved segment and that genes are adjacent in two genomes only by coincidence, we obtain m segments, the total number of homologous genes identified in the two genomes. All solutions lie somewhere between these two extremes. For an appropriate choice of weighting parameters, α, β, γ, and for all a, $c \leq a \leq m$, we wish to find the subgroupings of conserved syntenic genes into a segments so as to minimize

$$D = \sum_{i=1}^{a} D_i,$$

where D_i is a weighted measure of the compactness, density and integrity of segment i. Formally,

$$D_i = \gamma \max_{x,y \in i(1)} |x - y| + \alpha s[i(1)] + \gamma \max_{x,y \in i(2)} |x - y| + \alpha s[i(2)] - \beta r(i),$$

where $x_{\epsilon i}(j)$ refers to a gene (or its map coordinate) in segment i in species j, $r(i)$ indicates the number of homologous gene pairs in segment i and $s[i(j)]$ denotes the number of *other* segments with elements within the range of segment i in species j.

4.2 The Algorithm

Direct minimization of $D = \sum D_i$ is generally not feasible, because what is included in segment i impacts the quality of other segments and vice-versa. Instead we propose a rapid stepwise upper-bound algorithm and show sufficient conditions for it to calculate D exactly. An advantage of this method is that it constructs solutions for all a in one pass.

Our procedure starts with the extreme solution where $a = m$, then combines step by step genes syntenic in both genomes into conserved segments.

Basic to the algorithm is the notion of a rooted binary branching tree T_i with the leaves, or terminal nodes, associated with the m_i genes in conserved synteny i. This is illustrated in Figure 5.

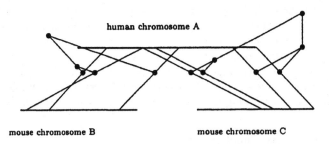

Fig. 5. Two rooted binary trees each representing successive solutions to the problem of identifying conserved segments within two conserved syntenies. Thin lines connect homologous genes in the two genomes. Note that the conserved syntenies overlap on the human chromosome and that the number of segments from the synteny on the right intervening between genes on the left changes as the trees are constructed from bottom up.

Each nonterminal node v denotes the formation of a segment from two smaller segments v_1, v_2 of distance $d(v_1, v_2) = D(v)$ apart. Note that d is a not a metric, and it is defined only for two segments v_1 and v_2 containing genes in the same synteny sets.

After precalculating all the distances d among the terminal nodes (segments consisting of single genes), we apply the following:

Algorithm conseg

Let m_k be the number of genes in the k-th conserved synteny. Set $a = m = \sum m_k$, the total number of homologous pairs of genes, and let seg to be the set of all these genes. For all k, set $S_k = -\beta m_k$. Initial construction step for T_k: Identify the terminal nodes with the m_k genes in the conserved synteny.

269

while there remains a conserved synteny with ≥ 2 segments in *seg*,
Find the two segments v_1 and v_2 that minimize $d(v_1, v_2)$.

Combine v_1, v_2 to form v. Add v to *seg*. Remove v_1 and v_2.
if v contains genes in the k-th synteny
Update T_k to indicate branching of v to v_1, v_2
Set $S_k = S_k + D(v) - D(v_1) - D(v_2)$.
endif
Set $a = a - 1$, and output configuration of the a segments in *seg*.
Recalculate all distances d given the decrease in number of segments in *seg*.
Set $D^* = \sum S_k$.

endwhile

A relatively literal implementation of this algorithm has worst-case performance in time cubic in m, the number of genes. Within the **while** loop, the distance update can take quadratic time (without any sophisticated data structures), though with small proportionality factor, and the loop itself must be executed $m - 1$ times. The search step is carried out at the same time as the update step. Improvement, possibly to quadratic performance, could be achieved by tracking which segments intervene in which other segments. With available data, however, there is little need for improved code.

The clustering procedure may seem a roundabout way of approaching the objective function, but to the extent that segments are disjoint, or overlap to a very limited extent, the following theorem [6] becomes pertinent:

Theorem 4. *For any a, the upper bound D^* achieved by the algorithm is equal to the objective D if no segment intervenes in any other segment by virtue of more than one gene.*

4.3 How Many Segments?

What value of a is the most reasonable? To answer this, we compare the number U_i of different human chromosomes represented among the a_i segments on a single mouse chromosome i, with the number u_i expected under a random hypothesis:

$$u_i = 22[1 - (\frac{21}{22})^{a_i}].$$

We chose the parameter values and a so that

$$\sum_{i=1}^{19} u_i = \sum_{i=1}^{19} U_i.$$

In our data set, these values are $a = 130$, $\alpha = 30$, $\gamma = 1$ and $\beta = 0.3$. There are 113 conserved syntenies in the data. Since we infer 130 segments, this means that about one conserved synteny per chromosome consists of more than one conserved segment, or that almost all the observed fragmentation of conserved syntenies is due to intrachromosomal movement and not interchromosomal events.

5   Gene Clumping and Non-uniform Densities

In Section 3.2, we used the value of $a = 130$ satisfying the criterion of Section 4.3 and 130 segments produced by the identification procedures in Section 4.2 as data for the maximum likelihood estimation of the total number, observed and unobserved, of segments. This was calculated making use of the exact values of (or, equivalently, the gaussian approximation to) $P(a, m, n)$, a valid procedure insofar as the basic Nadeau-Taylor model represents reality, with uniformly distributed breakpoints and uniformly and independently distributed genes. One check on this is the comparison in Figure 3 of the distribution predicted by the model $Q(r) = \Pi(r \,|r > 0)$ (dotted curve in the figure) with the $f(r), r = 1, ...$, the relative frequency of segments containing r genes, $r = 1, ...$

Based on data for 1423 genes and an analysis giving $a = 130$ segments, we find two major discrepancies. First, $f(1)$ is far greater than $Q(1)$, and second, $f(r)$ is systematically less than $Q(r)$ for r in the range [3,18]. To the extent the basic Nadeau-Taylor model needs refinement, we must rely less on Theorem 2 and maximum likelihood estimation based on it. Instead we use in this section a method which is most sensitive to a systematic discrepancy between $f(r)$ and $Q(r)$ over a range of values of r, namely a Kolmogorov-Smirnov approach. To estimate n, we simply choose the value which minimizes $\sup_r |F(r) - G(r)|$, where F and G are the cumulative distributions of f and Q, respectively. As is reflected in $Q(0)$ particularly and in the first few other inflated values of $Q(r)$ in Figure 3 (uninterrupted curve), compared to the maximum likelihood estimate of 141, the Kolmogorov-Smirnov-based estimate for n is 181, due to its sensitivity to the large $|F(1) - G(1)|$ discrepancy. (Indeed, $\sup_r |F(r) - G(r)| = |F(1) - G(1)| = f(1) - Q(1) = 0.095$.)

The excess observations accounting for the value of $f(1)$ may include a good proportion of experimental error, as we previously [6] noticed from changes in the data set over time for many of the chromosomal assignments involved. By removing the case $r = 1$ from the analysis (involving 27 of 130 observed segments), and conditioning both f and Q by $r \geq 2$, we obtain a better fit as seen in Figure 6. With the effect of f(1) removed, n is estimated at 129, greatly diminished from the exaggerated value of 181. The statistic $\sup_r |F(r) - G(r)|$ is dramatically reduced from 0.095 to 0.043. The range for which $f(r)$ is systematically less than $Q(r)$ is contracted to [12,18].

We undertook two approaches to modifying our basic model, relaxing the hypotheses of independence of gene distribution and uniformity of gene and breakpoint distributions [2].

Instead of distributing the genes one at a time according to the uniform distribution, we constructed a model where z genes, where z was fixed to be 2,3, or more, were positioned at the same point. (Thus, only $\frac{m}{z}$ points were sampled from the uniform.) This non-independence of gene distribution turned out to have little effect on the general shape of the predicted frequency curve, despite its effect on the first few values of r.

A second type of modified model divided the genes into two fractions and the breakpoints into two fractions and distributed the first fraction of genes among the first fraction of breakpoints and the rest of the genes among the remaining breakpoints.

The inhomogeneities of distribution rectify to some extent the discrepancies between the predictions and the observed results, both when data on $r = 1$ are

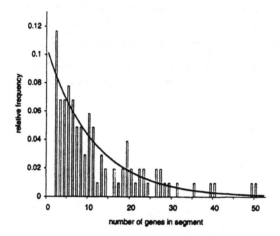

Fig. 6. Comparison of relative frequencies $f(r)/(1 - f(1))$ of segments containing $r \geq 2$ genes with predictions of Nadeau-Taylor model. Value of n in formula for Q (curve shown also conditioned for $r \geq 2$) is taken to be 129, as estimated by minimizing a Kolmogorov-Smirnov-type statistic. Values shown for $Q(0)$ and $Q(1)$, though [0,1] is outside the range of the conditional distribution, to permit a comparison of the estimated number of empty or single-gene segments with the predictions for $r \geq 2$. Three data points are off-scale, with $r = 54, 65$ and 83.

retained and when they are excluded. For example, when the genes are divided into two equal groups, and the breakpoints are divided unevenly, proportion α in one part of the genome and $1 - \alpha$ in the other, the best fit, as obtained by minimizing $\sup_r |F(r) - G(r)|$ with respect to α is illustrated in Figures. 7 and 8. In the case where $r = 1$ data are included, half the genes are distributed within a portion of the genome containing 20% of the 157 breakpoints and the other half among the other 80%. Note that 157 is a distinct reduction from the 181 needed in the homogeneous model, and the statistic of goodness-of-fit is reduced from 0.095 to 0.079. The fit of the model to the data is improved both for $r = 1$ and in the range [12,18]. In the case where the $r = 1$ segments are excluded, the best fit is with $n = 118$ and the split of the breakpoints is 29% vs. 71%. Here the improvement in $\sup_r |F(r) - G(r)|$ is from 0.043 to 0.036 as the fit is improved for $r = 2$ and in the range [12,18].

6 Discussion

The analytic insights of Nadeau and Taylor [3] and the prophetic accuracy of their estimation of the number of segments conserved between the mouse and human genomes have become increasingly relevant with the recent massive increases in the available genomic data, whether genetic maps, physical maps or complete sequences. Their work serves as a starting point for a variety of algorithmic, probabilistic, statistical and other applications of mathematical science.

6.1 The Original Approach of Nadeau and Taylor

In the intellectual climate of the early 80's, Nadeau and Taylor used $r \geq 2$ as a criterion for the existence of a conserved segment, in contradistinction to a model

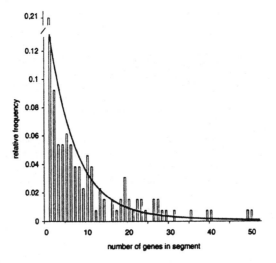

Fig. 7. Comparison of relative frequencies $f(r)$ of segments containing r genes with predictions of inhomogeneous genome model. Values of n and α are taken to be 157 and 0.2, respectively, as estimated by minimizing Kolmogorov-Smirnov-type statistic.

of random gene scrambling throughout the genome. Their analysis was based on the estimation of average segment length, in centimorgans, prior to the estimation of of the number of segments. This work involved a good number of mathematical assumptions and approximations that, while justifiable, turn out to be unnecessary within our formulation of the key assumptions of spatial homogeneity and independence of breakpoint and gene distributions in Sections 2 and 3.

6.2 The Distribution $P(a, m, n)$

When appropriately formulated, the probabilistic model fundamental to the Nadeau-Taylor theory derives from a classical occupancy problem related to statistical mechanics ([1], p. 62). As such, it makes no reference to the linear nature of chromosomes, though considerations of order are central to the identification of segments in Section 4, prior to statistical analysis.

6.3 Why So Few Segments?

The applications of our method in this paper were all based on the estimate of a in Section 4.3. This estimate of 130, contrasting with the 140-185 segments seen elsewhere in the literature may be considered low for reasons definitional, methodological, or biological.

The criterion in Section 4.3 is designed to estimate the number of reciprocal translocations based on the total number of conserved syntenies detected on each chromosome, and is not influenced by how fragmented each of these syntenies may be. This choice follows from our goal specified in Section 4 of recovering the history of translocation and ignoring the effects of intrachromosomal rearrangement. It is not, however, a fundamental aspect of our methodology; we could have chosen a somewhat larger value of a in the hope that the **conseg** algorithm would identify segments created by inversions and intrachromosomal transposition as well as translocation, for example, while excluding multiple counts of single segments due

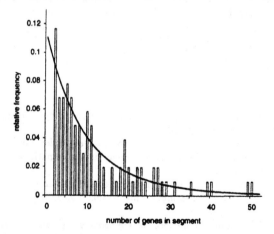

Fig. 8. Comparison of relative frequencies $f(r)/(1 - f(1))$ of segments containing $r \geq 2$ genes with predictions of inhomogeneous genome model. Values of n and α are taken to be 118 and 0.29, respectively, as estimated by minimizing Kolmogorov-Smirnov-type statistic.

simply to small mapping errors. This new value of a and the corresponding n_r could have equally well served to draw Figures 3 and 6-8, and to do the calculations in Sections 3.2 and 5.

Another explanation of the small estimate of a is the rather simple formula used in Section 4.3. A more detailed analysis of how segments are distributed, symmetric with respect to the two organisms, and using likelihood techniques, could have resulted in a larger value of a, though not very much so. This is a direction for future research.

A final type of explanation would depend on the cellular mechanisms, as yet unassessed, resulting in the fixation of a chromosomal aberration such as reciprocal translocation. These explanations might invoke differences in chromosome size or differential tendencies among chromosomes for synteny preservation, fusion, fission and translocation. For the time being these considerations remain purely speculative, but they have the greatest potential for revising and deepening our analysis of conserved segments.

6.4 The Study of Inhomogeneities

In our study of the fit of the distribution Π, or its version conditioned on $r \geq 1$, to the relative frequency f of segment sizes, the greatest discrepancy would seem to be for $r = 1$, which is most likely a reflection of error in the identification of homologous genes or other experimental error in chromosome assignment. Nevertheless, when this source of error is removed, there is clear evidence that allowing inhomogeneity in breakpoint and gene distributions offers a closer fit to the data. A refinement of our model of inhomogeneity, and associated statistical tests, are potential directions for combined empirical and theoretical research.

References

1. W. Feller. *An Introduction to Probability Theory and its Applications, Vol.1. 3d ed.* New York: John Wiley and Son, 1968.
2. I. Marchand. *Généralisations du modèle de Nadeau et Taylor sur les segments chromosomiques conservés.* MSc thesis, Département de mathématiques et de statistique, Université de Montréal. 1997.
3. J.H. Nadeau and B.A. Taylor Lengths of chromosomal segments conserved since divergence of man and mouse. *Proceedings of the National Academy of Sciences USA*, 81: 814-818, 1984.
4. M.-N. Parent. *Estimation du nombre de segments vides dans le modèle de Nadeau et Taylor sur les segments chromosomiques conservés.* MSc thesis, Département de mathématiques et de statistique, Université de Montréal. 1997.
5. D. Sankoff and V. Ferretti. Karotype distributions in a stochastic model of reciprocal translocation. *Genome Research* 6, 1-9, 1996.
6. D. Sankoff, V. Ferretti and J.H. Nadeau. Conserved segment identification. *RECOMB 97. Proceedings of the First Annual International Conference on Computational Molecular Biology.* New York: ACM Press, 1997, pp. 252-256.
7. D. Sankoff and J.H. Nadeau. Conserved synteny as a measure of genomic distance. *Discrete Applied Mathematics* 71, 247-257, 1996.

Iterative versus Simultaneous Multiple Sequence Alignment

Andreas Dress

University of Bielefeld
dress@Mathematik.Uni-Bielefeld.DE

Abstract. Due to unsurmountable time requirements, straight-forward dynamical-programming approaches to multiple sequence alignment were abandoned a long time ago in favour of iterative procedures. Some of those have been elaborated to great perfection, using many years of experience regarding the optimal adjustment of biologically relevant aligment parameters.

The situation changed somewhat with the availability of "MSA" – a program which is based on some clever "branch bound" procedures to reduce, by several orders of magnitude, the enormous search space one has to cope with. This way, MSA is able to compute optimal multiple sequence aligments with regard to appropriately defined "sum of pairs scores" for, say, up to six protein sequences of modest length (up to, say, 300 amino acids) and sufficiently high sequence homology.

Still, many biologically important sequence families turned out to be by far too complex to be aligned simultaneously by MSA. Here, the new DIVIDE & CONQUER multiple sequence alignment algorithm "DCA" can now be invoked to find (almost) optimal simultaneous alignments in situations where formerly the construction of such alignments was out of question.

In the lecture, the workings of DCA will be explained shortly, some applications to biologically relevant data sets will be discussed and the potential of this new approach regarding some basic problems in sequence alignment will be evaluated.

Modern Comparative Lexicostatistics

Joseph B. Kruskal

Bell Labs, Lucent Technologies
Murray Hill, NJ 07974
kruskal@research.bell-labs.com

Abstract. The problem most often dealt with in comparative lexico-statistics is to reconstruct a family tree for a family of dialects by comparing their lexicons (in a carefully chosen manner). A second problem (often distinguished by the name glottochronology) is to estimate the time at which branchings of the tree occurred. The fundamental data have this form: For a specified meaning, is the word in Dialect A cognate or not cognate to the word in Dialect B. This determination must be made by a highly-skilled linguist who has extensive knowledge of the dialect family, and is of course subject to error like any other measurement process.

Earlier work in comparative lexicostatistics treated the replacement rates for different meanings as equal, although many authors have pointed out the likelihood and effects of varying replacement rates. More recent work has dispensed with this equality assumption. Replacement rates have been explicitly estimated (by maximum likelihood) for hundreds of meanings in three different language families, and the rates have been used to estimate branching times in a tree of 84 Indoeuropean dialects.

Author Index

Abdeddaïm, Saïd, 167
Arvestad, Lars, 180

Ben-Dor, Amir, 247
Berman, Piotr, 40

Chen, Ting, 206
Cho, Yookun, 28
Crochemore, Maxime, 116
Cryan, Mary, 130

Das, Gautam, 12
Dress, Andreas, 275

Ferretti, Vincent, 262
Fleischer, Rudolf, 12
Fraenkel, Aviezri S., 76

Gasieniec, Leszek, 12, 90
Goldberg, Leslie Ann, 130
Gunopulos, Dimitris, 12

Indyk, Piotr, 90

Kärkkäinen, Juha, 12
Karpinski, Marek, 40
Kim, Dong Kyue, 28
Kitajima, João Paulo, 102
Klein, Shmuel T., 65
Kruskal, Joseph B., 276
Krysta, Piotr, 90
Kurtz, Stefan, 52

Lancia, Guiseppe, 247
Larmore, Lawrence L., 40
Lee, Jee-Soo, 28

Marchand, Isabelle, 262
Miyazaki, Masamichi, 1
Myers, Gene, 52

Navarro, Gonzalo, 102

Ohlebusch, Enno, 191

Parent, Marie-Noelle, 262
Park, Kunsoo, 28
Paterson, Mike, 76
Perone, Jennifer, 247
Phillips, Cynthia A., 130
Plandowski, Wojciech, 40

Ravi, R., 247
Ribeiro-Neto, Berthier A., 102
Richter, Thorsten, 150
Rytter, Wojciech, 40

Sagot, Marie-France, 224
Sankoff, David, 262
Shinohara, Ayumi, 1
Simpson, Jamie, 76
Skiena, Steven S., 206

Takeda, Masayuki, 1
Tran, Nicholas, 83

Vérin, Renaud, 116
Viari, Alain, 224

Ziviani, Nivio, 102

Lecture Notes in Computer Science

For information about Vols. 1–1191

please contact your bookseller or Springer-Verlag

Vol. 1192: M. Dam (Ed.), Analysis and Verification of Multiple-Agent Languages. Proceedings, 1996. VIII, 435 pages. 1997.

Vol. 1193: J.P. Müller, M.J. Wooldridge, N.R. Jennings (Eds.), Intelligent Agents III. XV, 401 pages. 1997. (Subseries LNAI).

Vol. 1194: M. Sipper, Evolution of Parallel Cellular Machines. XIII, 199 pages. 1997.

Vol. 1195: R. Trappl, P. Petta (Eds.), Creating Personalities for Synthetic Actors. VII, 251 pages. 1997. (Subseries LNAI).

Vol. 1196: L. Vulkov, J. Waśniewski, P. Yalamov (Eds.), Numerical Analysis and Its Applications. Proceedings, 1996. XIII, 608 pages. 1997.

Vol. 1197: F. d'Amore, P.G. Franciosa, A. Marchetti-Spaccamela (Eds.), Graph-Theoretic Concepts in Computer Science. Proceedings, 1996. XI, 410 pages. 1997.

Vol. 1198: H.S. Nwana, N. Azarmi (Eds.), Software Agents and Soft Computing: Towards Enhancing Machine Intelligence. XIV, 298 pages. 1997. (Subseries LNAI).

Vol. 1199: D.K. Panda, C.B. Stunkel (Eds.), Communication and Architectural Support for Network-Based Parallel Computing. Proceedings, 1997. X, 269 pages. 1997.

Vol. 1200: R. Reischuk, M. Morvan (Eds.), STACS 97. Proceedings, 1997. XIII, 614 pages. 1997.

Vol. 1201: O. Maler (Ed.), Hybrid and Real-Time Systems. Proceedings, 1997. IX, 417 pages. 1997.

Vol. 1203: G. Bongiovanni, D.P. Bovet, G. Di Battista (Eds.), Algorithms and Complexity. Proceedings, 1997. VIII, 311 pages. 1997.

Vol. 1204: H. Mössenböck (Ed.), Modular Programming Languages. Proceedings, 1997. X, 379 pages. 1997.

Vol. 1205: J. Troccaz, E. Grimson, R. Mösges (Eds.), CVRMed-MRCAS'97. Proceedings, 1997. XIX, 834 pages. 1997.

Vol. 1206: J. Bigün, G. Chollet, G. Borgefors (Eds.), Audio- and Video-based Biometric Person Authentication. Proceedings, 1997. XII, 450 pages. 1997.

Vol. 1207: J. Gallagher (Ed.), Logic Program Synthesis and Transformation. Proceedings, 1996. VII, 325 pages. 1997.

Vol. 1208: S. Ben-David (Ed.), Computational Learning Theory. Proceedings, 1997. VIII, 331 pages. 1997. (Subseries LNAI).

Vol. 1209: L. Cavedon, A. Rao, W. Wobcke (Eds.), Intelligent Agent Systems. Proceedings, 1996. IX, 188 pages. 1997. (Subseries LNAI).

Vol. 1210: P. de Groote, J.R. Hindley (Eds.), Typed Lambda Calculi and Applications. Proceedings, 1997. VIII, 405 pages. 1997.

Vol. 1211: E. Keravnou, C. Garbay, R. Baud, J. Wyatt (Eds.), Artificial Intelligence in Medicine. Proceedings, 1997. XIII, 526 pages. 1997. (Subseries LNAI).

Vol. 1212: J. P. Bowen, M.G. Hinchey, D. Till (Eds.), ZUM '97: The Z Formal Specification Notation. Proceedings, 1997. X, 435 pages. 1997.

Vol. 1213: P. J. Angeline, R. G. Reynolds, J. R. McDonnell, R. Eberhart (Eds.), Evolutionary Programming VI. Proceedings, 1997. X, 457 pages. 1997.

Vol. 1214: M. Bidoit, M. Dauchet (Eds.), TAPSOFT '97: Theory and Practice of Software Development. Proceedings, 1997. XV, 884 pages. 1997.

Vol. 1215: J. M. L. M. Palma, J. Dongarra (Eds.), Vector and Parallel Processing – VECPAR'96. Proceedings, 1996. XI, 471 pages. 1997.

Vol. 1216: J. Dix, L. Moniz Pereira, T.C. Przymusinski (Eds.), Non-Monotonic Extensions of Logic Programming. Proceedings, 1996. XI, 224 pages. 1997. (Subseries LNAI).

Vol. 1217: E. Brinksma (Ed.), Tools and Algorithms for the Construction and Analysis of Systems. Proceedings, 1997. X, 433 pages. 1997.

Vol. 1218: G. Păun, A. Salomaa (Eds.), New Trends in Formal Languages. IX, 465 pages. 1997.

Vol. 1219: K. Rothermel, R. Popescu-Zeletin (Eds.), Mobile Agents. Proceedings, 1997. VIII, 223 pages. 1997.

Vol. 1220: P. Brezany, Input/Output Intensive Massively Parallel Computing. XIV, 288 pages. 1997.

Vol. 1221: G. Weiß (Ed.), Distributed Artificial Intelligence Meets Machine Learning. Proceedings, 1996. X, 294 pages. 1997. (Subseries LNAI).

Vol. 1222: J. Vitek, C. Tschudin (Eds.), Mobile Object Systems. Proceedings, 1996. X, 319 pages. 1997.

Vol. 1223: M. Pelillo, E.R. Hancock (Eds.), Energy Minimization Methods in Computer Vision and Pattern Recognition. Proceedings, 1997. XII, 549 pages. 1997.

Vol. 1224: M. van Someren, G. Widmer (Eds.), Machine Learning: ECML-97. Proceedings, 1997. XI, 361 pages. 1997. (Subseries LNAI).

Vol. 1225: B. Hertzberger, P. Sloot (Eds.), High-Performance Computing and Networking. Proceedings, 1997. XXI, 1066 pages. 1997.

Vol. 1226: B. Reusch (Ed.), Computational Intelligence. Proceedings, 1997. XIII, 609 pages. 1997.

Vol. 1227: D. Galmiche (Ed.), Automated Reasoning with Analytic Tableaux and Related Methods. Proceedings, 1997. XI, 373 pages. 1997. (Subseries LNAI).

Vol. 1228: S.-H. Nienhuys-Cheng, R. de Wolf, Foundations of Inductive Logic Programming. XVII, 404 pages. 1997. (Subseries LNAI).

Vol. 1230: J. Duncan, G. Gindi (Eds.), Information Processing in Medical Imaging. Proceedings, 1997. XVI, 557 pages. 1997.

Vol. 1231: M. Bertran, T. Rus (Eds.), Transformation-Based Reactive Systems Development. Proceedings, 1997. XI, 431 pages. 1997.

Vol. 1232: H. Comon (Ed.), Rewriting Techniques and Applications. Proceedings, 1997. XI, 339 pages. 1997.

Vol. 1233: W. Fumy (Ed.), Advances in Cryptology — EUROCRYPT '97. Proceedings, 1997. XI, 509 pages. 1997.

Vol 1234: S. Adian, A. Nerode (Eds.), Logical Foundations of Computer Science. Proceedings, 1997. IX, 431 pages. 1997.

Vol. 1235: R. Conradi (Ed.), Software Configuration Management. Proceedings, 1997. VIII, 234 pages. 1997.

Vol. 1236: E. Maier, M. Mast, S. LuperFoy (Eds.), Dialogue Processing in Spoken Language Systems. Proceedings, 1996. VIII, 220 pages. 1997. (Subseries LNAI).

Vol. 1238: A. Mullery, M. Besson, M. Campolargo, R. Gobbi, R. Reed (Eds.), Intelligence in Services and Networks: Technology for Cooperative Competition. Proceedings, 1997. XII, 480 pages. 1997.

Vol. 1239: D. Sehr, U. Banerjee, D. Gelernter, A. Nicolau, D. Padua (Eds.), Languages and Compilers for Parallel Computing. Proceedings, 1996. XIII, 612 pages. 1997.

Vol. 1240: J. Mira, R. Moreno-Díaz, J. Cabestany (Eds.), Biological and Artificial Computation: From Neuroscience to Technology. Proceedings, 1997. XXI, 1401 pages. 1997.

Vol. 1241: M. Akşit, S. Matsuoka (Eds.), ECOOP'97 – Object-Oriented Programming. Proceedings, 1997. XI, 531 pages. 1997.

Vol. 1242: S. Fdida, M. Morganti (Eds.), Multimedia Applications, Services and Techniques – ECMAST '97. Proceedings, 1997. XIV, 772 pages. 1997.

Vol. 1243: A. Mazurkiewicz, J. Winkowski (Eds.), CONCUR'97: Concurrency Theory. Proceedings, 1997. VIII, 421 pages. 1997.

Vol. 1244: D. M. Gabbay, R. Kruse, A. Nonnengart, H.J. Ohlbach (Eds.), Qualitative and Quantitative Practical Reasoning. Proceedings, 1997. X, 621 pages. 1997. (Subseries LNAI).

Vol. 1245: M. Calzarossa, R. Marie, B. Plateau, G. Rubino (Eds.), Computer Performance Evaluation. Proceedings, 1997. VIII, 231 pages. 1997.

Vol. 1246: S. Tucker Taft, R. A. Duff (Eds.), Ada 95 Reference Manual. XXII, 526 pages. 1997.

Vol. 1247: J. Barnes (Ed.), Ada 95 Rationale. XVI, 458 pages. 1997.

Vol. 1248: P. Azéma, G. Balbo (Eds.), Application and Theory of Petri Nets 1997. Proceedings, 1997. VIII, 467 pages. 1997.

Vol. 1249: W. McCune (Ed.), Automated Deduction – CADE-14. Proceedings, 1997. XIV, 462 pages. 1997. (Subseries LNAI).

Vol. 1250: A. Olivé, J.A. Pastor (Eds.), Advanced Information Systems Engineering. Proceedings, 1997. XI, 451 pages. 1997.

Vol. 1251: K. Hardy, J. Briggs (Eds.), Reliable Software Technologies – Ada-Europe '97. Proceedings, 1997. VIII, 293 pages. 1997.

Vol. 1252: B. ter Haar Romeny, L. Florack, J. Koenderink, M. Viergever (Eds.), Scale-Space Theory in Computer Vision. Proceedings, 1997. IX, 365 pages. 1997.

Vol. 1253: G. Bilardi, A. Ferreira, R. Lüling, J. Rolim (Eds.), Solving Irregularly Structured Problems in Parallel. Proceedings, 1997. X, 287 pages. 1997.

Vol. 1254: O. Grumberg (Ed.), Computer Aided Verification. Proceedings, 1997. XI, 486 pages. 1997.

Vol. 1255: T. Mora, H. Mattson (Eds.), Applied Algebra, Algebraic Algorithms and Error-Correcting Codes. Proceedings, 1997. X, 353 pages. 1997.

Vol. 1256: P. Degano, R. Gorrieri, A. Marchetti-Spaccamela (Eds.), Automata, Languages and Programming. Proceedings, 1997. XVI, 862 pages. 1997.

Vol. 1258: D. van Dalen, M. Bezem (Eds.), Computer Science Logic. Proceedings, 1996. VIII, 473 pages. 1997.

Vol. 1259: T. Higuchi, M. Iwata, W. Liu (Eds.), Evolvable Systems: From Biology to Hardware. Proceedings, 1996. XI, 484 pages. 1997.

Vol. 1260: D. Raymond, D. Wood, S. Yu (Eds.), Automata Implementation. Proceedings, 1996. VIII, 189 pages. 1997.

Vol. 1261: J. Mycielski, G. Rozenberg, A. Salomaa (Eds.), Structures in Logic and Computer Science. X, 371 pages. 1997.

Vol. 1262: M. Scholl, A. Voisard (Eds.), Advances in Spatial Databases. Proceedings, 1997. XI, 379 pages. 1997.

Vol. 1263: J. Komorowski, J. Zytkow (Eds.), Principles of Data Mining and Knowledge Discovery. Proceedings, 1997. IX, 397 pages. 1997. (Subseries LNAI).

Vol. 1264: A. Apostolico, J. Hein (Eds.), Combinatorial Pattern Matching. Proceedings, 1997. VIII, 277 pages. 1997.

Vol. 1265: J. Dix, U. Fuhrbach, A. Nerode (Eds.), Logic Programming and Nonmonotonic Reasoning. Proceedings, 1997. X, 453 pages. 1997. (Subseries LNAI).

Vol. 1266: D.B. Leake, E. Plaza (Eds.), Case-Based Reasoning Research and Development. Proceedings, 1997. XIII, 648 pages. 1997 (Subseries LNAI).

Vol. 1267: E. Biham (Ed.), Fast Software Encryption. Proceedings, 1997. VIII, 289 pages. 1997.

Vol. 1268: W. Kluge (Ed.), Implementation of Functional Languages. Proceedings, 1996. XI, 284 pages. 1997.

Vol. 1269: J. Rolim (Ed.), Randomization and Approximation Techniques in Computer Science. Proceedings, 1997. VIII, 227 pages. 1997.

Vol. 1270: V. Varadharajan, J. Pieprzyk, Y. Mu (Eds.), Information Security and Privacy. Proceedings, 1997. XI, 337 pages. 1997.

Vol. 1271: C. Small, P. Douglas, R. Johnson, P. King, N. Martin (Eds.), Advances in Databases. Proceedings, 1997. XI, 233 pages. 1997.